Digital Signal Processing

THE OXFORD SERIES IN ELECTRICAL AND COMPUTER ENGINEERING

ADEL S. SEDRA, Series Editor

Digital Signal Processing

Spectral Computation and Filter Design

Chi-Tsong Chen

State University of New York at Stony Brook

New York   Oxford

OXFORD UNIVERSITY PRESS

2001

Oxford University Press

Oxford   New York
Athens   Auckland   Bangkok   Bogotá   Buenos Aires   Calcutta
Cape Town   Chennai   Dar es Salaam   Delhi   Florence   Hong Kong   Istanbul
Karachi   Kuala Lumpur   Madrid   Melbourne   Mexico City   Mumbai
Nairobi   Paris   São Paulo   Shanghai   Singapore   Taipei   Tokyo   Toronto   Warsaw

and associated companies in
Berlin   Ibadan

Published by Oxford University Press, Inc.
198 Madison Avenue, New York, New York, 10016
http://www.oup-usa.org

Oxford is a registered trademark of Oxford University Press

Library of Congress Cataloging-in-Publication Data

Chen, Chi-Tsong.
    Digital signal processing : spectral computation and filter design / Chi-Tsong Chen.
        p.   cm.
    Includes bibliographical references and index.
    ISBN 0-19-513638-1
    1. Signal processing—Digital techniques.   2. Electric filters—Design and construction.
    3. Spectral sequences (Mathematics)   I. Title.

TK5102.9 .C476   2000
621.382'2—dc21   00-027881

Printing (last digit): 10 9 8 7 6 5 4 3 2 1

Printed in the United States of America
on acid-free paper

In memory of my father and mother

CONTENTS ━━━━━━━━━━━━━━━

PART 2 Digital Filter Design

5 Linear Time-Invariant Lumped Systems   191

6 Ideal and Some Practical Digital Filters   252

PREFACE ▬▬▬▬▬▬▬▬▬▬▬▬▬▬▬▬▬▬▬▬▬▬▬▬

This text is intended for use in a first course on digital signal processing (DSP), typically in senior or first-year graduate level. It may also be useful to engineers and scientists who use digital computers to process measured data. Some elementary knowledge on signals and systems is helpful but is not necessary. An attempt has been made to make this text as self-contained as possible.

As an introductory text, we discuss only two major topics: computer computation of frequency contents of signals and design of digital filters. These two topics follow logically a basic text or course on *Signals and Systems*, which now often covers both continuous-time (CT) and discrete-time (DT) cases. To be self-contained, we start from scratch. Chapter 1 introduces CT signals, DT signals, and digital signals. We discuss why CT signals are now widely processed digitally and how they are converted into digital signals. We also discuss why we study in this text, as in every other DSP text, only CT and DT signals even though digital signals are the signals processed on digital computers. In Chapter 2, we first discuss the frequency of CT and DT sinusoidal signals. We then introduce CT and DT Fourier series (frequency components) for periodic signals and establish their relationships. We then use the fast Fourier transform (FFT) to compute their frequency components.

Chapter 3 extends CT and DT Fourier series to CT and DT Fourier transforms (frequency spectra) that can reveal frequency contents of aperiodic signals as well. We establish the sampling theorem and discuss the effects of truncation of signals on frequency spectra. In Chapter 4, we show that computer computation of DT Fourier transform (DTFT) leads naturally to the discrete Fourier transform (DFT), which differs from the DT Fourier series only by a fraction. We then introduce a version of FFT, an efficient way of computing DFT. We use FFT to compute frequency spectra of DT and CT signals and to compute DT and CT signals from their frequency spectra. This completes the discussion of spectral computation of signals.

The second part of this text discusses the design of digital filters. We introduce in Chapter 5 the class of digital filters (linear, time-invariant, lumped, causal, and stable) to be designed and the mathematical tools (convolutions, impulse responses, z-transform, and transfer functions) to be used. We then introduce the concept of frequency responses for stable systems. Chapter 6 discusses specifications of digital filters and how to use poles and zeros to shape magnitude responses of simple digital filters. Chapter 7 introduces various methods to design finite-impulse-response (FIR) digital filters. In Chapter 8, we first discuss the reasons for not designing directly infinite-impulse-response (IIR) digital filters. We then introduce analog prototype filters, which are lowpass filters with 1 rad/s as their passband or stopband edge frequency. All other analog and all digital frequency-selective filters can then be obtained by using frequency transformations.

The last chapter discusses a number of block diagrams for digital filters and some problems due to finite-word-length implementation. This chapter is essentially independent of Chapters 7 and 8 and may be studied right after Chapter 6.

Although most topics in this text are standard, our presentation and emphasis are significantly different from other existing DSP texts. They are discussed in the following.

- Most DSP texts discuss systems and signals intertwined. This text discusses in the first part only signals and their spectral computation. Thus the discussion can be more focused. It is also more efficient for those who wish to learn how to use FFT to analyze measured data.
- This text shows that the frequency of a DT sinusoidal sequence cannot be defined from its fundamental period as in the CT case. It is then defined from a CT sinusoid and justified using a physical argument. Thus this text uses the same notation to denote frequencies in CT and DT signals as opposed to different notations used in most DSP texts. We do use different notations when the analog frequency range $(-\infty, \infty)$ is compressed into the digital frequency range $(-\pi, \pi]$ in bilinear transformations.
- Assuming the reader to have had Fourier analysis of CT signals, most DSP texts cover only Fourier analysis of DT signals. Because signals processed by DSP are mostly CT, this text covers Fourier analyses of both CT and DT signals. The discussion of the CT part, however, is not exhaustive. It is introduced to the extent to show the differences between CT and DT Fourier analyses and to establish their relationships. We establish sampling theorems for pure sinusoids, periodic signals, and general signals.
- Most DSP texts assume the sampling period T to be 1 in DT Fourier analysis. Although the equations involved are simpler, they become more complex in relating spectra of CT signals and their sampled sequences. This text does not assume $T = 1$ and discusses how to select T in spectral computation of CT signals from their time samples. In digital filter design, however, we can select T to be 1 or any other value such as π in MATLAB,[1] as discussed in Section 5.3.2.
- Many DSP texts introduce DFT after the Fourier transforms and discuss it as an independent mathematical entity. Our discussion of DFT is brief because it is essentially the same as the DT Fourier series. It is introduced as a computational tool.
- Most DSP texts introduce first the two-sided z-transform and then reduce it to the one-sided z-transform. This text does not need the two-sided transform. Thus we concentrate on the one-sided z-transform and give reasons for forgoing the concept of the region of convergence, which is a difficult concept and rarely needed in application.
- Fourier analysis of DT systems are covered in most DSP texts. It is not covered in this text because Fourier analysis of DT systems is less general and more complex than the z-transform (Section 5.5.3).
- Most DSP texts use exclusively negative-power transfer functions. This text uses both negative-power and positive-power transfer functions because the latter is more convenient in introducing the concepts of properness, degrees, poles, and zeros. Furthermore, both forms are used in MATLAB.

---

[1] MATLAB is a trademark of the MathWorks, Inc.

- The bilinear transformation introduced in many DSP texts normalize the frequency range by assuming $T = 1$ and yet introduce the factor $T/2$ where T may be different from 1. This is inconsistent. We introduce an arbitrary factor and show that the design is independent of the factor. This is more logical and can also justify the inconsistence.

In addition, this text discusses a number of topics not found in most DSP texts. We list some of them in the following:

- Give a formal definition of the frequency of sinusoidal sequences and give a precise frequency range $(-\pi/T, \pi/T]$ for DT signals (Section 2.2).
- Establish a simplified version of the sampling theorem for periodic signals and then use FFT to compute frequency components of CT periodic signals from their sampled sequences. Discuss how to use the MATLAB functions fftshift, ceil, floor and a newly defined function shift to plot FFT computed frequency components in $(-\pi/T, \pi/T]$ or $[-\pi/T, \pi/T)$ (Sections 2.6 and 2.7).
- Discuss the nonuniqueness of inverse DFT and a method of determining the location of a time signal computed from frequency samples of its frequency spectrum (Section 4.7).
- Use FFT to compute the inverse z-transform (Section 5.4.2).
- Discuss steady-state and transient responses of digital filters, and give an estimated time for a transient response to die out (Section 5.7.1). Compare two ways of introducing the concept of frequency responses. We argue that although the concept can be more easily introduced by assuming an input to be applied from $-\infty$, two concepts of practical importance may often be concealed (Section 5.7.3).
- Give a mathematical justification of using an antialiasing analog filter in digital signal processing (Section 6.7.1).
- Introduce a discrete least squares method to design FIR filters. The method, although very simple, does not seem to appear anywhere in the literature. The method is more flexible than the method of frequency sampling and leads naturally to the MATLAB function firls (Sections 7.6 and 7.6.1).
- Introduce an analog bandstop transformation that yields better bandstop filters than the ones generated by MATLAB (Section 8.4).
- Although FFT is very efficient in computing the convolution of two long sequences, we give possible reasons for not using FFT in the MATLAB function conv (Section 9.4.1).

All terminology in this text is carefully defined. For example, both the Fourier series and Fourier transforms of periodic signals are often called discrete frequency spectra in the literature. This text reserves *frequency spectra* exclusively for the Fourier transforms, and calls Fourier series *frequency components*. We attempt to make all discussion mathematically correct; however, we will not be constrained by pure mathematics (Section 3.9 and the footnote in Section 6.3). As a text intended for practical application, the discussion is not necessarily rigorous. For example, we use the terms "very close" and "practically zero" loosely. Because of the information explosion, we all have less time to study a subject area. Thus the discussion in this text is not exhaustive; we concentrate only on concepts and results that, in our opinion, are essential in practical application.

MATLAB is an integral part of this text. We use *while loops* and *if-else-end structures* to develop MATLAB programs. However, our emphasis is not on MATLAB but rather on basic ideas and procedures in digital signal processing. Thus this text lists only essential MATLAB functions in most programs and skips functions that adjust shapes and sizes of plots, and draw horizontal and vertical coordinates. It is recommended that the reader repeat each program. Even though he may not obtain identical results, all essential information will be generated. Clearly any other DSP package can also be used.

All numerical examples in the text and most problems at the end of each chapter are very simple and can be solved analytically by hand. The reader is urged to do so. After obtaining results by hand, one can then compare them with computer-generated results. This is the best way of learning a topic and a computational tool. Once mastering the subject and tool, we can then apply the tool to real-world problems. A solutions manual, in which complete programs are listed for problems that use MATLAB, is available from the publisher.

This text is a complete restructure and rewriting of *One-Dimensional Digital Signal Processing*, published in 1979. We deleted the discussion of the two-sided z-transform, its region of convergence, and the topics such as error analyses, Wiener FIR, and IIR filters, that require statistical methods. All aforementioned sections, except Section 6.6.1, are new in this text.

I am indebted to many people in developing this text. First I like to thank Professors Petar Djuric, Adrian Leuciuc, John Murray, Nam Phamdo, Stephen Rappaport, Kenneth Short, and Armen Zemanian, my colleagues at Stony Brook. I went to them whenever I had any question or doubt. Many people reviewed a earlier version of this text. Their comments prompted me to rewrite many sections. I thank them all.

I am grateful to my editor Peter C. Gordon for his confidence in and enthusiasm about this project. Many people at Oxford University Press including Christine D'Antonio, Jacki Hartt, Anita Sung, and Jasmine Urmeneta, were most helpful in this undertaking. Finally I thank my wife, Bih-Jau, for her support.

Chi-Tsong Chen

Digital Signal Processing

CHAPTER 1

Introduction

1.1 Continuous-Time (CT), Discrete-Time (DT), and Digital Signals

Anything that bears information can be considered a signal. For example, speech, music, interest rates, the speed of an automobile are signals. This type of signal depends on one independent variable, namely, time and is called the one-dimensional signal. Pictures, x-ray images, and sonograms are also signals. They depend on two independent spatial variables, and are called two-dimensional signals. Temperature, wind speed, and air pressure are four-dimensional signals because they depend on the geographical location (longitude and latitude), altitude, and time. However, if we study the temperature at a fixed location, then the temperature becomes a one-dimensional signal, a function of time. This text studies only one-dimensional signals, and the independent variable is time.

A one-dimensional signal can be represented by $x(t)$. For every t, the signal is required to assume a unique value; otherwise the signal is not well defined. In mathematics, $x(t)$ is called a *function* because it has a unique value for every t. Thus there is no difference between signals and functions, and they will be used interchangeably.

Consider the temperature $x(t)$ at Stony Brook as shown in Fig. 1.1(a). It is defined at every time instant and is called a continuous-time (CT) signal. A CT signal is also called an *analog* signal because its waveform is often analogous to that of the physical variable. Other examples of analog signals are the waveform of the potential at household electric outlets in Fig. 1.1(b), and the electrocardiogram (EKG) waveforms in Fig. 1.1(c). The vast majority of signals encountered in practice are continuous-time or analog signals. Note that a continuous-time signal is not necessarily a continuous function of time as in Fig. 1.1(d), which is used as a time basis in televisions.

A signal is called a discrete-time (DT) signal if the signal is defined at discrete time instants, as shown in Fig. 1.2. Most discrete-time signals arise from sampling continuous-time signals. For example, if the temperature $x(t)$ in Fig. 1.1(a) is measured and recorded every hour, then the resulting signal is as shown in Fig. 1.2(a) and will be denoted by[1]

---

[1] We use A:=B to denote that A, by definition, equals B. We use A=:B to denote that B, by definition, equals A.

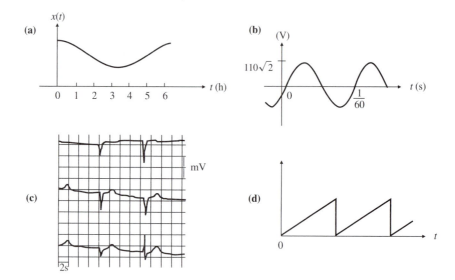

Figure 1.1 Examples of continuous-time signals: (a) Temperature. (b) Household electric potential. (c) EKG. (d) Saw-tooth waveform.

$$x[n] := x(nT) = x(t)|_{t=nT}$$

where T is called the *sampling period* and n is an integer ranging from $-\infty$ to ∞ called the *time index*. We call $x(nT)$ the *sampled signal* or *sampled sequence* of $x(t)$. The instants at which signals appear are called *sampling instants*. Figure 1.2(b) shows another sample of $x(t)$ with sampling period $T = 0.5$. Note the horizontal coordinates t and n.

If we use an analog thermometer to measure the temperature, then the reading can assume any value in a continuous range. If we use a digital thermometer that displays integer numbers, then the reading can only be an integer. In this case, the amplitude is said to be *quantized*. More generally, if the amplitude of a signal can assume only a value from a finite set of numbers, the amplitude is said to be *discretized* or *quantized*. The set of values is generally of equal distance as shown in Fig. 1.3. The values that the amplitude can assume are called *quantization levels*, and the distance between two adjacent quantization levels is called the *quantization step*. A digital thermometer that displays only integer numbers has quantization step 1; if it displays up to the first decimal point, then the quantization step is 0.1.

A signal is called a digital signal if its time is discretized and its amplitude is quantized as shown in Fig. 1.3. A digital computer can accept only sequences of numbers (discrete-time), and the numbers are limited by the number of bits used. Therefore, all signals processed on digital computers are digital signals. In summary, we have introduced the following three types of signals:

• Continuous-time (CT) or analog signals: continuous in time and continuous in amplitude
• Discrete-time (DT) signals: discretized in time and continuous in amplitude
• Digital signals: discretized in time and quantized in amplitude

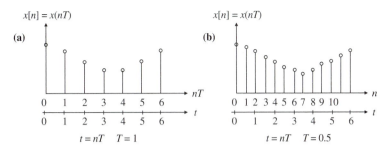

Figure 1.2 Discrete-time signals: (a) Samples of Fig. 1.1.(a) with $T = 1$. (b) Samples of Fig. 1.1(a) with $T = 0.5$.

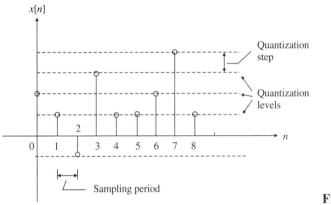

Figure 1.3 Digital signal.

The term *continuous in amplitude* means that the amplitude can assume any value in the continuous range from $-\infty$ to ∞.

1.1.1 Plotting of DT Signals

If a DT signal is obtained from a CT signal by sampling, then it can be expressed as

$$x[n] = x(t)|_{t=nT} = x(nT)$$

where $T > 0$ is the sampling period and n is the time index and can assume only integers. Because T is fixed and n is a variable, it is logical to plot the DT signal with time index as the horizontal coordinate. Such a plot, however, will not reveal an important information of the signal, as we shall demonstrate next.

Consider the DT signal $\sin(\pi n T_1)$, with $T_1 = 0.2$, plotted in Figs. 1.4(a) and (aa). The former plots the amplitude versus *time in seconds* and the latter the amplitude versus *time index*. Strictly speaking, the time index is unitless; however, it is often associated with sample. Figures 1.4(b) and (bb) are the corresponding plots of $\sin(\pi n T_2)$ with $T_2 = 0.1$. We can see from Figs. 1.4(a) and (b) that they are samples, with different sampling periods, of the same CT sinusoid. However, we cannot see this fact from Figs. 1.4(aa) and (bb) because the sampling periods do not appear explicitly in the plots. We plot in Figs. 1.4(c) and (cc) the DT sinusoid $\sin(2\pi n T_1)$ versus time and time index. The plots in Figs. 1.4(aa) and (cc) are identical, but their actual DT sinusoids are different. Thus we conclude that plots of DT signals versus time index will not reveal actual waveforms of signals. To see actual waveforms, we must actually or mentally replot them.

The first part of this text studies mainly waveforms or frequency contents of signals. Thus most DT signals in Chapters 2 through 4 will be plotted with respect to time. In computer processing, DT signals can be considered, as we will discuss in Chapter 5, as streams of numbers (independent of sampling period). Thus most DT signals, from Chapter 5 on, will be plotted with respect to time index.

1.2 Representation of Digital Signals

Messages sent by telegraph are digital signals. They are written using 26 alphabetical characters, 10 numerals, and symbols such as commas and periods. These symbols certainly can be represented by different voltage levels, for example, A by 10 V (volts), B by 9.5 V, C by 9 V, and so forth. This type of representation, however, is not used because it is susceptible to noise, shifting of power supply, and any other disturbances. In practice, these symbols are coded as sequences of dashes and dots or, equivalently, ones and zeros. The symbol 0 can be represented by a voltage from 0 to 0.8 V. The symbol 1 can be represented by a voltage from 2.0 to 5.0 V. The precise voltage of each symbol is not important, but it is critical that its value lies within one of the two allowable ranges.[2] On the compact disc, 1 and 0 are represented by dimples as shown in Fig. 1.5. The leading and trailing edges of a dimple represent a 1; no change represents a 0. Each bit occupies approximately a distance of 1 micrometer (10^{-6} meter). This type of representation of 1 and 0 is much less susceptible to noise. Furthermore, it is easy to implement. Therefore, digital signals are always coded by ones and zeros in physical implementation. This is called *binary coding*.

There are many types of binary coding. In this section, we discuss the simplest one, called *sign-and-magnitude* coding. Consider the following sequences of a_i:

$$
\begin{array}{cccccc}
a_0 & a_1 & a_2 & a_3 & a_4 & a_5 \\
\text{Sign bit} & 2^1 & 2^0 & 2^{-1} & 2^{-2} & 2^{-3}
\end{array}
$$

Each a_i can assume only either 1 or 0 and is called a binary digit or bit. There are six bits in the sequence. The left-most bit is called the sign bit. The sequence represents a positive number if $a_0 = 0$ and a negative number if $a_0 = 1$. The remaining five bits represent the magnitude of the number. For example, we have

$$110011 \longleftrightarrow -(1 \times 2^1 + 0 \times 2^0 + 0 \times 2^{-1} + 1 \times 2^{-2} + 1 \times 2^{-3}) = -2.375$$

---

[2] If the value lies outside the two allowable ranges, the telegraph must make a choice of either 1 or 0, and an error may occur. In this case, an error message can be sent, and the message be retransmitted.

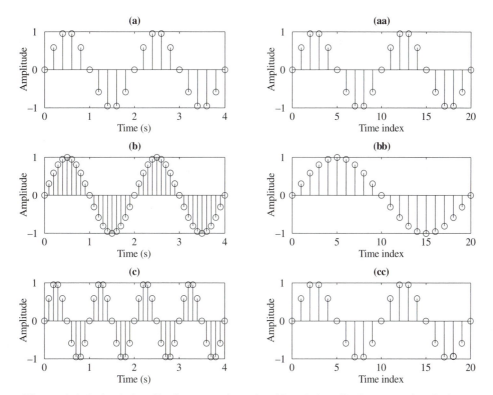

Figure 1.4 (a, b, c) Amplitude versus time. (aa, bb, cc) Amplitude versus time index.

Figure 1.5 Dimples on compact discs.

and

$$0\ 0\ 1\ 1\ 1\ 0 \longleftrightarrow +(0 \times 2^1 + 1 \times 2^0 + 1 \times 2^{-1} + 1 \times 2^{-2} + 0 \times 2^{-3}) = 1.75$$

The left-most bit, excluding the sign bit, is the most significant bit (MSB) and the right-most bit is the least significant bit (LSB). If we use 10 bits to represent the decimal part of a number, then the LSB represents $2^{-10} = 1/1024 = 0.0009765$, which yields the quantization step. Figure 1.6 shows the conversion of a CT signal into a 4-bit digital signal. The digital signal shown in Fig. 1.6(c) is said to be pulse-code modulated (PCM). The width of a pulse can be in the order of nanoseconds (10^{-9} seconds), and the sampling period can be in the order of microseconds (10^{-6} seconds), or we can take one million samples in one second. The waveform of the digital signal in Fig. 1.6(c) does not resemble in any way the waveform of the physical variable; thus we may call digital signals nonanalog signals.

Signals encountered in practice are mostly CT signals. To process an analog signal digitally, it must first be discretized in time to yield a discrete-time signal and then quantized in amplitude to yield a digital signal. Therefore, in actual digital signal processing, we deal exclusively with digital signals.

An analytical study of digital signals, however, is difficult, because quantization is not a linear process. To simplify the discussion, we use decimal numbers to illustrate this point. Suppose every number is to be rounded to the nearest integer (that is, the quantization step is 1); then we have

$$Q(2.6 + 2.7) = Q(5.3) = 5 \neq Q(2.6) + Q(2.7) = 3 + 3 = 6$$

and

$$Q(2.6 \times 2.7) = Q(7.02) = 7 \neq Q(2.6) \times Q(2.7) = 3 \times 3 = 9$$

where Q stands for quantization. Because of these nonlinear phenomena, analytical study of digital signals is complicated. There are, however, no such problems in studying discrete-time signals. For this reason, in analysis and design, all digital signals will be considered as discrete-time signals. In actual processing or implementation, all discrete-time signals must be converted into digital signals. In quantization, if the amplitude of a discrete-time signal does not fall exactly on a quantization level, then the value must be approximated by a quantization level either by truncation or rounding. In either case, errors will occur. Such errors are called *quantization errors*. In general, quantization errors are studied separately in digital signal processing. Such a study is important in specialized hardware that uses a small number of bits such as 4 or 8 bits. On digital computers and DSP processors that have 16 or 32 bits, quantization errors are very small and can often be simply ignored. For convenience, we use digital signals and discrete-time signals interchangeably with the understanding that *all DT signals must be transformed into digital signals in implementation and all digital signals are considered as DT signals in analysis and design*.

1.3 A/D and D/A Conversions[3]

When an analog signal is to be processed, stored, or transmitted digitally, it must be converted into a digital signal. To recover the original analog signal, the digital signal must be converted back into an analog signal. This section discusses how these conversions are carried out.

We discuss first the digital-to-analog (D/A) converter, or DAC. Figure 1.7 shows the basic structure of a 4-bit D/A converter. It consists of 4 latches, 4 transistor switches, a number of

---

[3] This section may be skipped without loss of continuity.

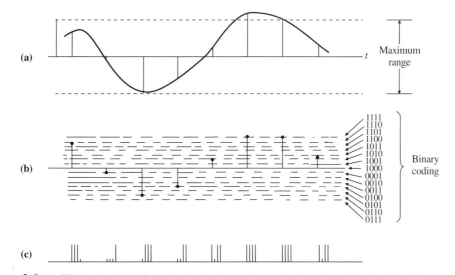

Figure 1.6 (a) CT signal. (b) Its discretization in time and quantization in amplitude. (c) Its binary-coded digital signal.

precision resistors, an operational amplifier, and a precision voltage reference. The 4-bit input denoted by $\{x_1\ x_2\ x_3\ x_4\}$, latched in the register, controls the closing or opening of the switches. The switch is closed if $x_i=1$, open if $x_i=0$. The input remains unchanged until the next string of x_i arrives. Thus, strictly speaking, the input is a stepwise continuous-time signal with amplitude expressed in binary form. The signal v_o in Fig. 1.7 denotes the output voltage of the operational amplifier. Because the current and voltage at the inverting terminal are zero as shown, we have

$$i_f = \frac{v_o}{R}; \quad i_1 = \frac{x_1 E}{2R}; \quad i_2 = \frac{x_2 E}{4R}; \quad i_3 = \frac{x_3 E}{8R}; \quad i_4 = \frac{x_4 E}{16R}$$

and

$$i_f = i_1 + i_2 + i_3 + i_4$$

Thus we have

$$v_o = R\left(\frac{x_1 E}{2R} + \frac{x_2 E}{4R} + \frac{x_3 E}{8R} + \frac{x_4 E}{16R}\right)$$

$$= (2^{-1}x_1 + 2^{-2}x_2 + 2^{-3}x_3 + 2^{-4}x_4)E$$

For example, if $x_1 x_2 x_3 x_4 = 1011$ and $E = 10$, then

$$v_o = (1 \times 2^{-1} + 0 \times 2^{-2} + 1 \times 2^{-3} + 1 \times 2^{-4}) \times 10$$

$$= (0.5 + 0.125 + 0.0625) \times 10 = 6.875$$

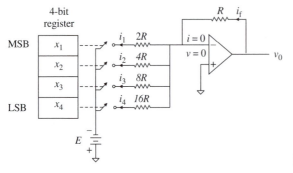

Figure 1.7 4-bit D/A converter.

The output of the operational amplifier will hold this value until the next set of binary numbers arrives. Thus the output of the D/A converter is stepwise, as shown in Fig. 1.8. This discontinuous signal can be smoothed by passing it through an analog low-pass filter as shown.

The D/A converter in Fig. 1.7 illustrates only the basic idea of the conversion. Because the values of the resistors in Fig. 1.7 are widely spread out, these resistors are difficult to fabricate in integrated form. Therefore, different arrangements of resistors are used in actual D/A converters. The interested reader is referred to Ref. 9.

Next we discuss the analog-to-digital (A/D) converter or ADC. An analog signal can be discretized by using a switch or sampler as shown in Fig. 1.9(a). The switch is closed for a short period of time every T seconds to yield the signal $\bar{x}$ shown. This signal is then applied to an A/D converter as shown in Fig. 1.10(a). The A/D converter consists of an operational amplifier that acts as a comparator, a D/A converter, a counter and output registers, and control logic. In the conversion, the counter starts to drive the D/A converter. The output $\hat{x}$ of the converter is compared with $\bar{x}$. The counter stops as soon as $\hat{x}$ exceeds $\bar{x}$, as shown in Fig. 1.10(b). The value of the counter is then transferred to the output registers and is the digital representation of the analog signal. This completes the A/D conversion.

We see from Fig. 1.10(b) that the A/D conversion takes a small amount of time, called the *conversion time*, to achieve the conversion. A 12-bit A/D converter may take 1 $\mu s = 10^{-6}$ s to complete a conversion. Because of this conversion time, if an analog signal changes very rapidly, the value converted may not be the value intended for conversion. This problem can be resolved by using a sample-and-hold circuit, as shown in Fig. 1.9(b). The circuit consists of a transistor switch, a capacitor with a large capacitance, and two voltage followers. The voltage followers are used to shield the capacitor to eliminate the loading problem. The switch is controlled by control logic. When the switch is closed, the input voltage $x(t)$ charges the capacitor rapidly to the input voltage. When the switch is open, the capacitor voltage remains roughly constant. Thus the output of the sampled-and-hold circuit is stepwise as shown. Using this circuit, the problem due to the conversion time can be eliminated. Therefore, a sample-and-hold circuit is often used, either internally or externally, with an A/D converter.

Inputs of A/D converters are electrical signals, usually ranging from 0 to 10 volts or from −5 to +5 volts. To process nonelectrical signals such as sound (acoustic pressure), temperature, displacement, or velocity, these signals must be transformed, using transducers,

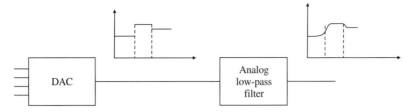

Figure 1.8 Smoothing the output of a D/A converter.

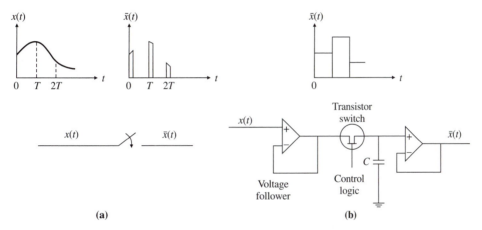

(a) **(b)**

Figure 1.9 (a) Sampler. (b) Sample-and-hold circuit.

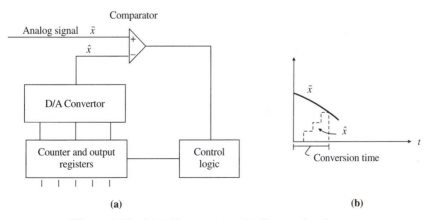

(a) **(b)**

Figure 1.10 (a) A/D converter. (b) Conversion time.

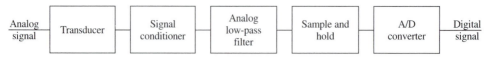

Figure 1.11 Conversion of analog signals into digital signals.

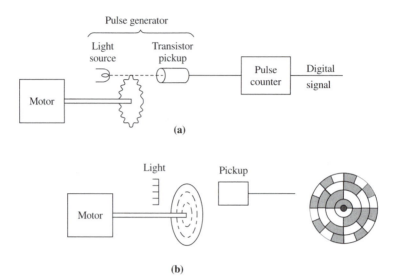

Figure 1.12 (a) Digital speed transducers. (b) Digital position transducer.

into electrical signals. A *transducer* is a device that transforms a signal from one form to another. Microphones, thermocouples, potentiometers, and tachometers are transducers; they transform acoustic pressure, temperature, displacement, and velocity into electrical voltage signals. Therefore, the transformation of a physical analog signal into a digital signal may require the setup shown in Fig. 1.11. The left-most block is a transducer. The output of the transducer may be in millivolt levels and must be amplified to drive the A/D converter. To improve the accuracy of the digital representation of the analog signal, the analog signal may also require scaling so that it will utilize the full range of the converter. Some transducers may exhibit nonlinearities and can be compensated in conversion. These operations (matching signal levels, scaling to improve accuracy, and compensation for nonlinearities) are called *signal conditioning*. After conditioning, the signal is generally applied to an analog low-pass filter. This filter will eliminate high-frequency noise that may be introduced by transducers. It is also introduced to reduce the effect of aliasing (to be discussed in later chapters). Therefore, the analog low-pass filter is also called an *antialiasing filter*. Thus the conversion of a physical analog signal to a digital signal generally requires the five operations shown in Fig. 1.11. We mention that in addition to analog transducers, there are digital transducers that can transform an analog signal directly into a digital signal. Examples of such transducers are shown in Fig. 1.12. They transform, respectively, the angular velocity and angular position of the motor shaft into digital signals. In these cases, the intermediate steps in Fig. 1.11 become unnecessary. At present, data acquisition devices are widely available commercially for achieving the operations in Fig. 1.11. Thus conversions of analog signals to digital signals or conversely can be readily achieved in practice.

1.4 Comparison of Digital and Analog Techniques

Digital techniques have become increasingly popular and have replaced, in many applications, analog techniques. We discuss some of the reasons in the following.

- *Digital techniques are less susceptible to noise and disturbance.* In the transmission and processing of analog signals, any noise or disturbance will affect the signals. Digital signals are coded in 1s and 0s, which are represented by ranges of voltages; therefore, small noise, disturbance, or perturbation in power supply may not affect the representation. Thus digital techniques are less susceptible to noise and disturbance. This reliability can further be improved by using error-detecting and error-correcting codes. On the compact disc, by using the so-called cross-interleaved Reed-Solomon code, it is possible to correct 2400-bit long errors (corresponding to 2-mm scratches). However, the number of bits must be increased considerably. For example, a portion of the audio signal represented originally by 192 bits now requires 588 bits on the disc. Another example is the transmission of the pictures of the Mars, taken by a spacecraft, to the ground station on the earth. After traveling over 200×10^6 kilometers, the received signal has a power level in the order of 10^{-18} watts. If the signal is transmitted by analog techniques, the received signal will be severely corrupted by noise and it is not possible to reconstruct the pictures. However, the pictures of the Mars have been transmitted successfully to Earth by using digital techniques.
- *The precision in digital techniques is much higher than that in analog techniques.* In analog display, the accuracy of the reading is often limited. Generally, we can achieve only an accuracy of 1% of the full scale. In digital display, the accuracy can be increased simply by increasing the number of bits used. In analog systems, it is difficult or very expensive to have a number of components with identical value. For example, if we buy ten 1-kΩ resistors, probably the resistances of the ten resistors will be all different and none exactly equals 1 kΩ. Even simple analog voltmeters require constant resetting in their use. Digital systems have no such problem; they can always be exactly reproduced.
- *The storage of digital signals is easier than that of analog signals.* Digital signals can be easily stored in shift registers, memory chips, floppy disks, or compact discs for as long as needed without loss of accuracy. These data can be retrieved for use in a few microseconds. This easy storage and rapid access is an important feature of digital techniques. The only convenient way to store analog signals is to tape or film them. Their retrieval is not as convenient and fast as in digital techniques.
- *Digital techniques are more flexible and versatile than analog techniques.* In digital display, digital signals can easily be frozen, expanded, or manipulated. A digital system can be easily altered by resetting its parameters. To change an analog system, we must physically replace its components. Using time multiplexing, a digital system can be used to process a number

of digital signals: for example, if the sampling period of a digital signal is 0.05 s and if the processing of the signal requires only 0.005 s. Then the digital system will be free in the next 0.045 s and can be used to process other signals. Another example of this type of arrangement is the digital transmission of human voices. On a telephone line, the voice is sampled 8000 times per second; each sample is coded by using eight bits. Thus the transmission of one channel of voices requires the transmission of 64,000 or 64K bits per second. The existing telephone line can transmit 1544K bits per second. Therefore, a single telephone line can be used to transmit 24 voice-band channels. If we use fiber optics, then the number of channels is even larger.

• *Personal computers are now widely available; so is computer software.* This fact speeds considerably the spreading of digital techniques.

From the preceding discussion, it is not surprising to see the avalanche of digital techniques. In many applications digital techniques have almost completely replaced analog techniques such as in analog oscilloscopes and analog computers.

1.5 Applications of Digital Signal Processing

Digital techniques have been advancing rapidly in recent years and have found many applications in almost every field of technology. This section discusses some of them.

The human voice can be transmitted digitally as shown in Fig. 1.13. The voice is sampled and coded by a series of 8-bit words. After transmission, the digital signal is converted back to analog signal by using a D/A converter. This type of transmission is much less sensitive to cross talk, distortion, and noise than direct transmission. Furthermore, the faded signal can be readily amplified without any distortion. Digital techniques can also be used to increase the capacity of speech transmission. By analyzing the frequency spectrum of speech, it is possible to extract key features of the speech. At the receiving end, we use these features to synthesize a voice that resembles the original one. The voice generated by this process will lose the individual distinction, but the information required to generate it will be less. Thus the capacity of the transmission can be increased. Digital processing is also useful in speech recognition and in vocal communication between men and computers.

Once music or a picture is transformed and stored in digital form, it is possible to carry out many manipulations. For example, the waveform in Fig. 1.14(a) is a section of sound track that contains two clicks and dropout due to faulty recording. It is possible to remove the clicks and reconstruct the missing sound as shown in the lower waveform in Fig. 1.14(b). Once a picture is stored digitally, it can be easily enlarged, enhanced, or sharpened. It is also possible to remove a divorced husband from a family portrait. These techniques are now available commercially.

Vibratory motions of the earth due to earthquakes, volcanic eruptions, or manmade explosions are called *seismic waves*. Figure 1.15 shows a typical setup of its measurement. A study of these waves may reveal the interior structure of the earth, sand-shale ratio, or existence of oil deposits. For example, by comparing the characteristics or the frequency spectra of of the seismic waves of a region with those of a known oil deposit, we may determine the possible existence of oil deposits in the region.

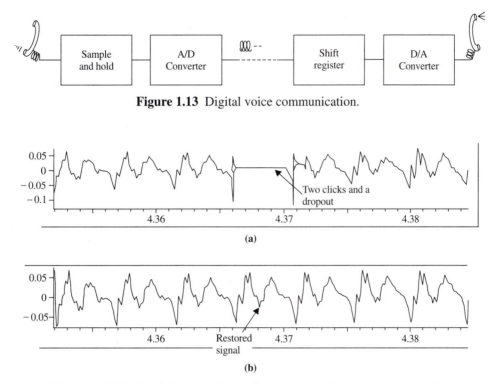

Figure 1.13 Digital voice communication.

Figure 1.14 (a) Sound track with missing sound. (b) Restored sound track.

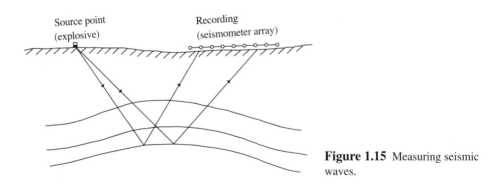

Figure 1.15 Measuring seismic waves.

Vibration study is also important in tall buildings, airplane wings, suspension bridges, mechanical systems, or printed circuit boards. Figure 1.16(a) shows a setup to measure the frequency spectrum of the vibration, called the *frequency response*, after a printed circuit board is hit by an impact of a very short duration (hammer blow). This frequency response such as the one in Fig. 1.16(b) will reveal the natural response of the board. If the board is attached to a device that may be subjected to some external force, then it is important that the frequency response

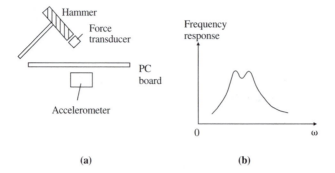

Figure 1.16 (a) Testing a PC board. (b) Measured structural frequency response.

of the board should disjoint from the frequency spectrum of the external force. Otherwise, the external force may induce resonance and the circuit board may crack.

Measuring frequency responses is also useful in production lines. For example, loudspeakers can be inspected visually for defects. An alternative and more reliable method is to use the setup in Fig. 1.17(a). The output of the loudspeaker excited by a voltage signal of a very short duration is picked up by a microphone and passes through an A/D converter. The FFT of the output yields the frequency response of the loudspeaker and can be displayed as shown in Fig. 1.17(b). From the display, whether the speaker passes the test can be determined.

Digital techniques are now used in cameras, automobiles, telephone answering machines, and many other fields. Consider the control of the air/fuel mixture of an engine shown in Fig. 1.18(a). A richer mixture will increase the power; a leaner mixture will reduce air pollution. The sensors monitor the rotational speed and smoothness (acceleration and deceleration) of the flywheel. These signals are processed and then control the mixture. A four-cycle engine may rotate from 600 rpm (revolution per minute) during idling to over 6000 rpm in high speed. If the sensor measures twice per revolution, then it must measure 100 or more times per second. This will be difficult to achieve using analog techniques. The fingerprint verifier shown in Fig. 1.18(b) is again out of the reach of analog techniques.

1.6 Scope of the Book

In this introductory chapter, we showed why CT signals are now widely processed digitally. We also showed how CT signals are sampled into digital signals by using A/D converters. Before using an A/D converter, we must select a sampling period T. How to select a sampling period is an important issue in DSP and will be discussed in this text. Once a CT signal is converted into a digital signal, all operations are carried out in binary form with quantized amplitudes. As discussed earlier, quantization is not a linear operation, and its study is complicated. To simplify analysis and design, all digital signals will be considered as DT signals. Thus we study in this text only CT and DT signals. The remainder of this text is divided into two parts.

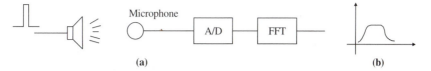

Figure 1.17 (a) Testing a speaker. (b) Its frequency response.

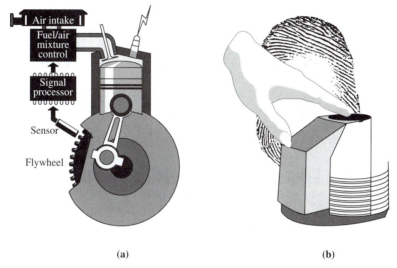

Figure 1.18 (a) Digital control of carburetors. (b) Fingerprint verifier.

1.6.1 Spectral Computation

This part consists of Chapters 2 through 4 and discusses the computation of frequency contents of CT and DT signals using the fast Fourier transform (FFT). As a preview, consider the sentence "The lathe is a big tool" recorded in Fig. 1.19(a). It is a CT signal and will be denoted as $x(t)$.[4] Let us sample the signal with sampling period $T = 1/8000 = 0.000125$ s. Because the sentence lasts about 2s, we need at least $2/0.000125 = 16000$ samples. For computational efficiency, as we will discuss in Chapter 4, we select the number of samples as $N = 2^{14} = 16384$. Then the program that follows

Program 1.1
```
T=1/8000;N=2^(14);D=2*pi/(T*N);
n=0:N-1;
```

---

[4] This signal was provided by Professor Nam Phamdo. It is interpolated from a digital signal. The discussion still holds if $x(t)$ is measured and stored in analog form.

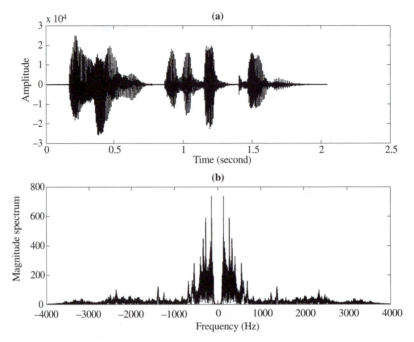

Figure 1.19 (a) Time signal. (b) Frequency content.

```
x=x(n*T);
X=T*fft(x);
m=-N/2:N/2-1;
plot(m*D/(2*pi),abs(fftshift(X)))
```

generates the plot in Fig. 1.19(b). The plot reveals the frequency content, called the magnitude spectrum, of the sentence. From the plot, we see that the frequency content of the sentence is limited roughly to [200, 3800] in Hz.

The plot shown in Fig. 1.19(b) is an even function of frequency; thus we may plot only the positive frequency part. This can be achieved by changing the last two lines of Program 1.1 as

```
Program 1.2
T=1/8000;N=2^(14);D=2*pi/(T*N);
n=0:N-1;
x=x(n*T);
X=T*fft(x);
m=0:N/2;
plot(m*D,abs(X(m+1)))
```

The result is plotted in Fig. 1.20. The plot is the same as the positive frequency part of Fig. 1.19(b) except that the frequency is in units of rad/s instead of Hz.

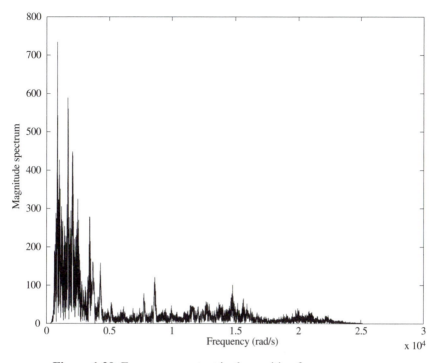

Figure 1.20 Frequency content in the positive frequency range.

These two programs are generic and can be applied to any CT signal. Most programs in this text are simple variations of these programs. Every function in the programs is available in the student edition of MATLAB, and the reader is urged to try them out. They are simple and involve only a number of concepts and results we will develop in this part. The programs are listed here as an enticement for the reader to go through this part.

We start from rudimentary concepts in Chapter 2. After defining the frequency for DT sinusoidal sequences, we show that its frequency is limited to the Nyquist frequency range $(-\pi/T, \pi/T]$, which is dictated entirely by the sampling period T. We then show that most CT and DT periodic signals can be expressed as CT and DT Fourier series. The Fourier series coefficients of CT and DT periodic signals are called *frequency components* and appear only at discrete frequencies. Their only difference is that the frequency range of CT signals is $(-\infty, \infty)$, whereas the frequency range of DT signals is $(-\pi/T, \pi/T]$. A direct consequence of this fact is that if all *significant* frequency components of a CT signal lie inside $[-W, W]$, and if T is selected so that $W < \pi/T$, then the frequency components of a CT periodic signal can be computed from its sampled sequence. We then develop programs similar to Programs 1.1 and 1.2 to compute and to plot CT and DT Fourier series. We also discuss how to select a sampling period T in actual computation.

Chapter 3 extends the Fourier series for periodic signals to the Fourier transforms for periodic as well as aperiodic signals. The CT and DT Fourier transforms appear at every frequency and are called *frequency spectra*. Again the frequency range of CT signals is $(-\infty, \infty)$, and the frequency range of DT signals is $(-\pi/T, \pi/T]$. The major differences between frequency components (Fourier series) and frequency spectra (Fourier transform) are (1) the former are sequences of numbers appeared only at discrete frequencies and the latter are mostly continuous functions appeared at all frequencies; (2) the former reveal distribution of power in frequencies and the later reveal distribution of energy in frequencies; and (3) the former are defined for only periodic signals and the latter for periodic and aperiodic signals. Thus the CT and DT Fourier transforms are more general than the corresponding CT and DT Fourier series. If all *significant* frequency spectrum of a CT signal lies inside $[-W, W]$, and if T is selected so that $W < \pi/T$, then the frequency spectrum of the CT signal can be computed from its sampled sequence. This chapter also discusses the effects of truncation of signals on frequency spectra.

Chapter 4 develops the discrete Fourier transform (DFT) and FFT for computer computation of frequency spectra. Spectral computation of aperiodic signals is more complex than computing Fourier series, and its discussion will be built on the discussion in Chapter 2. Programs 1.1 and 1.2 will be fully explained by the end of this chapter.

The signals selected for illustration in this part are all very simple, and their frequency components or spectra can be obtained analytically. Therefore, it is possible to compare FFT computed results with analytical results and gain confidence in using FFT. What we learn in this part will form a solid foundation for studying real-world signals.

1.6.2 Digital Filter Design

The second part of this text consists of Chapters 5 through 9 and discusses the design of digital filters to process CT or DT signals. In this part, we assume implicitly that all CT signals have been sampled, using an appropriate sampling period T, into DT signals. Thus the input of a digital filter is a sequence of numbers. The digital filter will then generate as its output a sequence of numbers. How fast each number of the output is computed depends mainly on the speed of the hardware used and the complexity of the filter. It has nothing to do with the sampling period. Thus in this part we may assume the sampling period T to be 1 or any other value. For example, all digital filters designed in MATLAB assume $T = 0.5$ if we use units of Hz or $T = \pi$ if we use units of rad/s.

Chapter 5 discusses the class of digital filters to be designed. Unlike the first part, where we discuss first CT signals and then DT signals, here we discuss first DT systems and then CT systems for two reasons. The mathematics involved in developing discrete-time impulse responses, convolutions, and transfer functions for DT systems is relatively simple. Although the same concepts and procedure can be applied to CT systems, the mathematics is much more complex. Furthermore, the concept of impulse responses, which is essential in the DT case, is not needed in the CT case. We need only the concepts of transfer functions and stability for CT systems. Thus the discussion for CT systems is brief.

After introducing the necessary mathematical background in Chapter 5, we discuss in Chapter 6 the design of digital filters. We first discuss general issues and then use poles and zeros to design simple digital filters to process CT signals. Chapters 7 and 8 introduce specific design techniques. Chapter 9 discusses structures of digital filters. Chapter 9 is essentially independent of Chapters 7 and 8 and may be studied right after Chapter 6.

PART 1

SPECTRAL COMPUTATION

CHAPTER

CT and DT Fourier Series—Frequency Components

2.1 Introduction

Signals encountered in practice are mostly CT signals and can be denoted as $x(t)$, where t is a continuum. Although some signals such as stock markets, savings account, and inventory are inherently discrete time, most DT signals are obtained from CT signals by sampling and can be denoted as $x[n] := x(nT)$, where T is the sampling period and n is the time index and can assume only integers. Both $x(t)$ and $x[n]$ are functions of time and are called the *time-domain* description. In signal analysis, we study frequency contents of signals. In order to do so, we must develop a different but equivalent description, called the *frequency-domain* description. From the description, we can more easily detect existence of periodicities of signals and determine the distribution of power in frequencies.

In digital processing of a CT signal $x(t)$, the first step is to select a sampling period T and then to sample $x(t)$ to yield $x(nT)$. It is clear that the smaller T is, the closer $x(nT)$ is to $x(t)$. However, a smaller T also require more computation. Thus an important task in DSP is to find the largest possible T so that all information (if not possible, all essential information) of $x(t)$ is retained in $x(nT)$. Without the frequency-domain description, it is not possible to find such a sampling period. Thus computing the frequency content of signals is a first step in digital signal processing.

The frequency-domain description is developed from complex-valued exponential functions. Therefore, even though all signals encountered in practice are real valued, we still have to deal with complex-valued functions. We study in this chapter first sinusoidal signals and then general periodic signals. Furthermore, they are assumed to be defined for all t and n in $(-\infty, \infty)$. This will simplify the discussion, as we will explain in Section 3.2.1.

2.1.1 Frequency of CT Sinusoids

Consider the CT sinusoids

$$x_1(t) = \sin \omega_0 t; \qquad x_2(t) = \cos \omega_0 t$$

defined for all t. They have frequency ω_0, in radians per second, or frequency $f := \omega_0/2\pi$, in Hz (cycles per second). They have period $P := 2\pi/\omega_0 = 1/f$, in seconds. Using the two functions, we can build a *complex exponential function* as

$$e^{j\omega_0 t} = \cos \omega_0 t + j \sin \omega_0 t \qquad (2.1)$$

where $j := \sqrt{-1}$. It is a complex-valued function with real part $\cos \omega_0 t$ and imaginary part $\sin \omega_0 t$. The function $e^{j\omega_0 t}$ can be plotted as shown in Fig. 2.1(a) with time as the horizontal coordinate. It is a three-dimensional plot and is difficult to visualize. Because

$$|e^{j\omega_0 t}| = \sqrt{(\cos \omega_0 t)^2 + (\sin \omega_0 t)^2} = 1$$

for all t and ω_0, the function can also be plotted as a vector rotating around a unit circle as shown in Fig. 2.1(b) with t as a parameter on the plot. By convention, the vector rotates counterclockwise if $\omega_0 > 0$, and clockwise if $\omega_0 < 0$. We will encounter both positive and negative frequencies in using complex exponentials. A negative-frequency complex exponential rotates merely in the opposite direction of a positive-frequency complex exponential.

2.2 Frequency and Frequency Range of Sinusoidal Sequences

The CT sinusoid $\sin \omega_0 t$ is periodic for every ω_0 in $(-\infty, \infty)$ and will repeat itself every $P = 2\pi/\omega_0$ seconds. Its counterpart in the DT case is much more complex and requires careful development, as we will do in this section.

Consider a discrete-time sequence $x[n] := x(nT)$, where n is an integer ranging from $-\infty$ to ∞, and $T > 0$ is the *sampling period*. In this text, variables inside brackets are limited to integers. The sequence $x[n]$ is said to be periodic if there exists an integer N such that

$$x[n] = x[n + N]$$

for all integer n. In other words, the sequence will repeat itself every N samples or every $P := NT$ seconds, and is said to have period N. If $x[n]$ is periodic with period N, then it is also periodic with period $2N, 3N, \ldots$. The smallest such integer is called the *fundamental period* or, simply, the period, unless stated otherwise.

The CT sinusoid $x(t) = \sin \omega_0 t$ is periodic for every ω_0. However, not every DT sinusoid of the form

$$x[n] := x(nT) = \sin \omega_0 nT$$

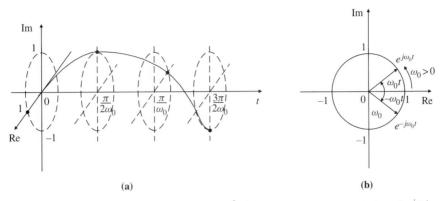

Figure 2.1 (a) Three-dimensional plot of $e^{j\omega_0 t}$. (b) Two-dimensional plot of $e^{j\omega_0 t}$.

is periodic for every ω_0 and every T. Indeed, if $x[n]$ is periodic, then there exists an integer N such that

$$\sin \omega_0 nT = \sin \omega_0 (n+N)T = \sin \omega_0 nT \cos \omega_0 NT + \cos \omega_0 nT \sin \omega_0 NT$$

This holds for every integer n if and only if

$$\cos \omega_0 NT = 1 \quad \text{and} \quad \sin \omega_0 NT = 0$$

which imply

$$\omega_0 NT = k2\pi \quad \text{or} \quad \omega_0 T = \frac{2k}{N}\pi \tag{2.2}$$

for some integer k. Note that $2k/N$, a ratio of two integers, is a rational number. Thus we conclude that $\sin \omega_0 nT$ is periodic if and only if $\omega_0 T$ is a rational-number multiple of π. Similarly, we can show that $\cos \omega_0 nT$ and, consequently, $e^{j\omega_0 nT}$ are periodic if and only if $\omega_0 T$ is a rational-number multiple of π. For example, $\sin 2n$ and e^{j2n} are not periodic because $\omega_0 T = 2$ is not a rational-number multiple of π. If we plot the vectors e^{j2n} for $n = 0, 1, 2, \ldots$, as shown in Fig. 2.2(a), then no vector will ever repeat itself. For example, no integer other than $n = 0$ will yield $e^{j2n} = 1$. On the other hand, $\sin 0.1\pi n$ and $\sin 0.3\pi n$ are periodic because $\omega_0 T = 0.1\pi$ and 0.3π are rational-number multiples of π.

If $\sin \omega_0 nT$ is periodic, what is its period? To answer this, we write (2.2) as

$$N = \frac{2k\pi}{\omega_0 T} \tag{2.3}$$

Then the smallest positive integer N meeting (2.3), for some integer k, is the fundamental period. This can be obtained by searching the smallest positive integer k to make $(2k\pi/\omega_0 T)$ an integer. For example, for $\sin 0.3\pi n$, we have

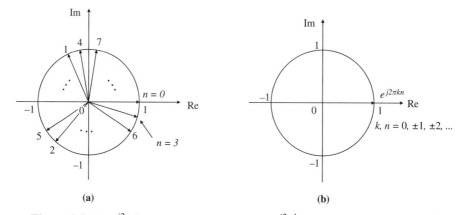

Figure 2.2 (a) e^{j2n} for $n = 0,\ 1,\ 2,\ \ldots$ (b) $e^{j2\pi kn}$ for all integers k and n.

$$N = \frac{2k\pi}{0.3\pi} = \frac{20k}{3}$$

If $k = 1$ or 2, then N is not an integer. If $k = 3$, then $N = 20$. This N is the smallest positive integer meeting (2.3) and, therefore, is the fundamental period of $\sin 0.3\pi n$. For $\sin 0.1\pi n$, we have

$$N = \frac{2k\pi}{0.1\pi} = 20k$$

which equals $N = 20$ for $k = 1$. Thus the fundamental period of $\sin 0.1\pi n$ is also $N = 20$. Figure 2.3 plots the two sequences $\sin 0.1\pi n$ and $\sin 0.3\pi n$ with $T = 1$. Indeed, they have the same fundamental period $N = 20$.

If we follow the CT case to define $f := 1/P = 1/NT$, in Hz, as the frequency of sinusoidal sequences, then $\sin 0.1\pi n$ and $\sin 0.3\pi n$ have the same frequency. This is not consistent with our perception of frequency because $\sin 0.3\pi n$ has a rate of change three times faster than $\sin 0.1\pi n$, as shown in Fig. 2.3. Thus the frequency of $\sin \omega_0 nT$ should not be defined as $1/NT$. In conclusion, $\sin \omega_0 nT$ may not be periodic. Even if it is periodic, its frequency should not be defined from its period as in the CT case.

If we cannot use the fundamental period to define the frequency of DT sinusoids, then how should we define it? The most natural way is to use their envelopes as shown in Fig. 2.3 with dashed lines. More specifically, a CT sinusoid $\sin \bar{\omega} t$ is called an *envelope* of $\sin \omega_0 nT$ if its sample with sampling period T equals $\sin \omega_0 nT$; that is,

$$\sin \omega_0 nT = \sin \bar{\omega} t|_{t=nT}$$

for all n. Note that if $\sin \bar{\omega} t$ is an envelope of $\sin \omega_0 nT$, so is $\sin[\bar{\omega} + k(2\pi/T)]t$ for any integer k. Indeed, because of

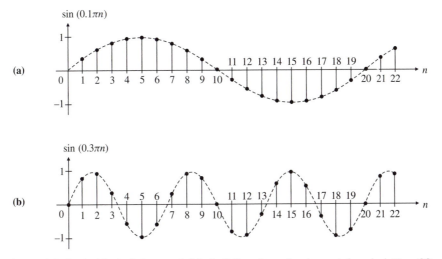

Figure 2.3 Both (a) $\sin 0.1\pi n$ and (b) $\sin 0.3\pi n$ have fundamental period $N = 20$.

$$\sin(2\pi nk) = 0 \quad \text{and} \quad \cos(2\pi nk) = 1 \qquad (2.4)$$

for any integers n and k (positive, negative, or zero), we have

$$\sin\left(\bar{\omega} + \frac{2\pi k}{T}\right)t\Big|_{t=nT} = \sin(\bar{\omega}nT + 2\pi nk)$$
$$= \sin \bar{\omega}nT \cos(2\pi nk) + \cos \bar{\omega}nT \sin(2\pi nk) = \sin \bar{\omega}nT \qquad (2.5)$$

Thus every DT sinusoid $\sin \omega_0 nT$ has infinitely many CT sinusoidal envelopes. In order to define the frequency of $\sin \omega_0 nT$ through its envelopes, we must use an envelope that is unique in some sense. The envelope $\sin \bar{\omega}t$ with the smallest $|\bar{\omega}|$ will be called the *primary envelope*. The frequency of the primary envelope will then be defined as the frequency of $\sin \omega_0 nT$. This is stated as a definition.

Definition 2.1 The frequency of DT $e^{j\omega_0 nT}$ ($\sin \omega_0 nT$ or $\cos \omega_0 nT$), periodic or not, is defined as the frequeny of CT $e^{j\bar{\omega}t}$ ($\sin \bar{\omega}t$ or $\cos \bar{\omega}t$) with the smallest $|\bar{\omega}|$ such that the sample of $e^{j\bar{\omega}t}$, with sampling period T, equals $e^{j\omega_0 nT}$.

From (2.5), we see that the $\bar{\omega}$ that has the smallest magnitude must lie in the range $-\pi/T \le \bar{\omega} \le \pi/T$. If $\bar{\omega} = \pi/T$ and $\bar{\omega} = -\pi/T$, then the two $\bar{\omega}$ have the same magnitude. In this case, we adopt the former. Thus the frequency range of DT sinusoids becomes

$$-\frac{\pi}{T} < \omega \le \frac{\pi}{T} \quad \text{(in rad/s)} \qquad (2.6)$$

denoted as $(-\pi/T, \pi/T]$ (note the left parenthesis and right bracket). This will be called the *Nyquist frequency range*. If we define $f := \omega/2\pi$ and $f_s := 1/T$, where f_s is called the *sampling frequency* or *sampling rate*, then the Nyquist frequency range is

$$-0.5f_s < f \leq 0.5f_s \quad \text{(in Hz)} \tag{2.7}$$

If $T = 1$, then the range becomes $(-\pi, \pi]$ (in rad/s) or $(-0.5, 0.5]$ (in Hz). We see that the frequency range of DT sinusoidal sequences is determined entirely by the sampling frequency or sampling period.

Let us consider again $\sin(0.1\pi n)$ with sampling period $T = 1$. It has envelopes $\sin(0.1\pi t)$, $\sin(2.1\pi t)$, $\sin(-1.9\pi t)$, ... Clearly, 0.1π has the smallest magnitude; thus the frequency of $\sin(0.1\pi n)$ is 0.1π (in rad/s). The sinusoidal sequence $\sin(0.3\pi n)$ has envelopes $\sin 0.3\pi t$, $\sin(2.3\pi t)$, $\sin(-1.7\pi t)$, ... Clearly, 0.3π has the smallest magnitude. Thus $\sin(0.3\pi n)$ with $T = 1$ has frequency 0.3π rad/s. In conclusion, Definition 2.1 appears to be an acceptable definition.

Before discussing how to compute the frequency of DT sinusoidal sequences, we discuss a property of $e^{j\omega_0 nT}$ that includes $\sin \omega_0 nT$ and $\cos \omega_0 nT$ as special cases. For CT complex exponentials, we have

$$e^{j\omega_1 t} \neq e^{j\omega_2 t} \quad \text{if } \omega_1 \neq \omega_2$$

However, two DT complex exponentials $e^{j\omega_1 nT}$ and $e^{j\omega_2 nT}$ may denote the same sequence even if $\omega_1 \neq \omega_2$. Because

$$e^{j2\pi kn} = 1$$

for all integers k and n, as shown in Fig. 2.2(b), we have

$$e^{j(\omega_0 + \frac{2\pi k}{T})nT} = e^{j\omega_0 nT} e^{j2\pi kn} = e^{j\omega_0 nT}$$

Thus, for $T = 1$, the following complex exponential sequences

$$e^{j3.2\pi n} \qquad e^{j1.2\pi n} \qquad e^{-j0.8\pi n} \qquad e^{-j2.8\pi n}$$

all denote the same sequence. So do the following sequences

$$e^{j8n} \qquad e^{j1.72n} \qquad e^{-j4.56n} \qquad e^{-j10.84}$$

where we have used $2\pi = 2 \times 3.14 = 6.28$. In general, we have

$$e^{j\omega_1 nT} = e^{j\omega_2 nT} \qquad \text{if } \omega_1 = \omega_2 \ (\text{mod } 2\pi/T) \tag{2.8}$$

where mod stands for modulo and $\omega_1 = \omega_2 \ (\text{mod } 2\pi/T)$ means that ω_1 and ω_2 may differ by $2\pi/T$ or its integer multiples; that is,

$$\omega_1 - \omega_2 = k(2\pi/T)$$

for some integer k (positive, negative, or zero). Thus, unlike analog $e^{j\omega_0 t}$, which has a unique representation on the ω axis from $-\infty$ to ∞ as shown in Fig. 2.4(a), the ω_0 in $e^{j\omega_0 nT}$ can be represented by every dot shown in Fig. 2.4(b). Thus ω_0 in $e^{j\omega_0 nT}$ is often said to be periodic with period $2\pi/T$. In order to have a unique representation, we must restrict the frequency to a range of $2\pi/T$. This is consistent with our selection of (2.6) or (2.7) as the frequency range. Thus the frequency of $e^{j\omega_0 nT}$ equals $\bar{\omega}$, where $\bar{\omega}$ equals ω_0 modulo $2\pi/T$ and lies inside the Nyquist frequency range.

◆ **Example 2.1**

Consider $\sin 4.4\pi n$. If the sampling period is $T = 0.1$, then its Nyquist frequency range is $(-\pi/0.1, \ \pi/0.1] = (-10\pi, \ 10\pi]$. First we write $\sin 4.4\pi n = \sin 44\pi nT$. Because 44π is outside the range, it must be reduced to the range by subtracting $2\pi/T = 20\pi$ or its multiple. Clearly, we have

$$44\pi = 24\pi = 4\pi \qquad (\text{mod } 20\pi)$$

Thus the frequency of $\sin 4.4\pi n$ with sampling period $T = 0.1$ is 4π rad/s or 2 Hz.

◆ **Example 2.2**

Consider $\sin 5n = \sin 50nT$ with sampling period $T = 0.1$. Its frequency must lie inside $(-10\pi, \ 10\pi] = (-31.4, \ 31.4]$. Because

$$50 = (50 - 62.8) = -12.8 \qquad (\text{mod } 2\pi/0.1 = 62.8)$$

the frequency of $\sin 5n$ with $T = 0.1$ is -12.8 rad/s. Note that $\sin 5n$ is not periodic, but its frequency is still defined and is a negative frequency.

We recapitulate what has been discussed so far. The DT sinusoid $\sin \omega_0 nT$ may or may not be periodic. Even if it is periodic, its frequency should not be defined from its fundamental period as in CT sinusoids. The frequency of $\sin \omega_0 nT$ is defined as the frequency of CT sinusoid $\sin \bar{\omega} t$, where $\bar{\omega}$ equals ω_0 modulo $2\pi/T$ and lies inside $(-\pi/T, \ \pi/T]$. This implies that if $|\omega_0| < \pi/T$ or $T < \pi/|\omega_0|$, then

$$\text{freq. of DT } \sin \omega_0 nT = \text{freq. of CT } \sin \omega_0 t = \omega_0$$

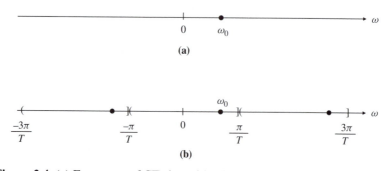

Figure 2.4 (a) Frequency of CT sinusoids. (b) Frequencies of DT sinusoids.

Thus the same notation ω will be used to denote the frequencies of CT and DT sinusoids. The frequency range of CT sinusoids is $(-\infty, \infty)$. The frequency range of DT sinusoids is $(-\pi/T, \pi/T]$. If T is small, the range can be very large.

To conclude this section, we mention that the Nyquist frequency range can be selected as $(-\pi/T, \pi/T]$ or $[-\pi/T, \pi/T)$ but not $[-\pi/T, \pi/T]$. The last one has a redundancy.

2.3 Frequencies of CT Sinusoids and Their Sampled Sequences

This section studies the relationship between the frequency of a CT sinusoid and the frequency of its sampled sequence. Consider a CT complex exponential $x(t) = e^{j\omega_0 t}$ with frequency ω_0 in rad/s or $f_0 = \omega_0/2\pi$ in Hz. Its sampled sequence, with sampling period T, is given by

$$x(nT) = e^{j\omega_0 t}|_{t=nT} = e^{j\omega_0 nT} = e^{j2\pi f_0 nT}$$

The frequency of $x(nT)$, as discussed in the preceding section, must lie inside the Nyquist frequency range

$$(-0.5f_s, \ 0.5f_s] \ \text{(in Hz)}$$

where $f_s := 1/T$. If f_0 lies inside the range, then the frequency of $e^{jf_0 2\pi nT}$ is f_0. If f_0 is outside the range, then f_0 must be reduced, by subtracting or adding f_s or its multiples, to lie inside the range as shown in Table 2.1. Thus the frequency of $e^{j\omega_0 t}$ and the frequency of its sampled sequence $e^{j\omega_0 nT}$ are related as shown in Fig. 2.5.

The relationship in Fig. 2.5 can also be established using a physical argument. Suppose we use a camcorder to shoot a wheel with a mark A on its rim. To simplify the discussion, we assume that the camcorder shoots one frame per second or, equivalently, the sampling period is $T = 1$ and the sampling frequency is $f_s = 1/T = 1$. Then n in $e^{jf_0 2\pi nT}$ denotes the nth frame on the tape. If $f_0 = 0 \pmod{f_s = 1}$ or $f_0 = 0, \pm 1, \pm 2, \ldots,$—no matter how many *complete* turns the wheel rotates in either direction—the mark A appears stationary as shown

Table 2.1 Frequency Aliasing

Freq. of $e^{j2\pi f_0 t} = f_0 =$	Freq. of $e^{j2\pi f_0 nT} =$
$0.6 f_s$	$-0.4 f_s$
f_s	0
$1.2 f_s$	$0.2 f_s$
$1.5 f_s$	$0.5 f_s$
$-0.6 f_s$	$0.4 f_s$
$-1.2 f_s$	$-0.2 f_s$
$-2.5 f_s$	$+0.5 f_s$
$-0.5 f_s < f_0 \leq 0.5 f_s$	f_0

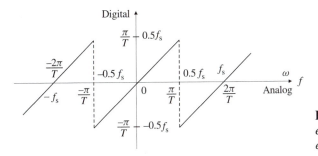

Figure 2.5 Frequencies of $e^{j\omega_0 t} = e^{2\pi f_0 t}$ and its sampled sequence $e^{j\omega_0 nT} = e^{j2\pi f_0 n/f_s}$.

in Fig. 2.6(a). Thus the frequency is 0 to our perception even though the wheel rotates many complete turns between frames. If the wheel rotates in the counterclockwise direction 3/4, 7/4, 11/4, ..., turns,[1] the mark A will appear as shown in Fig. 2.6(b). Thus the wheel rotates, to our perception, clockwise 1/4 turns even though it is actually rotating in the counterclockwise direction. Thus the frequency is $-1/4$ Hz. Proceeding forward, the relationship in Fig. 2.5 can be established.

Figure 2.6(c) shows $e^{j\omega_0 n} = e^{j2\pi f_0 n}$ with $\omega_0 = 2$ or $f_0 = 1/\pi = 0.318$ for $n = 0, 1, 2, \ldots$. The sequence never repeats itself and is not periodic; however, the mark A still rotates with frequency 0.318 Hz. Thus it is well justified to define frequency for nonperiodic complex exponential sequences.

Given a CT $e^{j2\pi f_0 t}$, the frequency of its sampled sequence is given by

$$\text{Freq. of } e^{j2\pi f_0 nT} = \begin{cases} f_0 & \text{if } |f_0| < 0.5 f_s = 0.5/T \\ |f_0| & \text{if } f_0 = \pm 0.5 f_s \\ \bar{f}_0 & \text{if } |f_0| > 0.5 f_s \end{cases} \tag{2.9}$$

---

[1] Rotation in the counterclockwise direction will be considered as positive rotation and in the clockwise direction, negative rotation.

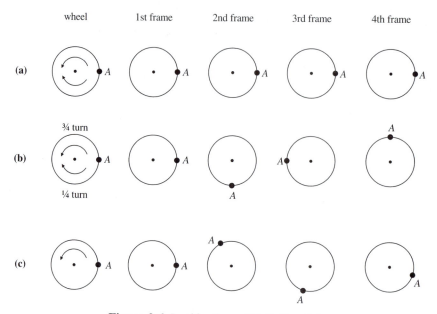

Figure 2.6 Justification of Definition 2.1.

where $\bar{f}_0$ lies inside $(-0.5f_s, \ 0.5f_s]$ and equals f_0 modulo f_s. We see that if

$$f_s > 2|f_0|$$

then the frequency of $e^{j2\pi f_0 t}$ equals the frequency of its sampled sequence. If $f_s \leq 2|f_0|$, then the frequency of its sampled sequence is different from the frequency of $e^{j2\pi f_0 t}$ or f_0. Such a frequency is called an *aliased frequency*. Thus time sampling may cause frequency aliasing if the sampling frequency is not sufficiently large or, equivalently, its sampling period $T = 1/f_s$ is not sufficiently small. This is a basic property and will be discussed further later.

2.3.1 Applications

Consider $e^{j2\pi f_0 t}$ with frequency f_0 in Hz. If the sampling frequency is chosen as $f_s = f_0$, then its sampled sequence $e^{jf_0 2\pi t}$ has frequency 0, or appears stationary, as shown in Fig. 2.6(a). This property can be used in a number of practical applications.

Consider a wheel that has an etched line, as shown in Fig. 2.7. When the wheel is spinning at a high speed, it is not possible to see the line. Let us aim a timing gun at the line. A timing gun is a device that periodically emits pulses of light and acts as a sampler. If the frequency of emitting light is the same as the frequency of the wheel, the etched line will appear stationary and become visible. This technique was used to set mechanically the ignition time in automobiles before the advent of electronic ignition.

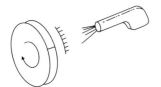

Figure 2.7 Timing gun.

A stroboscope is a device that periodically emits flashes of light and acts as a sampling system. The frequency of emitting flashes can be adjusted. If the frequency of flashing (sampling frequency) is the same as the rotational speed of an object, the object will appear stationary. If the sampling frequency is slightly smaller than the rotational frequency, the object will appear to rotate slowly in the opposite direction of the actual rotation of the object. If the sampling frequency is slightly larger, the object will appear to rotate slowly in the same direction. This phenomenon is often exhibited in science museums and may appear in the rotation of wagon wheels in old movies. The stroboscope can be used to observe vibration or rotation of mechanical objects and to measure rotational speed.

2.3.2 Recovering a Sinusoid from Its Sampled Sequence

Given a complex exponential $e^{j\omega_0 t} = e^{j2\pi f_0 t}$, we showed in the preceding section that its frequency f_0 can be determined from its sampled sequence if $f_s > 2|f_0|$. In this section, we shall establish a more general statement for the sinusoid

$$x(t) = A\sin(\omega_0 t + \theta) \qquad (2.10)$$

with fundamental period $P := 2\pi/\omega_0$. We shall show that if $f_s = 2|f_0| = |\omega_0|/\pi$ or $T = P/2$ and if its sampled sequence $x(nT)$ is not identically zero, then only ω_0 can be determined uniquely from the sampled sequence but not A and θ. However, if $f_s > 2|f_0| = |\omega_0|/\pi$ or $T < P/2$, then $x(t)$ can be recovered from its sampled sequence $x(nT)$ in the sense that A, ω_0, and θ can uniquely be determined from $x(nT)$.[2] Before proceeding, we plot in Fig. 2.8(a) the sampling of a CT sinusoid with a sampling period larger than $P/2$. In this case, there exists a sinusoid with a smaller frequency as shown with dotted line that has the same sampled values. See Problem 2.3. Thus if $T > P/2$, it is not possible to recover from the sampled sequence the original sinusoid.

The sampled sequence of (2.10) with sampling period T is

$$x(nT) = A\sin(\omega_0 nT + \theta) \qquad (2.11)$$

which yields, for $n = 0, 1, -1,$

---

[2] The remainder of this section may be skipped without loss of continuity. In this section, we have assumed that the model in (2.10) is known, and the problem reduces to parameter identification of the model. If the model is not known, the problem is more complex.

$$x(0) = A \sin \theta \qquad\qquad (2.12)$$

$$x(T) = A \sin(\omega_0 T + \theta) = A(\sin \omega_0 T \cos \theta + \cos \omega_0 T \sin \theta) \qquad (2.13)$$

$$x(-T) = A \sin(-\omega_0 T + \theta)$$

$$= A(-\sin \omega_0 T \cos \theta + \cos \omega_0 T \sin \theta) \qquad\qquad (2.14)$$

If $T = P/2$, that is, the sampling period equals half of the fundamental period of (2.10), then we have

$$x(nT) = \begin{cases} a & n \text{ even} \\ -a & n \text{ odd} \end{cases}$$

for some constant a, as shown in Fig. 2.8(b). Clearly, if $a = 0$, or the sampled sequence is identically zero, there is no way to recover $x(t)$ from the sampled sequence. Now we assume $a \neq 0$ and consider $x(0) = a$ and $x(T) = x(-T) = -a$. Adding (2.13) and (2.14) and then substituting (2.12), we obtain

$$-2a = 2A \cos \omega_0 T \sin \theta = 2a \cos \omega_0 T$$

or

$$\cos \omega_0 T = -1$$

Thus we have $\omega_0 = \pi/T$ or $-\pi/T$. Because ω_0 is required to lie inside $(-\pi/T, \pi/T]$, we select $\omega_0 = \pi/T$. Substituting $\omega_0 T = \pi$ into $x(nT)$ yields the same equation, $A \sin \theta = a$, for all n. There are many solutions in $A \sin \theta = a$ such as $A = a, \theta = 0$ or $A = 2a, \theta = 30°$. Thus if $f_s = 2|f_0|$ or $T = P/2$, we cannot recover $x(t)$ uniquely from $x(nT)$.

If $f_s > 2|f_0|$ or $T < P/2$, generally we have $x(0) \neq x(T) \neq x(-T)$ as shown in Fig. 2.8(c). Adding (2.13) and (2.14) and then substituting (2.12), we obtain

$$\cos \omega_0 T = \frac{x(T) + x(-T)}{2x(0)} =: b \qquad\qquad (2.15)$$

From this equation, we can obtain a positive and a negative ω_0 in $(-\pi/T, \pi/T]$ as

$$\omega_0 = \frac{\cos^{-1} b}{T} \qquad\qquad (2.16)$$

However, both will lead to the same result as we will demonstrate in the example that follows shortly. Once an ω_0 is computed, we can write (2.13) as

$$A \cos \theta = \frac{x(T) - x(0) \cos \omega_0 T}{\sin \omega_0 T}$$

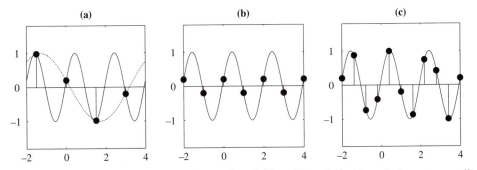

Figure 2.8 (a) Sampling with $T > P/2$. (b) $T = P/2$. (c) $T < P/2$. All vertical axes are amplitude, and all horizontal axes are time in seconds.

Its ratio with (2.12) yields

$$\frac{A\sin\theta}{A\cos\theta} = \tan\theta = \frac{x(0)\sin\omega_0 T}{x(T) - x(0)\cos\omega_0 T} =: c \qquad (2.17)$$

From this equation we can compute a unique θ in $(-\pi/2, \ \pi/2]$ and then compute A from (2.12).

◆ **Example 2.3**

Consider (2.10) with unknown A, ω, and θ. Find the three unknowns from the following measured data

$$x(-T) = -1.5136, \quad x(0) = 1.6829, \quad x(T) = -1.8186$$

with $T = 0.1$.

   We use (2.15) to compute $b = -0.990$. Then (2.16) yields $\omega_0 = \pm 30$. First we select $\omega_0 = 30$ and compute c in (2.17) as $c = -1.5574$. Thus we have $\theta = \tan^{-1} c = -1$. We then use (2.12) to compute A as $A = x(0)/\sin\theta = -2$. Thus the CT sinusoid that generates the sequence is

$$x(t) = -2\sin(30t - 1)$$

   If we select $\omega_0 = -30$ in (2.16), then we will obtain $\theta = 1$ and $A = 2$. Thus the CT sinusoid is

$$x(t) = 2\sin(-30t + 1)$$

which is the same as the one computed using a positive ω_0.

In conclusion, for a sinusoid with period P or frequency $f_0 = 1/P$, we can recover the sinusoid from its sampled sequence with sampling period T if $T < P/2$ or $f_s = 1/T > 2f_0$. This is in fact a special case of the *Nyquist sampling theorem*, which states that if the highest-frequency component of a CT signal $x(t)$ is f_{max}, then $x(t)$ can be recovered from its sampled sequence $x(nT)$ if the sampling frequency f_s is larger than $2f_{max}$. This will formally be established in the next chapter.

2.4 Continuous-Time Fourier Series (CTFS)

This section will extend the discussion for CT sinusoids to CT periodic signals. We will show that most periodic CT signal can be expressed as a linear combination of complex exponential functions. Although it is also possible to express it as a linear combination of sine and cosine functions, the use of complex exponential functions can be more easily extended to aperiodic signals. However, in using complex exponential functions, we must use both positive and negative frequencies and complex coefficients such as

$$
\begin{aligned}
\sin(3t + \pi/4) &= \frac{e^{j(3t+\pi/4)} - e^{-j(3t+\pi/4)}}{2j} \\
&= \frac{e^{j\pi/4}}{2j}e^{j3t} - \frac{e^{-j\pi/4}}{2j}e^{-j3t} \\
&= (0.3536 - j0.3536)e^{j3t} + (0.3536 + j0.3536)e^{-j3t}
\end{aligned}
$$

A CT signal, real or complex valued, is said to be periodic with period P if

$$
x(t) = x(t + P) \tag{2.18}
$$

for all t in $(-\infty, \infty)$, where P is a real positive constant. If (2.18) holds, then

$$
x(t) = x(t + P) = x(t + 2P) = \cdots = x(t + kP)
$$

for all t and for every positive integer k. Therefore if $x(t)$ is periodic with period P, it is also periodic with period $2P$, $3P$, ... The smallest such P is called the *fundamental* period or, simply, the period unless stated otherwise. The *fundamental frequency* is then defined as

$$
\omega_0 := \frac{2\pi}{P} \quad \text{(in rad/s)}; \quad f_0 := \frac{1}{P} \quad \text{(in Hz)}
$$

Let us consider the set of complex exponentials

$$
\phi_m(t) := e^{jm\omega_0 t} = \cos m\omega_0 t + j \sin m\omega_0 t, \quad m = 0, \pm 1, \pm 2, \ldots \tag{2.19}
$$

where m is an integer, ranging from $-\infty$ to ∞. Because the frequency of $e^{jm\omega_0 t}$ is $m\omega_0$, the integer m is called the *frequency index*. There are infinitely many $\phi_m(t)$.

We give some examples of $\phi_m(t)$. For $m = 0$, the function equals 1 for all t and is called a dc signal. For $m = 1$, the function is $\cos \omega_0 t + j \sin \omega_0 t$ and has fundamental period $P := 2\pi/\omega_0$. For $m = 2$, the function is $\cos 2\omega_0 t + j \sin 2\omega_0 t$ and has fundamental period $2\pi/2\omega_0 = P/2$. Thus the function $e^{j2\omega_0 t}$ will repeat itself twice in every time interval P. In general, $e^{jm\omega_0 t}$ is periodic with fundamental period P/m and will repeat itself m times in every time interval P. Thus all $\phi_m(t)$ are periodic with (not necessarily fundamental) period P and are said to be *harmonically related*. There are infinitely many harmonically related complex exponentials.

Because every time interval P contains *complete* cycles of $e^{jm\omega_0 t}$, we have

$$\int_{<P>} e^{jm\omega_0 t} dt := \int_{t_0}^{t_0+P} e^{jm\omega_0 t} dt = \int_{t_0}^{t_0+P} (\cos m\omega_0 t + j \sin m\omega_0 t) dt$$

$$= \begin{cases} P & \text{if } m = 0 \\ 0 & \text{if } m \neq 0 \end{cases} \tag{2.20}$$

for any t_0. The notation $<P>$ is used to denote *any* time interval P. If $m = 0$, the integrand becomes 1 and its integration over P equals P. If $m \neq 0$, the positive part and negative part of $\cos m\omega_0 t$ and $\sin m\omega_0 t$ cancel out completely over P; thus the integration is 0. This establishes (2.20). Of course, (2.20) can also be established by direct integration.

The set of complex exponentials $\phi_m(t)$, for all integers m, has a very important property. Let us use an asterisk to denote complex conjugation. For example, we have

$$\phi_m^*(t) = (e^{jm\omega_0 t})^* = (\cos m\omega_0 t + j \sin m\omega_0 t)^*$$

$$= \cos m\omega_0 t - j \sin m\omega_0 t = e^{-jm\omega_0 t}$$

Then we have

$$\int_{t=<P>} \phi_m(t) \phi_k^*(t) dt = \int_{t_0}^{t_0+P} e^{jm\omega_0 t} e^{-jk\omega_0 t} dt = \int_{<P>} e^{j(m-k)\omega_0 t} dt$$

$$= \begin{cases} P & \text{if } m = k \\ 0 & \text{if } m \neq k \end{cases} \tag{2.21}$$

for any t_0. This follows directly from (2.20) and is called the *orthogonality* property of the set $\phi_m(t)$.

Now we are ready to show that most CT periodic signal $x(t)$ with period P can be expressed as a *linear combination* of $\phi_m(t)$ as

$$x(t) = \sum_{m=-\infty}^{\infty} c_m e^{jm\omega_0 t} \quad \text{(synthesis eq.)} \tag{2.22}$$

with $\omega_0 = 2\pi/P$ and

$$c_m = \frac{1}{P} \int_{<P>} x(t) e^{-jm\omega_0 t} dt$$

$$= \frac{1}{P} \int_{-P/2}^{P/2} x(t) e^{-jm\omega_0 t} dt \quad \text{(analysis eq.)} \tag{2.23}$$

for $m = 0, \pm1, \pm2, \ldots$, and m is called the *frequency index* because it is associated with frequency $m\omega_0$. The set of two equations in (2.22) and (2.23) is called the *continuous-time complex exponential Fourier series*, *CT Fourier series*, or *CTFS*. In this Fourier series, both positive and negative frequencies must be used. If we use only positive frequencies, then most real-valued periodic functions cannot be expressed in the series.[3] As we will discuss shortly, the coefficients c_m reveal the frequency content of the signal. Thus (2.23) is called the *analysis equation*. The function $x(t)$ can be constructed from c_m by using (2.22). Thus (2.22) is called the synthesis equation.

To establish (2.23), we multiply (2.22) by $e^{-jk\omega_0 t}$ and then integrate it over a time interval P to yield

$$\int_{t=<P>} x(t) e^{-jk\omega_0 t} dt = \int_{t=<P>} \left(\sum_{m=-\infty}^{\infty} c_m e^{jm\omega_0 t} e^{-jk\omega_0 t} \right) dt$$

$$= \sum_{m=-\infty}^{\infty} c_m \left(\int_{t=<P>} e^{j(m-k)\omega_0 t} dt \right) \tag{2.24}$$

where we have changed the order of integration and summation.[4] Note that in the equation, the integer k is fixed and m ranges over all integers. The orthogonality property in (2.21) implies that the integral in the last large parentheses in (2.24) equals 0 if $m \neq k$ and P if $m = k$. Thus (2.24) reduces to

$$\int_{<P>} x(t) e^{-jk\omega_0 t} dt = c_k P$$

This becomes (2.23) after renaming the index k to m.

Not every CT periodic signal can be expressed as a Fourier series. Sufficient conditions for $x(t)$ to be so represented are

1. $x(t)$ is absolutely integrable over one period; that is,

$$\int_0^P |x(t)| dt \leq M < \infty$$

and

---

[3] If we use the set of sine and cosine functions as a basis, then we need only positive frequencies.
[4] The condition of uniform convergence is needed to permit such a change of order. We will not be concerned with this mathematical condition and will proceed intuitively.

2. $x(t)$ has a finite number of discontinuities and a finite number of maxima and minima in one period.

These are called the *Dirichlet conditions*. Except for some mathematically contrived functions such as the ones in Problem 2.22, most CT periodic signals meet these conditions; thus these conditions will not be discussed further. If a periodic signal meets the Dirichlet conditions, the infinite summation in (2.22) will converge to a value at every t. The value equals $x(t)$ if $x(t)$ is continuous at t. If $x(t)$ is not continuous at t, that is, $x(t+) \neq x(t-)$, then we have

$$\frac{x(t+) + x(t-)}{2} = \sum_{m=-\infty}^{\infty} c_m e^{jm\omega_0 t} \tag{2.25}$$

Thus the infinite summation converges to the midpoint at every discontinuity.

In the literature, the set of c_m is often called the *discrete or line frequency spectrum*. We will not use this terminology, as we will explain in the next chapter. Instead, we call c_m the *CTFS coefficients* or the *frequency components*.

Before giving examples, we discuss some properties of c_m. CTFS coefficients are in general complex-valued functions of m or $m\omega_0$. Let $c_m = \alpha_m + j\beta_m$, where α_m is the real part of c_m and β_m is the imaginary part. Then the magnitude and phase of c_m are

$$|c_m| = \sqrt{\alpha_m^2 + \beta_m^2} \quad \text{and} \quad \measuredangle c_m = \tan^{-1} \frac{\beta_m}{\alpha_m}$$

Let us express c_m as

$$c_m = \frac{1}{P} \int_{-P/2}^{P/2} x(t)(\cos m\omega_0 t - j \sin m\omega_0 t) dt$$

If $x(t)$ is a real-valued signal, then we have

$$\alpha_m = \operatorname{Re} c_m = \frac{1}{P} \int_{-P/2}^{P/2} x(t) \cos m\omega_0 t \, dt = \alpha_{-m} \tag{2.26}$$

and

$$\beta_m = \operatorname{Im} c_m = -\frac{1}{P} \int_{-P/2}^{P/2} x(t) \sin m\omega_0 t \, dt = -\beta_{-m} \tag{2.27}$$

where we have used $\cos(-\theta) = \cos\theta$ and $\sin(-\theta) = -\sin\theta$. Thus the real part of c_m is even and the imaginary part of c_m is odd. Such c_m is said to be *conjugate symmetric*. Thus if $x(t)$ is real, then its CTFS coefficients are located at discrete frequencies $m(2\pi/P)$ with $m = 0, \pm 1, \pm 2, \ldots$, and are conjugate symmetric ($c_m = c_{-m}^*$).

In applications, we often compute the magnitude and phase of c_m. If $x(t)$ is real, then we have

$$|c_{-m}| = \sqrt{\alpha^2_{-m} + \beta^2_{-m}} = \sqrt{\alpha^2_m + (-\beta_m)^2} = |c_m|$$

and

$$\angle c_{-m} = \tan^{-1}(\beta_{-m}/\alpha_{-m}) = -\tan^{-1}(\beta_m/\alpha_m) = -\angle c_m$$

Thus we conclude that if $x(t)$ is a real-valued signal, then the magnitude and real part of c_m are even, and the phase and imaginary part of c_m are odd.

A remark is in order regarding the phase or angle of c_m. Throughout this text, two angles are considered to be the same if they differ by 2π radians or 360 degrees. For example, if $c_0 = -1$, then we have $\angle c_0 = 180°$ or $\angle c_0 = -180°$. With this convention, we now discuss the condition $\angle c_m = -\angle c_{-m}$. In general, if a signal $x(t)$ or a sequence $x[n]$ is odd in the sense $x(t) = -x(-t)$ or $x[n] = -x[-n]$, then we must have $x(0) = 0$ or $x[0] = 0$. In other words, oddness generally requires the function to be zero at the origin. However, the condition $\angle c_0 = -\angle c_0$ does not require $\angle c_0 = 0$. For example, if c_0 is real and negative, then we have $\angle c_0 = 180°$ or $\angle c_0 = -180°$, and $\angle c_0 = -\angle c_0$ (mod 360°). Thus in discussing the oddness of the plot of $\angle c_m$, we must exclude the point at $m = 0$ as the next example illustrates.

◆ **Example 2.4**

Consider the signal

$$x(t) = -1.2 + 0.8 \sin 0.6t - 1.6 \cos 1.5t \qquad (2.28)$$

plotted in Fig. 2.9(a). It is a CT periodic signal with fundamental frequency $\omega_0 = 0.3$ and fundamental period $P = 2\pi/0.3 = 20.93$. For this signal, there is no need to use (2.23) to compute its CT Fourier series. We use the following Euler formula

$$\sin \omega t = \frac{e^{j\omega t} - e^{-j\omega t}}{2j}; \qquad \cos \omega t = \frac{e^{j\omega t} + e^{-j\omega t}}{2}$$

to express it as

$$x(t) = -0.8e^{-j5\omega_0 t} + 0.4je^{-j2\omega_0 t} - 1.2 - 0.4je^{j2\omega_0 t} - 0.8e^{j5\omega_0 t}. \qquad (2.29)$$

This is the CT Fourier series of the periodic signal in (2.28). Its coefficients are

$$c_{-5} = -0.8, \quad c_{-2} = 0.4j, \quad c_0 = -1.2, \quad c_2 = -0.4j, \quad c_5 = -0.8 \qquad (2.30)$$

and the rest of the c_m are zero. Their magnitudes are plotted in Fig. 2.9(b) with frequency (rad/s) as its horizontal coordinate. If we plot them with respect to frequency index m, then the plot will not reveal the actual frequency. We see that $|c_m|$ is even. The phases of $\angle c_m$ for $m = 0, -5$, and 5 can be 180 or -180

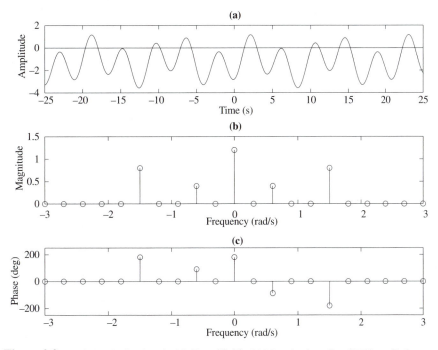

Figure 2.9 (a) CT periodic signal with $P = 20.93$. (b) Magnitudes of its CTFS coefficients. (c) Corresponding phases.

degrees. If they are all chosen as 180°, then the phase plot is not odd. If we select $\angle c_0 = \angle c_{-5} = 180°$ and $\angle c_5 = -180°$, then the phase plot excluding $m = 0$ is odd, as shown in Fig. 2.9(c).

If $x(t)$ is real, then its frequency components have even magnitudes and odd phases. If $x(t)$ is, in addition, even [$x(t) = x(-t)$], then its frequency components are real and even. Indeed, because $x(t)$ is even and $\sin m\omega_0 t$ is odd, their product $x(t) \sin m\omega_0 t$ is odd. Thus the integration in (2.27) is zero, and we have $c_m = \text{Re } c_m = \alpha_m = \alpha_{-m} = c_{-m}$. Thus if $x(t)$ is real and even, so is the set of c_m. Values of real c_m are called *amplitudes*. Their phases are 0 if $c_m \geq 0$, and π or $-\pi$ if $c_m < 0$.

◆ **Example 2.5**

Consider the CT periodic signal with period $P = 4$ shown in Fig. 2.10(a). It is a train of boxcars with width $2a = 2$ and height 1. Its fundamental frequency is $\omega_0 = 2\pi/P$. To express the signal in the CT Fourier series, we compute

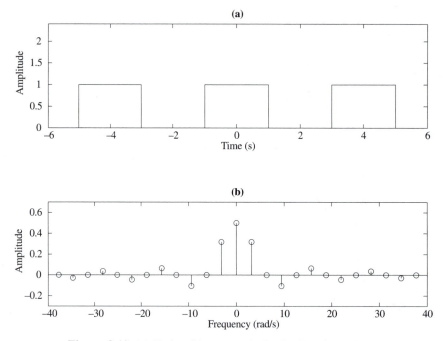

Figure 2.10 (a) Train of boxcars. (b) Its CT Fourier series.

$$c_0 = \frac{1}{P} \int_{-P/2}^{P/2} x(t)dt = \frac{1}{P} \int_{-a}^{a} 1dt = \frac{2a}{P} \tag{2.31}$$

and

$$c_m = \frac{1}{P} \int_{-P/2}^{P/2} x(t)e^{-jm\omega_0 t} dt = \frac{1}{P} \int_{-a}^{a} e^{-jm\omega_0 t} dt$$

$$= \frac{1}{-jm\omega_0 P} \left(e^{-jm\omega_0 a} - e^{jm\omega_0 a} \right) = \frac{2\sin m\omega_0 a}{m\omega_0 P} \tag{2.32}$$

Thus the CT Fourier series of the train of boxcars is

$$x(t) = \sum_{m=-\infty}^{\infty} c_m e^{jm\omega_0 t}$$

with c_m in (2.31) and (2.32). Because $x(t)$ is real and even, so is c_m. The set is plotted in Fig. 2.10(b) for $P = 4$ and $a = 1$.

◆ Example 2.6 Sampling Function

Consider the sequence of impulses shown in Fig. 2.11. The impulse at $t = 0$ can be denoted by $\delta(t)$; the ones at $t = \pm T$, by $\delta(t \mp T)$ and so forth. See the appendix. Thus the sequence can be expressed as

$$r(t) = \sum_{k=-\infty}^{\infty} \delta(t - kT) \tag{2.33}$$

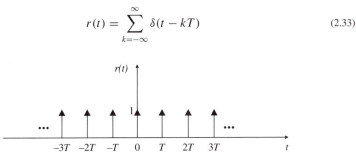

Figure 2.11 Impulse sequence.

It is periodic with fundamental period $P = T$ and fundamental frequency $\omega_0 = 2\pi/T$. Its CT Fourier series is

$$r(t) = \sum_{m=-\infty}^{\infty} c_m e^{jm\omega_0 t} \tag{2.32}$$

with, using (2.23) and the sifting property of the impulse in (A.6),

$$c_m = \frac{1}{T} \int_{-T/2}^{T/2} \delta(t) e^{-jm\omega_0 t} \, dt = \frac{1}{T} e^{-jm\omega_0 t} \Big|_{t=0} = \frac{1}{T}$$

for all m. Thus we have

$$r(t) = \sum_{k=-\infty}^{\infty} \delta(t - kT) = \sum_{m=-\infty}^{\infty} \frac{1}{T} e^{jm\omega_0 t} \tag{2.35}$$

The function in (2.33) is called the *sampling function* for reasons to be given in Chapter 3.

Let us recapitulate the key points here. Most periodic CT signals can be expressed in complex exponential Fourier series with frequency components c_m located at discrete frequencies $m\omega_0 = 2m\pi/P$, where P is the fundamental period and the frequency index m ranges from $-\infty$ to ∞. Thus the frequency range of CT periodic signals is $(-\infty, \infty)$.

If $c_{\pm M} \neq 0$ and

$$c_m = 0 \quad \text{for all } |m| > M$$

the signal is said to be *bandlimited* to M or $M\omega_0$. For example, the periodic signal in Example 2.4 is bandlimited to 1.5 rad/s or $M = 5$, the periodic signals in Examples 2.5 and 2.6 are not bandlimited.

2.4.1 Distribution of Average Power in Frequencies

This section discusses the physical significance of CTFS coefficients. Suppose $x(t)$ is a voltage signal. If it is applied to a 1-Ω resistor, then the current passing through it is $x(t)/1$ and the power dissipated is $x(t)x(t)$. Thus the total energy provided by $x(t)$ from $-\infty$ to ∞ is

$$E = \int_{-\infty}^{\infty} x^2(t)dt$$

where we have assumed implicitly that $x(t)$ is real. If $x(t)$ is complex, then the total energy is defined as

$$E := \int_{-\infty}^{\infty} x(t)x^*(t)dt = \int_{-\infty}^{\infty} |x(t)|^2 dt \tag{2.36}$$

It is clear that every nonzero periodic signal has $E = \infty$. Thus it is meaningless to discuss total energy for periodic signals. Instead, we discuss its average power defined by

$$P_{\text{av}} = \lim_{L \to \infty} \frac{1}{2L} \int_{-L}^{L} |x(t)|^2 dt \tag{2.37}$$

For a periodic signal with period P, (2.37) reduces to

$$P_{\text{av}} = \frac{1}{P} \int_0^P x(t)x^*(t)dt = \frac{1}{P} \int_0^P |x(t)|^2 dt \tag{2.38}$$

This average power can be computed directly in the time domain. It can also be computed from its Fourier series, as we will develop next. Substituting (2.22) into (2.38) and interchanging the order of integration and summation, we obtain

$$P_{\text{av}} = \frac{1}{P} \int_0^P \left(\sum_{-\infty}^{\infty} c_m e^{jm\omega_0 t} \right) x^*(t)dt = \sum_{-\infty}^{\infty} c_m \left(\frac{1}{P} \int_0^P x(t)e^{-jm\omega_0 t} dt \right)^*$$

The term inside the large parentheses is the c_m in (2.23). Thus we have

$$P_{\text{av}} = \frac{1}{P} \int_0^P |x(t)|^2 dt = \sum_{m=-\infty}^{\infty} c_m c_m^* = \sum_{m=-\infty}^{\infty} |c_m|^2 \tag{2.39}$$

This is called *Parseval's formula*. It states that the average power equals the sum of squared magnitudes of c_m. Because $|c_m|^2$ is the power at frequency $m\omega_0$, from c_m we can see the distribution of power over frequencies. Note that average power depends only on the magnitudes of c_m and is independent of the phases of c_m.

◆ **Example 2.7**

Consider the train of boxcars shown in Fig. 2.10 with $P = 4$ and $a = 1$. Find the percentage of the average power lying inside the frequency range $[-5, \ 5]$ in rad/s.

The total average power of the signal can be computed directly in the time domain as

$$P_{av} = \frac{1}{P} \int_{-a}^{a} 1^2 dt = \frac{1}{P}[a - (-a)] = \frac{1+1}{4} = 0.5$$

To find the average power inside the range $[-5, \ 5]$, we must use the Fourier series coefficients computed in Example 2.5. Because $\omega_0 = 2\pi/P = \pi/2$ and $5/\omega_0 = 3.1$, we conclude that c_m, for $m = -3, -2, \ldots, 2, 3$, lie inside the range. From (2.31) and (2.32), we have $c_0 = 0.5$ and

$$c_{-1} = \frac{2\sin[-(\pi/2)1]}{-(\pi/2)4} = \frac{2(-1)}{-2\pi} = \frac{1}{\pi}$$

Likewise, we can compute $c_1 = 1/\pi, c_{\pm 2} = 0$, and $c_{\pm 3} = -1/3\pi$. Thus the average power in $[-5, \ 5]$ is

$$2\left(\frac{-1}{3\pi}\right)^2 + 2\left(\frac{1}{\pi}\right)^2 + (0.5)^2 = 0.4752$$

It is 0.4752/0.5=0.95= 95% of the total average power. Because most power is located at low frequencies, the train is a low-frequency signal.

2.4.2 Are Phases Important?

The average power of CT periodic signals depends only on the magnitudes of CTFS coefficients and is independent of their phases. Does this imply that phases are not important? To answer this, we consider the two periodic signals

$$x_1(t) = \sin t - \cos 2t; \quad x_2(t) = \sin t + \sin 2t$$

They have fundamental frequency $\omega_0 = 1$ and fundamental period $P = 2\pi/\omega_0 = 6.28$, as shown in Figs. 2.12(a) and (d). The magnitudes of their CTFS coefficients are plotted in Figs. 2.12(b) and (e), and their phases in Figs. 2.12(c) and (f) (Problem 2.10). We see that their CTFS

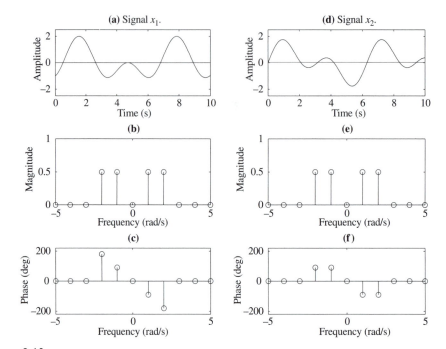

Figure 2.12 Two CT periodic signals in (a) and (d) with same CTFS magnitudes in (b) and (e) but different phases in (c) and (f).

coefficients have the same magnitudes but different phases. Clearly the two signals have the same average power.

Even though their CTFS coefficients have the same magnitudes, their waveforms, as shown in Figs. 2.12(a) and (d), are quite different. This is due to their different phases. In other words, different phases will generate different waveforms. Thus phases are important in image and video signals. However, in some application, we may disregard phases. For example, hearing is caused by the vibration of eardrums, which, in turn, is excited by power. Because power is independent of phases, our perception of sound is insensitive to phases or "phase dead." See Ref. 10 (p. 270). Thus in frequency-domain speech processing, we often disregard phases. In conclusion, we *may* disregard phases in audio signals but *cannot* in video signals or image processing.

Figure 2.12 also shows the usefulness of the frequency-domain analysis. Although the time-domain and frequency-domain representations carry exactly the same amount of information, the latter is much more revealing of its frequency content than the former. Thus the frequency-domain study of signals is important.

The CT Fourier series has a number of useful properties. Because they will not be used in the remainder of this chapter and because they are essentially the same as those to be introduced in the next chapter, they will not be discussed here.

2.5 Discrete-Time Fourier Series (DTFS)

In this section we develop the DT counterpart of the CT Fourier series. The development will follow closely the continuous-time case.

Consider a DT signal $x[n] = x(nT)$. It is said to be periodic with period N in samples, or $P = NT$ in seconds, if

$$x[n] = x[n + N]$$

for all integers n in $(-\infty, \infty)$. If it is periodic with period N, then it is also periodic with period $2N, 3N, \ldots$. The smallest such N is called the *fundamental period*.

Consider the set of complex exponential sequences

$$\phi_m[n] := e^{jm\omega_0 nT} = e^{j2\pi mn/N} \quad m = 0, \pm 1, \pm 2, \ldots \tag{2.40}$$

where

$$\omega_0 := \frac{2\pi}{NT}$$

Even though (2.40) is obtained from (2.19) by sampling or by replacing t by nT, the final form is independent of the sampling period T and is defined only by the given period N. Thus $\phi_m[n]$ can be defined without referring to the sampling period T. The reason for including T will be given later. Equation (3.40) has two integer indices n and m. The index n denotes the time instant nT and will be called the *time index*. Following (2.19), we called m the *frequency index*, whose meaning will be discussed as we proceed.

For each m, $\phi_m[n]$ is a complex-valued time sequence with n ranging from $(-\infty, \infty)$. As discussed in Section 2.2, *not* every DT complex exponential is periodic. However, for every m, the DT sequence defined in (2.40) is periodic with period (not necessarily fundamental period) N because

$$\phi_m[n + N] = e^{j2\pi m(n+N)/N} = e^{j2\pi mn/N} e^{j2\pi m} = \phi_m[n]$$

for all integer n.

Even though there are infinitely many complex exponential sequences in (2.40), there are only N distinct sequences. Because

$$\phi_{m+N}[n] = e^{j2\pi (m+N)n/N} = e^{j2\pi mn/N} e^{j2\pi n} = \phi_m[n] \tag{2.41}$$

we conclude that

$$\phi_m[n], \quad \phi_{m\pm N}[n], \quad \phi_{m\pm 2N}[n], \ldots$$

all denote the same sequence. Thus the set of sequences in (2.40) has only N distinct complex exponential sequences. By distinct, we mean that no sequence in the set can be expressed as a linear combination of the rest.[5] For example, if $N = 2$, the two distinct sequences are periodic extensions of $\{1 \ -1\}$ and $\{1 \ 1\}$ to all n with period 2. The first sequence has fundamental period 2, and the second sequence has fundamental period 1. Both are periodic with (not necessarily fundamental) period 2 and are harmonically related. If $N = 4$, the four distinct sequences are periodic extensions of $\{1 \ j \ -1 \ -j\}$, $\{1 \ -1 \ 1 \ -1\}$, $\{1 \ -j \ -1 \ j\}$, and $\{1 \ 1 \ 1 \ 1\}$ to all n with period 4. The first and third sequences have fundamental period 4, the second, 2 and the last, 1. They are harmonically related because they are all periodic with period 4. In contrast to infinitely many distinct CT complex exponential functions in (2.19), the set in (2.40) has only N distinct harmonically related complex exponential sequences. This is the major difference between CT and DT complex exponentials.

Let us discuss a general property of $\phi_m[n]$, which is the DT counterpart of (2.20). For every m, we have

$$\sum_{n=<N>} \phi_m[n] := \sum_{n=n_0}^{n_0+N-1} e^{j2\pi mn/N}$$

$$= \begin{cases} N & \text{for } m = 0 \ (\text{mod } N) \\ 0 & \text{for } m \neq 0 \ (\text{mod } N) \end{cases} \tag{2.42}$$

for any integer n_0, in particular for $n_0 = 0$ or $-N/2$ if N is even. The notation $n = <N>$ is used to denote any consecutive N time instants. The equality $m = 0 \ (\text{mod } N)$ means that the remainder of m divided by N is 0. Thus $m = 0 \ (\text{mod } N)$ if $m = 0, \pm N, \pm 2N, \ldots$. If $m = 0$ $(\text{mod } N)$, every term in the summation equals 1. Thus the sum of N terms is N. To show the case for $m \neq 0 \ (\text{mod } N)$, we use the formula

$$\sum_{n=0}^{N-1} r^n = 1 + r + r^2 + \cdots + r^{N-1} = \frac{1-r^N}{1-r} \tag{2.43}$$

to write

$$\sum_{n=0}^{N-1} \phi_m[n] = \sum_{n=0}^{N-1} \left(e^{j2\pi m/N}\right)^n = \frac{1 - e^{(j2\pi m/N)N}}{1 - e^{j2\pi m/N}} = \frac{1-1}{1 - e^{j2\pi m/N}}$$

If $m \neq 0 \ (\text{mod } N)$, we have $e^{j2\pi m/N} \neq 1$ and the denominator $(1 - e^{j2\pi m/N})$ is different from 0. Thus the preceding summation is 0. This establishes (2.42).

Let us use an asterisk to denote complex conjugation. For example, we have

$$\phi_k^*[n] = (e^{j2\pi kn/N})^* = e^{-j2\pi kn/N}$$

Then (2.42) implies

---

[5] To be more precise, the set is said to be *linearly independent*.

$$\sum_{n=<N>} \phi_m[n]\phi_k^*[n] = \sum_{n=<N>} e^{j2\pi mn/N} e^{-j2\pi kn/N} = \sum_{n=<N>} e^{j2\pi(m-k)n/N}$$

$$= \begin{cases} N & \text{if } m = k \ (\text{mod } N) \\ 0 & \text{if } m \neq k \ (\text{mod } N) \end{cases} \tag{2.44}$$

This is the *orthogonality* property of the set $\phi_m[n]$ and is the DT counterpart of (2.21).

With this preliminary, we are ready to introduce the discrete-time Fourier series. Let $x(nT)$ be periodic with period N. Define $\omega_0 = 2\pi/NT$. Then $x(nT)$ can be expressed as

$$x[n] := x(nT) = \sum_{m=<N>} c_{md} e^{jm\omega_0 nT}$$

$$= \sum_{m=<N>} c_{md} e^{j2\pi mn/N} \quad \text{(synthesis eq.)} \tag{2.45}$$

with

$$c_{md} = \frac{1}{N} \sum_{n=<N>} x(nT) e^{-jm\omega_0 nT}$$

$$= \frac{1}{N} \sum_{n=0}^{N-1} x[n] e^{-j2\pi mn/N} \quad \text{(analysis eq.)} \tag{2.46}$$

for all integer m, called the *frequency index*. The second subscript d denotes discrete time. The set of two equations is called the *DT Fourier series* or *DTFS*. The second equation is called the *analysis equation* because c_{md} reveals the frequency content of $x[n]$. The first equation is called the *synthesis equation* because $x[n]$ can be constructed from c_{md}. We call c_{md} DTFS coefficients or frequency components.

To establish (2.46), we multiply (2.45) by $e^{-j2\pi kn/N}$, for a fixed k, and then sum it over N consecutive terms to yield

$$\sum_{n=<N>} x(nT) e^{-j2\pi kn/N} = \sum_{n=<N>} \left(\sum_{m=<N>} c_{md} e^{j2\pi(m-k)n/N} \right)$$

$$= \sum_{m=<N>} c_{md} \left(\sum_{n=<N>} e^{j2\pi(m-k)n/N} \right) \tag{2.47}$$

where we have changed the order of summations. The orthogonality property in (2.44) implies that the summation inside the parentheses equals 0 if $m \neq k$, and N if $m = k$. Thus (2.47) reduces to

$$\sum_{n=<N>} x(nT) e^{-j2\pi kn/N} = c_{kd} N$$

This becomes (2.46) after renaming the integer k to m.

First we show that the set of DTFS coefficients c_{md} is periodic with period N or

$$c_{(m+N)d} = c_{md} \tag{2.48}$$

for all integer m. Indeed, we have, using $e^{-j2\pi n} = 1$ for all n,

$$c_{(m+N)d} = \frac{1}{N} \sum_{n=<N>} x(nT)e^{-j2\pi(m+N)n/N}$$

$$= \frac{1}{N} \sum_{n=<N>} x(nT)e^{-j2\pi mn/N} e^{-j2\pi n} = c_{md}$$

Thus there are only N distinct c_{md} and we sum only over N consecutive terms in (2.45). If we sum (2.45) over all integers m as in the CT case in (2.22), then the summation will diverge and the equation is meaningless. This is the major difference between (2.22) and (2.45) or between the CT and DT Fourier series. It is important to mention that the DT Fourier series can be defined without referring to the sampling period T. However, we have purposely inserted the middle term in (2.40), (2.45), and (2.46). By so doing, it becomes easier to relate the CT Fourier series of $x(t)$ and the DT Fourier series of its sampled sequence $x(nT)$.

Before giving some examples, we mention that if $x[n] = x(nT)$ is real, then we have $|c_{(-m)d}| = |c_{md}|$ and $\measuredangle\, c_{(-m)d} = -\measuredangle\, c_{md}$. If $x[n] = x(nT)$ is real and even, so is c_{md}. Their proofs are identical to the CT case and will not be repeated.

◆ **Example 2.8**

Consider the periodic sequence with period $N = 3$ shown in Fig. 2.13(a). It is plotted with respect to time (in seconds), not time index. The sampling period is $T = 0.5$ and we have $x[0] = x(0) = -2$, $x[1] = x(T) = 1$ and $x[2] = x(2T) = -0.6$. Let us compute its DTFS coefficients for $m = 0, 1, 2$:

$$c_{0d} = \frac{1}{3} \sum_{n=0}^{2} x[n]e^{-j2\pi \times 0 \times n/3} = \frac{1}{3}(x[0] + x[1] + x[2]) = \frac{-2+1-0.6}{3}$$

$$= -0.53 = 0.53e^{j180^o}$$

$$c_{1d} = \frac{1}{3} \sum_{n=0}^{2} x[n]e^{-j2\pi \times 1 \times n/3} = \frac{1}{3}\left(x[0] + x[1]e^{-j2\pi/3} + x[2]e^{-j4\pi/3}\right)$$

$$= \frac{1}{3}[-2 + (-0.5 - j0.886) - 0.6(-0.5 + j0.886)]$$

$$= -0.73 - j0.46 = 0.87e^{-j147.8^o}$$

$$c_{2d} = \frac{1}{3} \sum_{n=0}^{2} x[n] e^{-j2\pi \cdot 2 \cdot n/3} = \frac{1}{3} \left(x[0] + x[1] e^{-j4\pi/3} + x[2] e^{-j8\pi/3} \right)$$

$$= \frac{1}{3} [-2 + (-0.5 + j0.886) - 0.6(-0.5 - j0.886)]$$

$$= -0.73 + j0.46 = 0.87 e^{j147.8^o}$$

Their magnitudes and phases are located at $m\omega_0 = 2\pi m/NT = 4\pi m/3$ in rad/s, for $m = 0, 1, 2$, as shown in Figs. 2.13(b) and 2.13(c) with solid dots. They are then extended periodically to all m, with hollow dots, as shown. Note that they are plotted with respect to frequency, not frequency index. Because $x(nT)$ is real, the magnitude plot is even. Note that the phase of $c_{0d} = -0.53$ can be selected as 180° or -180°. The phase plot, excluding $\not\prec c_{0d}$, is odd; the reason for excluding $\not\prec c_{0d}$ is the same as in the CT case. We mention that the sampling period T does not appear explicitly in computing c_{md}. It is used only to determine the discrete frequencies $m\omega_0 = m(2\pi/NT)$ in rad/s. Note that the scales of the horizontal axes in Figs. 2.13(b) and 2.13(c) are radians per second.

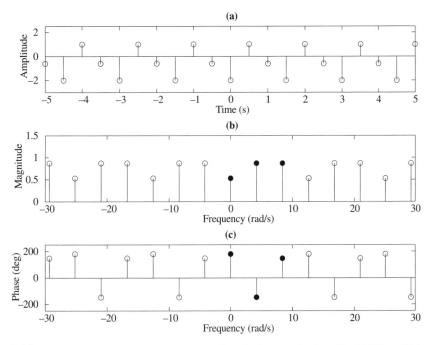

Figure 2.13 (a) DT periodic sequence with period $N = 3$. (b) Magnitudes of its DTFS coefficients. (c) Corresponding phases. Both (b) and (c) are periodic with period $N = 3$ or $N\omega_0 = 2\pi/T = 4\pi$ (rad/s).

◆ **Example 2.9**

Consider the periodic sequence shown in Fig. 2.14(a) with $T = 1$, $x[0] = 2.5$, $x[1] = -0.4$, $x[2] = 1$, and $x[3] = -2$. It is periodic with period $N = 4$. We compute its DTFS coefficients as

$$c_{0d} = \frac{1}{4} \sum_{n=0}^{3} x[n]e^{-j2\pi \times 0 \times n/4} = 0.25(2.5 - 0.4 + 1 - 2) = 0.275$$

$$c_{1d} = 0.25 \left(2.5 - 0.4e^{-j2\pi \times 1 \times 1/4} + e^{-j2\pi \times 1 \times 2/4} - 2 \times e^{-j2\pi \times 1 \times 3/4}\right)$$

$$= 0.375 - 0.4j = 0.55e^{-j46.8^o}$$

$$c_{2d} = 0.25 \left(2.5 - 0.4e^{-j2\pi \times 2 \times 1/4} + e^{-j2\pi \times 2 \times 2/4} - 2e^{-j2\pi \times 2 \times 3/4}\right) = 1.475$$

$$c_{3d} = 0.25 \left(2.5 - 0.4\, e^{-j2\pi \times 3 \times 1/4} + e^{j2\pi \times 3 \times 2/4} - 2e^{-j2\pi \times 3 \times 3/4}\right)$$

$$= 0.375 + 0.4j = 0.55e^{j46.8^o}$$

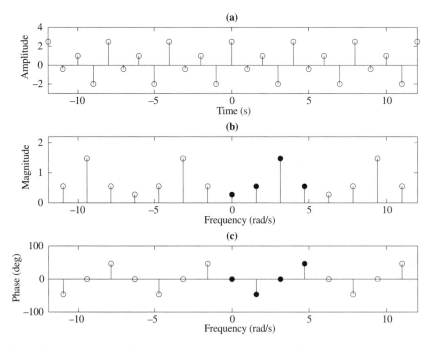

Figure 2.14 (a) DT periodic sequence with period $N = 4$. (b) Magnitudes of its DTFS coefficients. (c) Corresponding phases. Both (b) and (c) are periodic with period $N = 4$ or $N\omega_0 = 2\pi/T = 2\pi$ (rad/s).

Their magnitudes and phases are located at $m\omega_0 = 2\pi m/4 = 0.5\pi m$, in rad/s, for $m = 0, 1, 2, 3$, and are plotted in Figs. 2.14(b) and (c) with solid dots. They are then extended periodically, with period $N = 4$, to all m with hollow dots as shown. Because $x(nT)$ is real, the periodic extension of c_{md} has even magnitude and odd phase as shown.

2.5.1 Range of Frequency Index m

Consider a periodic sequence with period N. The sequence is completely specified by any N consecutive $x[n]$, and we can use any N consecutive $x[n]$ to compute its DTFS. Its DTFS is also periodic with period N, and we can compute any N consecutive c_{md}. Most often, we use $x[n], n = 0, 1, \ldots, N - 1$, to compute c_{md} for $m = 0, 1, \ldots, N - 1$, as in Examples 2.8 and 2.9. Thus it is fair to say that in DTFS, we deal with only two sets of N numbers. The set $x[n]$ is always real valued; the set c_{md} is mostly complex valued. We use different indices because they carry different physical meanings. The time index n denotes time instant nT. Now we discuss the meaning of the frequency index m.

The frequency index m in c_{md} denotes the discrete "frequency" $m\omega_0 = 2m\pi/NT$ or, to be more precise, is associated with the complex exponential sequence $\phi_m[n] = e^{j2\pi mn/N} = e^{jm\omega_0 nT}$. We now discuss the frequency of $\phi_m[n]$. As discussed in Section 2.2, if $m\omega_0 = 2\pi m/NT$ lies inside the Nyquist frequency range

$$(-\pi/T, \ \pi/T] \tag{2.49}$$

then the frequency of $\phi_m[n]$ is $m\omega_0$; otherwise, $m\omega_0$ is not the frequency of $\phi_m[n]$. Thus it is desirable to select those m so that $m\omega_0 = 2\pi m/NT$ lie inside (2.49). Then the frequency index m denotes the discrete frequency $2m\pi/NT$ in rad/s. Therefore, instead of using any N consecutive c_{md}, we shall use those c_{md} with frequency in (2.49) to denote the DT Fourier series.

Before proceeding, we introduce some notation. We use $\overline{b}$ to denote the least integer equal to or larger than b. That is, the overbar rounds a number upward to an integer. For example, we have $\overline{-3.1} = -3$ and $\overline{3.1} = 4$. We use colon (:) between two integers to denote the integer range from the first integer to the second integer. For example, $m = -1 : 3$ means $m = -1, 0, 1, 2, 3$.

Let us consider the range

$$m = \overline{-(N-1)/2} : \overline{(N-1)/2} \tag{2.50}$$

for any integer N. For example, if $N = 8$, then

$$m = \overline{-3.5} : \overline{3.5} = -3{:}4 = -3, -2, -1, 0, 1, 2, 3, 4$$

If $N = 9$, then

$$m = \overline{-4} : \overline{4} = -4 : 4 = -4, -3, -2, -1, 0, 1, 2, 3, 4$$

The number of indices in (2.50) equals N. Excluding $m = 0$, (2.50) has equal number of positive and negative indices if N is odd, and one extra positive index if N is even. Note that (2.50) can also be expressed as

$$m = -(N-1)/2 : (N-1)/2$$

if N is odd and

$$m = -N/2 + 1 : N/2$$

if N is even.

Next we show that $m\omega_0 = 2m\pi/NT$ with m in (2.50) lies inside the Nyquist frequency range. The left-most frequencies on the frequency horizontal coordinate for N even and odd are, respectively, at

$$-\frac{(N-1)\pi}{NT} \quad \text{and} \quad \left(-\frac{N}{2}+1\right)\frac{2\pi}{NT} = -\frac{(N-2)\pi}{NT}$$

Clearly, they are larger than $-N\pi/NT$. The right-most frequencies for N even and odd are, respectively, at

$$\frac{(N-1)\pi}{NT} \quad \text{and} \quad \frac{\pi}{T}$$

The first one is less than $N\pi/NT$ and the second one equals π/T. Thus we conclude that $m\omega_0$ lies inside $(-\pi/T, \ \pi/T]$ for all m in (2.50).

Even though we are interested in those m in (2.50), it is more convenient to compute c_{md} with m ranging from 0 to $N-1$. Once these c_{md} are computed, we can readily obtain, by periodic extension, c_{md} with m in (2.50).

2.5.2 Time Shifting

This subsection discusses one property of the DT Fourier series that will be needed in the next section. Consider a DT periodic sequence $x[n] = x(nT)$ with period N and with DTFS coefficients c_{md}. Let n_0 be an integer. If n_0 is positive, then $x[n-n_0] = x((n-n_0)T)$ is the sequence of $x[n]$ shifted to the right n_0 samples. If n_0 is negative, then it is shifted to the left. Clearly, $x[n-n_0]$ is also periodic with period N and can be expressed in the DT Fourier series as

$$x[n-n_0] = \sum_{m=<N>} \bar{c}_{md} e^{j2\pi mn/N}$$

with

$$\bar{c}_{md} = \frac{1}{N} \sum_{n=<N>} x[n-n_0] e^{-j2\pi mn/N}$$

We express $\bar{c}_{md}$ as

$$\bar{c}_{md} = e^{-j2\pi mn_0/N} \frac{1}{N} \sum_{n=<N>} x[n-n_0]e^{-j2\pi m(n-n_0)/N}$$

$$= e^{-j2\pi mn_0/N} \left(\frac{1}{N} \sum_{k=<N>} x[k]e^{-j2\pi mk/N} \right)$$

where the summation inside the parentheses is the DTFS coefficient c_{md}. Thus we have

$$\bar{c}_{md} = e^{-j2\pi mn_0/N} c_{md} \tag{2.51}$$

which implies

$$|\bar{c}_{md}| = |e^{-j2\pi mn_0/N}||c_{md}| = |c_{md}| \tag{2.52}$$

and

$$\measuredangle \bar{c}_{md} = \measuredangle e^{-j2\pi mn_0/N} + \measuredangle c_{md} = \measuredangle c_{md} - 2\pi mn_0/N \tag{2.53}$$

for all m. Thus time shifting does not affect the magnitudes of DTFS coefficients, but introduces a linear phase (in terms of m) into the phases of the coefficients.

◆ **Example 2.10**

Consider a periodic sequence with period $N = 4$, $T = 1$, $\bar{x}[-1] = 2.5$, $\bar{x}[0] = -0.4$, $\bar{x}[1] = 1$, and $\bar{x}[2] = -2$. It is the shifting of the sequence in Example 2.9 three samples to the right or, equivalently, one sample to the left. Thus if we use $x[n]$ to denote the sequence in Example 2.9, and $\bar{x}[n]$ to denote the sequence here, then we have $\bar{x}[n] = x[n-3] = x[n+1]$. The DTFS coefficients of $\bar{x}[n]$ will be computed, by summing from $n = -1$ to 2, in the following:

$$\bar{c}_{0d} = \frac{1}{4} \sum_{n=-1}^{2} \bar{x}[n]e^{-j2\pi \times 0 \times n/4} = 0.25(2.5 - 0.4 + 1 - 2) = 0.275$$

$$\bar{c}_{1d} = 0.25 \left(2.5e^{-j2\pi \times 1 \times (-1)/4} - 0.4 + e^{-j2\pi \times 1 \times 1/4} - 2e^{-j2\pi \times 1 \times 2/4} \right)$$

$$= 0.4 + 0.375j = 0.55e^{j43.2^o}$$

$$\bar{c}_{2d} = 0.25 \left(2.5e^{-j2\pi \times 2 \times (-1)/4} - 0.4 + e^{-j2\pi \times 2 \times 1/4)} - 2e^{-j2\pi \times 2 \times 2/4} \right)$$

$$= -1.475 = 1.475e^{j180^o}$$

$$\bar{c}_{3d} = 0.25 \left(2.5e^{-j2\pi \times 3 \ (-1)/4} - 0.4 + e^{-j2\pi \times 3 \times 1/4} - 2e^{-j2\pi \times 3 \times 2/4} \right)$$

$$= 0.4 - 0.375j = 0.55e^{-j43.2^o}$$

Comparing with c_{md} in Example 2.9, we have $|\bar{c}_{md}| = |c_{md}|$, for all m, and

$$\angle \bar{c}_{0d} = \angle c_{0d} = 0$$

$$\angle \bar{c}_{1d} = 43.2^o = \angle c_{1d} + 360^o \times 1 \times 1/4 = -46.8^o + 90^o$$

$$\angle \bar{c}_{2d} = 180^o = \angle c_{2d} + 360^o \times 2 \times 1/4 = 0^o + 180^o$$

$$\angle \bar{c}_{3d} = -43.2^o = \angle c_{3d} + 360^o \times 3 \times 1/4$$

$$= 46.8^o + 270^o = 316.8^o \quad (\text{mod } 360^o)$$

This verifies (2.51). Note that we have used degrees here instead of radians as in (2.53).

2.6 FFT Computation of DTFS Coefficients

From Examples 2.8 through 2.10, we saw that computing DT Fourier series (DTFS) by hand is straightforward but very tedious. Therefore, it is better to delegate its computation to a digital computer. Although (2.46) can be directly programmed on a computer, Ref. 8, entitled "An algorithm for the machine computation of complex Fourier series," developed an efficient way of computing it. The algorithm together with its large number of variations are now collectively called the fast Fourier transform (FFT), which we will introduce in Chapter 4. The FFT is a workhorse in DSP and is widely available. Its employment is also very simple, as we will demonstrate in the remainder of this chapter.

Consider a set of N numbers $\{x[n], n = 0 : N - 1\}$. The FFT is an efficient method of computing

$$X[m] = \sum_{n=0}^{N-1} x[n]e^{-jn(2\pi/N)m} \tag{2.54}$$

for $m = 0, 1, \ldots, N - 1$. Comparing (2.54) and (2.46) yields immediately

$$c_{md} = \frac{X[m]}{N} \tag{2.55}$$

Thus the DT Fourier series can be computed using FFT. We will use the FFT in MATLAB to carry out all computation in this text. In MATLAB, if x is an N-point sequence, the function fft(x) or fft(x,N) generates N data of (2.54), for $m = 0, 1, \ldots, N - 1$. Dividing those data by N yields the DTFS coefficients. For the sequence in Example 2.8, typing

```
x=[-2 1 -0.6];
c=fft(x)/3
```

yields

$$c = -0.5333 \quad -0.7333 - 0.4619i \quad -0.7333 + 0.4619i$$

which are the DT Fourier series coefficients computed in Example 2.8. Note that in MATLAB, both i and j denote $\sqrt{-1}$. For the sequence in Example 2.9, typing

```
x=[2.5 -0.4 1 -2];
c=fft(x/4)
```

yields

$$c = 0.275 \quad 0.375 - 0.4000i \quad 1.475 \quad 0.375 + 0.4000i$$

which are the same as those computed in Example 2.9. We see that the factor $1/N$ in (2.55) can be computed after computing FFT or be included in the input data as x/N. Note that if x is a column vector, then fft(x) yields a column vector. In conclusion, the DT Fourier series can be easily obtained by calling FFT in MATLAB or in any computer software package.

We mention an important difference, other than the factor $1/N$, between DTFS and FFT. Consider a periodic sequence with period N.

- Its DTFS can be computed using any one period.
- If we use FFT, the input data must be from $n = 0$ to $N - 1$.

The reason is that the FFT is developed specifically for the equation in (2.54). If the input data of FFT is not from $n = 0$ to $N - 1$, then the result will be incorrect. For example, consider the sequence $\bar{x}[-1] = 2.5, \bar{x}[0] = -0.4, \bar{x}[1] = 1, \bar{x}[-2] = -2$ in Example 2.10. In computing its DT Fourier series, we used the four data directly as shown in the example by summing from $n = -1$ to 2. In employing FFT, if we use the four data directly, then FFT will compute the DT Fourier series of the sequence in Example 2.9 rather than the one in Example 2.10. To compute the DT Fourier series of the sequence in Example 2.10, we must use the data from 0 to 3. The value of the sequence at $n = 3$ is $\bar{x}[3] = \bar{x}[-1] = 2.5$. Thus typing

```
x=[-0.4 1 -2 2.5];
cb=fft(x/4)
```

will yield the four DT Fourier series coefficients computed in Example 2.10.

2.6.1 Rearranging the Output of the FFT

Using FFT to compute the DT Fourier series of a periodic sequence is very simple. We now discuss how to arrange the output of FFT to correspond the range of m in (2.50). Before proceeding, we discuss indexing in MATLAB. No matter how an array of N numbers is generated, its elements are indexed from 1 to N (not from 0 to $N - 1$) and are enclosed by parentheses. For example, if we type x=[2 6 -8], then we have x(1)=2, x(2)=6, x(3)=-8. If we type x(0) or x(4), then error messages will appear. Thus the MATLAB function X=fft(x/N) will generate N data

X(1) X(2) X(3) $\cdots$ X(N)

They are related to the DTFS coefficients c_{md} by

$$c_{md} = X(m+1) \quad \text{for } m = 0, 1, \ldots, N-1 \tag{2.56}$$

These N output data are to be located at $m(2\pi/NT)$ for $m = 0, 1, \ldots, N-1$ or inside the "frequency" range $[0, \ 2\pi/T)$. If we are interested in the Nyquist frequency range, then the range of m should be as shown in (2.50).

In MATLAB, the function ceil(b) rounds the number b upward to an integer or, equivalently, generates $\overline{b}$. Thus the range of m in (2.50) can be generated in MATLAB as

$$m = \text{ceil} \, (-(N-1)/2) : \ \text{ceil} \, ((N-1)/2)$$

The corresponding c_{md} can be generated by periodic extension of c_{md} with $m = 0, 1, \ldots, N-1$. However, they can be more easily obtained by rearranging c_{md} or X(m). First we group X(m) as

X(m)=[X(m1) X(m2)]

where m1 indicates the range of m from 0 to ceil$((N-1)/2)$ in c_{md}. In view of (2.56), the corresponding indices of X(m1) should be

m1=1:ceil((N-1)/2)+1=1:ceil((N-1)/2+1)=1:ceil((N+1)/2)=:1:p

where p:=ceil((N+1)/2). The remaining m2 clearly ranges from p+1 to N or

m2=p+1:N

Once X(m) is grouped as [X(m1) X(m2)], by interchanging the positions of X(m1) and X(m2) to [X(m2) X(m1)], we will obtain the DT Fourier series coefficients corresponding to the range in (2.50). The preceding rearrangement can be programmed in MATLAB as[6]

Program 2.1
function y=shift(X)
N=max(size(X));
p=ceil((N+1)/2);

---

[6] One advantage of using MATLAB is that the user can create his own function and store it as an M file. It then becomes a part of MATLAB.

```
m1=1:p;
m2=p+1:N;
        y=[X(m2) X(m1)];
```

The second line defines the function shift. The vector X is $1 \times N$; thus its maximum size is N. The rest of the program was explained above.

Let us summerize how to use MATLAB to compute and to plot the DT Fourier series of periodic sequences. Given a periodic sequence $x[n]$ with period N and sampling period T, the following program will plot the magnitude and phase of its DT Fourier series in the frequency range $(-\pi/T, \pi/T]$.

Program 2.2
```
N= ;T= ;D=2*pi/(N*T);
x=[x[0] x[1] x[2] ··· x[N-1]];
X=fft(x/N);
m=ceil(-(N-1)/2):ceil((N-1)/2);
w=m*D;
subplot(2.1,1)
stem(w, abs(shift(X))),title('(a)')
subplot(2.1,2)
stem(w,angle(shift(X))*180/pi),title('(b)')
```

Using this program, we plot in Figs. 2.15(a) and (b) the magnitude and phase of the DT Fourier series of the sequence in Example 2.8.[7] They generate the parts of Figs. 2.13(b) and (c) in the Nyquist frequency range $(-2\pi, 2\pi] = (-6.28, 6.28]$. Figures 2.15(c) and (d) generate the parts of Figs. 2.14(b) and (c) in the Nyquist frequency range $(-3.14, 3.14]$. Note that for N odd, there are same number of entries in the positive and negative frequencies. For N even, there is one extra entry at the positive frequency $\omega = \pi/T = 3.14$.

We summarize the procedure of using FFT to compute the DT Fourier series:

- Given a periodic sequence $x(nT)$ with period N. Let x denote the N data from $n = 0$ to $N - 1$. Then c=fft(x/N) generates its N DTFS coefficients at

$$\omega = m\frac{2\pi}{NT} \quad \text{for } m = 0, 1, \ldots, N - 1$$

Note that the sampling period T does not play any role in using FFT; it is used only to locate the output of FFT.

---

[7] The program will not reproduce exactly the plots in Figs. 2.15(a) and (b). We have added the axis command axis([-2*pi 2*pi 0 1.5]) to shape the size of Fig. 2.15(a), and the command plot([-2*pi 2*pi],[0 0]) to draw the horizontal coordinate in Fig. 2.15(b).

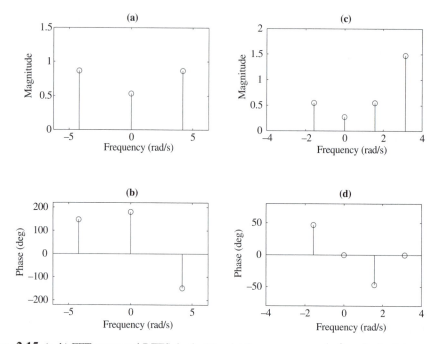

Figure 2.15 (a, b) FFT-computed DTFS, in the Nyquist frequency range $(-2\pi, \ 2\pi]$, of the sequence in Fig. 2.13(a). (c, d) FFT-computed DTFS, in the Nyquist frequency range $(-\pi, \ \pi]$, of the sequence in Fig. 2.14(a).

• If we apply the function shift, defined in Program 2.1, to the output of FFT, that is, X=shift(c)=shift(fft(x/N)), then its output yields the N DTFS coefficients, arranged from left to right, inside the Nyquist frequency range at

$$\omega = m\frac{2\pi}{NT} \quad \text{with } m = \overline{-(N-1)/2} : \overline{(N-1)/2}$$

If N is odd, all frequencies lie inside $(-\pi/T, \ \pi/T)$. If N is even, the right-most frequency is at the boundary π/T.

Before concluding this subsection, we discuss the MATLAB function fftshift. In fact, the function shift in Program 2.1 is modified from fftshift. To see the difference between shift and fftshift, we type shift([0 1 2 3 4]) and fftshift([0 1 2 3 4]), which yield, respectively,

[3 4 0 1 2]   and   [3 4 0 1 2]

and type shift([0 1 2 3 4 5]), and fftshift([0 1 2 3 4 5]) which yield

[4 5 0 1 2 3]   and   [3 4 5 0 1 2]

We see that if the number of entries N is odd, then shift=fftshift. If N is even, then fftshift shifts one extra entry to the left.

If we use fftshift instead of shift, then the output of fftshift(fft(x/N)) will locate at $2m\pi/NT$ with

$$m = \underline{-(N-1)/2} : \underline{(N-1)/2}$$

where the underline rounds a number downward to an integer and can be carried out in MATLAB by the command floor. The range can also be expressed as

$$m = -(N-1)/2 : (N-1)/2;$$

for N odd and

$$m = -N/2 : N/2 - 1;$$

for N even. We can readily verify that for N odd, all frequencies $2m\pi/NT$ lie inside $(-\pi/T,$ $\pi/T)$. For N even, the left-most frequency is at $-\pi/T$ and the rest inside $(-\pi/T, \pi/T)$. Thus the frequency range of fftshift is $[-\pi/T, \pi/T)$. Because we have adopted the frequency range $(-\pi/T, \pi/T]$, we should use shift. We mention that shift in Program 2.1 is developed for a row vector X. If X is a column vector, then shift(X) will yield an error message. This can be corrected by taking the transpose of X as shift(X'). The function fftshift in MATLAB is developed for general matrices. Therefore, fftshift(X) will generate a shifted X whether X is a column or a row vector.

To conclude this section, we mention that all shift and fftshift in this text can be interchanged. The Nyquist frequency range is $(-\pi/T, \pi/T]$ if we use shift and $[-\pi/T, \pi/T)$ if we use fftshift. The corresponding frequency indices m must be interchanged as listed in Table 2.2.

2.7 FFT Computation of CTFS Coefficients

This section discusses FFT computation of CTFS coefficients. Before proceeding, we must establish the relationship between the CTFS of a CT signal and the DTFS of its sampled sequence. Consider a CT periodic signal with fundamental period P and fundamental frequency $\omega_0 = 2\pi/P$. It can be expressed in CT Fourier series as

$$x(t) = \sum_{m=-\infty}^{\infty} c_m e^{jm\omega_0 t} \tag{2.57}$$

with

$$c_m = \frac{1}{P} \int_0^P x(t) e^{-jm\omega_0 t} dt \tag{2.58}$$

Table 2.2 Frequency Indices Corresponding to shift and fftshift

	shift	fftshift
N	m=ceil(-(N-1)/2):ceil((N-1)/2)	m=floor(-(N-1)/2):floor((N-1)/2)
N(even)	m=-N/2+1:N/2	m=-N/2:N/2-1
N(odd)	m=-(N-1)/2:(N-1)/2	m=-(N-1)/2:(N-1)/2

These frequency components are located at $m\omega_0$ and lie inside the frequency range $(-\infty, \infty)$. Now we take the sample of x(t). If we take N samples per period, then the sampling period is

$$T = \frac{P}{N}$$

and the resulting sampled sequence is $x(nT)$. It is clear that $x(nT)$ is periodic with period N (in samples) or $NT = P$ (in seconds) and $\omega_0 = 2\pi/NT$. If we select T first, and if P/T is not an integer, then $x(nT)$ is not periodic. Thus we first select N and then compute $T = P/N$. We express $x(nT)$ in DT Fourier series as

$$x(nT) = \sum_{m=0}^{N-1} c_{md} e^{jm\omega_0 nT} \tag{2.59}$$

with

$$c_{md} = \frac{1}{N} \sum_{n=0}^{N-1} x(nT) e^{-jm\omega_0 nT} \tag{2.60}$$

If m are selected as in (2.50), then these coefficients lie inside the Nyquist frequency range $(-\pi/T, \pi/T]$.

Let us establish the relationship between c_m and c_{md}. Equation (2.57) holds for all t, in particular, at $t = nT$. Thus we have

$$x(nT) = \sum_{m=-\infty}^{\infty} c_m e^{jm\omega_0 nT}$$

This infinite summation will be divided into sections, each section with N terms, as

$$x(nT) = \sum_{k=-\infty}^{\infty} \sum_{m=kN}^{kN+N-1} c_m e^{jm\omega_0 nT}$$

Let us introduce the new index $\bar{m} := m - kN$. Then the summation can be written as

$$x(nT) = \sum_{k=-\infty}^{\infty} \sum_{\bar{m}=0}^{N-1} c_{(\bar{m}+kN)} e^{j(\bar{m}+kN)\omega_0 nT}$$

which becomes, after using $e^{jkN\omega_0 nT} = e^{jk(2\pi)n} = 1$, for all integers k and n, interchanging the order of the two summations, and renaming $\bar{m}$ to m,

$$x(nT) = \sum_{\bar{m}=0}^{N-1} \sum_{k=-\infty}^{\infty} c_{(\bar{m}+kN)} e^{j\bar{m}\omega_0 nT}$$

$$= \sum_{m=0}^{N-1} \left(\sum_{k=-\infty}^{\infty} c_{(m+kN)} \right) e^{jm\omega_0 nT} \qquad (2.61)$$

Comparing (2.59) and (2.61) yields immediately

$$c_{md} = \sum_{k=-\infty}^{\infty} c_{(m+kN)} \qquad (2.62)$$

This relates the CTFS coefficients of $x(t)$ and the DTFS coefficients of its sampled sequence $x(nT)$. It is important to mention that the fundamental frequencies of $x(t)$ and $x(nT)$ are the same and equal $\omega_0 = 2\pi/p = 2\pi/NT$. The CTFS coefficients c_m of $x(t)$ are located at $m\omega_0$ with $m = \infty : \infty$; whereas the DTFS coefficients c_{md} of $x(nT)$ are located at $m\omega_0$ with $m = \overline{-(N-1)/2} : \overline{(N-1)/2}$.

Let us discuss the implication of (2.62). Suppose $\omega_0 = 1$ and c_m is real and even as shown in Fig. 2.16(a) with highest-frequency $M\omega_0 = 2\omega_0$. This signal $x(t)$ is bandlimited to $2\omega_0$ and has at most $(2M+1)$ nonzero frequency components. For any integer N, c_{m+kN} is the shifting of c_m to the left kN units of ω_0, and c_{m-kN} is the shifting of c_m to the right kN units of ω_0. Thus we conclude that the DT Fourier series of $x(nT)$ is the sum of all repetitive shiftings, with period N (in samples) or $N\omega_0$ (in rad/s), of the CT Fourier series of $x(t)$. We plot in Fig. 2.16(b) for $N = 7$. We see that repetitive shiftings of c_m do not overlap; thus we have

$$c_m = \begin{cases} c_{md} & \text{for } -3 \le m \le 3 \\ 0 & \text{for } |m| > 3 \end{cases} \qquad (2.63)$$

This establishes essentially the theorem that follows.

Theorem 2.1   Consider a CT periodic signal $x(t)$ with period P and bandlimited to $M\omega_0 = M(2\pi/P)$. Let $x(nT)$ be its sampled sequence with sampling period $T = P/N$ for some integer N. If $N > 2M$, then the CTFS coefficients c_m of $x(t)$ can be computed from the DTFS coefficients c_{md} of $x(nT)$ as

$$c_m = \begin{cases} c_{md} & \text{for } \overline{-(N-1)/2} \le m \le \overline{(N-1)/2} \\ 0 & \text{otherwise} \end{cases} \qquad (2.64)$$

where the overline rounds a number upward to an integer.

This theorem can be stated differently. The highest nonzero frequency component is $\omega_{\max} =$

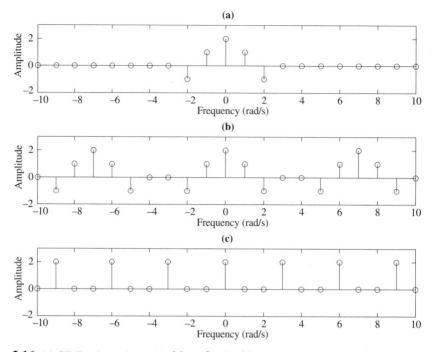

Figure 2.16 (a) CT Fourier series, with $M = 2$, of $x(t)$ with frequency range $(-\infty, \infty)$. (b) DT Fourier series of $x(nT)$ with $T = 2\pi/7$, which is periodic with period $N = 7$ and has the Nyquist frequency range $(-3.5, 3.5]$. (c) DT Fourier series of $x(nT)$ with $T = 2\pi/3$, which is periodic with period $N = 3$ and has the Nyquist frequency range $(-1.5, 1.5]$.

$M\omega_0 = M(2\pi/P)$ in rad/s, or $f_{\max} = M/P$ in Hz. The sampling frequency is $f_s = 1/T = N/P$. If $N > 2M$ or, equivalently, if

$$f_s > 2f_{\max}$$

then the CT Fourier series of $x(t)$ can be computed from the DT Fourier series of $x(nT)$. Once the CT Fourier series is obtained, the time signal $x(t)$ can be computed from (2.57). Thus Theorem 2.1 can also be stated as: If $f_s = N/P > 2f_{\max} = 2M/P$, then the CT periodic signal $x(t)$ can be recovered from its sampled sequence. This is a simplified version of the sampling theorem that we will develop in the next chapter.

◆ **Example 2.11**

Consider the CT periodic signal

$$x(t) = 1 + 2\cos t - \sin t$$

Its fundamental frequency is $\omega_0 = 1$ and its fundamental period is $P = 2\pi/\omega_0 = 2\pi$. Using Euler's formula, we have

$$x(t) = 1 + e^{jt} + e^{-jt} - \frac{e^{jt} - e^{-jt}}{2j}$$

$$= (1 - 0.5j)e^{-jt} + 1 + (1 + 0.5j)e^{jt}$$

Thus its CTFS coefficients are $c_{-1} = 1 - 0.5j$, $c_0 = 1$, $c_1 = 1 + 0.5j$, and $c_m = 0$ for all other m.

Next we compute the CTFS coefficients of $x(t)$ from its time samples. If we take N samples of $x(t)$ in $[0, \ P) = [0, \ 2\pi)$, then the sampling period is $T = 2\pi/N$ and the N time samples are $x(nT)$ for $n = 0, 1, \ldots, N-1$. The fundamental frequency of $x(nT)$ is $\omega_0 = 2\pi/(NT) = 2\pi/P = 1$, which is independent of N and is the same as the fundamental frequency of $x(t)$. If we denote the N values of $x(nT)$ by x, then shift(fft(x/N)) yields N DTFS coefficients of $x(nT)$ in the Nyquist frequency range $(-\pi/T, \pi/T] = (-N/2, N/2]$ in rad/s or at frequencies $\omega_0 = m\omega_0 = m$ with m in

$$m = \overline{-(N-1)/2} \ : \ \overline{(N-1)/2}$$

Consider the program

```
P=2*pi;N= ;T=P/N;n=0:N-1;
x=1+2*cos(n*T)-sin(n*T);
cd=shift(fft(x/N))
m=ceil(-(N-1)/2):ceil((N-1)/2)
```

For $N = 2$, the program generates cd=[1 2], which are to be located at m=[0 1]. For $N = 3$, we have cd=[1-0.5i 1 1+0.5i] and m=[-1 0 1]. Note that i and j both denote $\sqrt{-1}$ in MATLAB. We list in the following the results of the program for various N:

m or ω	$\ldots$	-3	-2	-1	0	1	2	3	$\ldots$
N=2					1	2			
N=3				$1 - 0.5i$	1	$1 + 0.5i$			
N=4				$1 - 0.5i$	1	$1 + 0.5i$	0		
N=5			0	$1 - 0.5i$	1	$1 + 0.5i$	0		
N=20	zeros(1, 6)	0	0	$1 - 0.5i$	1	$1 + 0.5i$	0	0	zeros(1, 7)
N=21	zeros(1, 7)	0	0	$1 - 0.5i$	1	$1 + 0.5i$	0	0	zeros(1, 7)

We see that if $N \geq 3$, the DTFS coefficiens of $x(nT)$ equal the CTFS coefficients of $x(t)$ in the Nyquist fequency range as shown above. This verifies Theorem 2.1.

For the signal in Example 2.11, if we know the signal to be bandlimited with $M = 1$, we may select $N = 3$ or $T = 2\pi/3 = 2.094$ and use FFT to compute three DTFS coefficients of $x(nT)$. The three coefficients are the three nonzero CTFS coefficients of $x(t)$ and we may then set to zero the remaining CTFS coefficients. If we selecte a larger N, FFT will generate three nonzero coefficients and the rest zero as shown. If we select $N = 1024$ or $T = 2\pi/1024 = 0.00614$, we will still obtain only three nonzero CTFS coefficients and the rest zero. Thus, for bandlimited signals, an unnecessarily large N or an unnecessarily small T will incur only more computation and will not generate new information except confirming that the signal is bandlimited.

We next give an example to illustrate a problem that may arise in computer computation.

◆ **Example 2.12**

Consider the CT periodic signal in (2.28). Its fundamental frequency is $\omega_0 = 0.3$, and its fundamental period is $P = 2\pi/0.3$. Its highest-frequency component is $1.5 = M\omega_0$ with $M = 5$. Let us select $N = 2M + 1 = 11$. Then the sampling period is $T = P/N = 2\pi/(11 \times 0.3)$. Computing the eleven-point DT Fourier series of $x(nT)$ by hand is complicated. We will use Program 2.2 with the first two lines replaced by

```
P=2*pi/0.3;N=11;T=P/N;D=2*pi/P;
n=0:N-1;
x=-1.2+0.8*sin(0.6*n*T)-1.6*cos(1.5*n*T);
```

Then the program will generate the plots in Figs. 2.17(a) and (b). We see that the magnitude plot is identical to the one in Fig. 2.9(b) for $|m| \leq 5$ or $|\omega| \leq 5\omega_0 = 1.5$. In the first sight, the phase plot in Fig. 2.17(b) is quite different from the one in Fig. 2.9(c). However, if we examine them more carefully, we will discover that if $c_m \neq 0$, then the corresponding phases in Fig. 2.17(b) denoted by solid dots and in Fig. 2.9(c) are the same except possibly modulo 360°. For example, the phases of c_0, $c_{\pm 2}$, and c_5 in both plots are the same. The phase of c_{-5} is 180° in Fig. 2.9(c) but -180° in Fig. 2.17(b). Thus we conclude that if $c_m \neq 0$, the phase of c_m can be computed from the phase of c_{md}.

If $c_m = 0$, we have automatically assigned its phase to be zero, as shown in Fig. 2.9(c). On a digital computer, the situation is more complex. For example, in the preceding computation, MATLAB generates

$$c_{1d} = 3.67 \cdot 10^{-16} - 6.13 \cdot 10^{-18}i; \quad c_{(-1)d} = -3.91 \cdot 10^{-16} + 5.29 \cdot 10^{-16}i$$

$$c_{3d} = -3.21 \cdot 10^{-16} - 7.11 \cdot 10^{-17}i; \quad c_{(-3)d} = 5.58 \cdot 10^{-17} + 8.46 \cdot 10^{-16}i$$

These infinitesimal values are due to computer roundoff errors[8] and can be consid-

---

[8] In theory, we should have $c_{id} = c_{(-i)d}^*$ for $i = 1, 3$. The computed c_{id} do not have the properties. Thus the nonzero infinitesimal values are due to computer roundoff errors.

ered as zero in practice. However, the ratios of their real and imaginary parts need not be small, and, consequently, their phases can be any value between $-180°$ and $180°$. Clearly, we should disregard these phases. If we disregard the phases corresponding to $c_{md} = 0$, then the part of Fig. 2.17(b) denoted by solid dots is the same, modulo $360°$, as the one in Fig. 2.9(c).

We repeat the computation for $N = 20$. The result is plotted in Figs. 2.17(c) and (d). We see that the magnitude plot is identical to the one in Fig. 2.9(b) or 2.17(a). If we disregard the phases corresponding to $c_{md} = 0$, the part of Fig. 2.17(d) denoted by solid dots is the same, modulo $360°$, as the one in Fig. 2.9(c) or 2.17(b).

We next discuss a method to resolve the problem of phases due to computational errors. In MATLAB, numbers are stored using 16 digits plus exponents. The magnitudes and phases of c_{md} in Program 2.2 are computed using stored data and computational errors will affect the computed phases. The MATLAB function **round** rounds a number to its nearest integer. Typing

Xr=round(X.∗10000)./10000

rounds **X** to four digits after the decimal point. Using rounded **Xr**, we will obtain the plots in Figs. 2.17(bb) and (dd). They are the same, modulo $360°$, as the plots in Fig. 2.9(c). This process eliminates the effect of computational rounding errors.

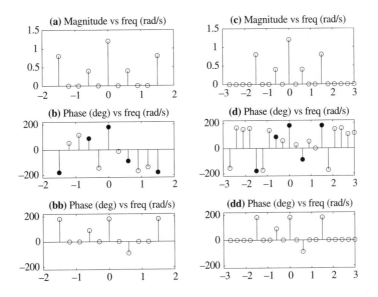

Figure 2.17 (a, b) CTFS from DTFS with $N = 11$. (c, d) With $N = 20$. (bb, dd) Using rounded DTFS.

From the preceding example, we see that the CT Fourier series of a bandlimited $x(t)$ can be computed from the DT Fourier series of its sampled sequence if the number of samples in one period is selected to be sufficiently large. In this case, repetitive shiftings of CTFS coefficients will not overlap and the DTFS equals the CTFS in the Nyquist frequency range. Note that they are always different outside the range because *the DTFS coefficients can be extended periodically but the CTFS coefficients are zero outside the range.*

To conclude this section, we mention that "folding" is also used in many DSP texts in place of "shifting." Using folding, the DT Fourier series becomes the sum of the repetitive *foldings* of the CT Fourier series. If a CT Fourier series is real and even, then folding and shifting will yield the same result. However, most CTFS coefficients encountered in practice have only even magnitudes but odd phases. In this case, folding will yield a correct magnitude plot but an incorrect phase plot. Thus we use shifting exclusively.

2.7.1 Frequency Aliasing due to Time Sampling

Consider again the CT periodic signal with its CTFS coefficients shown in Fig. 2.16(a). Now if we select $N = 3$ or sample the CT signal with sampling period $T = P/N = P/3$, then the DT Fourier series of the sampled sequence is the sum of repetitive shiftings, with period $N = 3$, of the CT Fourier series. The repetitive shiftings of the CT Fourier series will overlap with each other, and their sum is plotted in Fig. 2.16(c). We see that the resulting DT Fourier series is different from the original CT Fourier series in the Nyquist frequency range. This is called *frequency aliasing.* Thus if the sampling period is not sufficiently small, time sampling will introduce frequency aliasing, and we cannot compute the CTFS coefficients of $x(t)$ from its sampled sequence $x(nT)$ or, equivalently, cannot recover $x(t)$ from its sampled sequence $x(nT)$.

Consider a CT signal. Suppose it is not bandlimited or, equivalently, it has infinitely many nonzero CTFS coefficients. In this case, no matter how small the sampling period is chosen, frequency aliasing will always occur. However, if the CT signal has a finite average power, as is often the case in practice, then c_m will approach zero as $|m|$ approaches infinity. In this case, if N is sufficiently large, then we have

$$c_{m+kN} \approx 0 \tag{2.65}$$

or

$$\operatorname{Re} c_{m+kN} \approx 0 \quad \text{and} \quad \operatorname{Im} c_{m+kN} \approx 0$$

for $k \neq 0$ and for $\overline{-\gamma} \le m \le \overline{\gamma}$, where $\gamma = (N-1)/2$, and Re and Im denote, respectively, the real and imaginary parts. Let us write (2.62) as

$$c_{md} = \cdots + c_{m+N} + c_m + c_{m-N} + \cdots$$

Then (2.65) implies

$$c_m \approx \begin{cases} c_{md} & \text{for } \overline{-\gamma} \le m \le \overline{\gamma} \\ 0 & \text{otherwise} \end{cases} \tag{2.66}$$

In general, c_m is complex and we also have

$$|c_m| \approx \begin{cases} |c_{md}| & \text{for } \overline{-\gamma} \leq m \leq \overline{\gamma} \\ 0 & \text{otherwise} \end{cases} \tag{2.67}$$

$$\text{Re}[c_m] \approx \begin{cases} \text{Re}[c_{md}] & \text{for } \overline{-\gamma} \leq m \leq \overline{\gamma} \\ 0 & \text{otherwise} \end{cases} \tag{2.68}$$

and

$$\text{Im}[c_m] \approx \begin{cases} \text{Im}[c_{md}] & \text{for } \overline{-\gamma} \leq m \leq \overline{\gamma} \\ 0 & \text{otherwise} \end{cases} \tag{2.69}$$

However, Re $c_m \approx 0$ and Im $c_m \approx 0$ do not imply $\measuredangle c_m \approx 0$. Here we refer to actual values of c_m, not rounding errors discussed in Example 2.12. See Problems 2.6 and 2.19. Thus, in general, we have

$$\measuredangle c_m \neq \measuredangle c_{md} \quad \text{for all } m \tag{2.70}$$

In conclusion, given a CT periodic signal, if a sufficiently large number of samples is taken in one period, then its CTFS coefficients can be computed approximately from the DTFS of its sampled sequence, as we show in the next example.

◆ Example 2.13

Consider the sequence of boxcars with $P = 4$ and $a = 1$ shown in Fig. 2.10(a) and with its CTFS coefficients shown in Fig. 2.10(b) and repeated in Fig. 2.18 with hollow dots. It has infinitely many nonzero coefficients and is not bandlimited. Because the signal is real and even, so is the set of its CTFS coefficients. First we select $N = 4$ or take four samples in [0, 4). The boxcar equals 1 in [0, 1] and [3, 4), and equals 0 in (1, 3). If $N = 4$, then the sampling period is $4/N = 1$, and the four samples are given by $x = [1 \ 1 \ 0 \ 1]$. Then the program that follows

Program 2.3
```
N=4;P=4;T=P/N;D=2*pi/P;
x=[1 1 0 1];
X=fft(x/N);
m=ceil(-(N-1)/2):ceil((N-1)/2);
stem(m*D,shift(X),'fill')
```

computes and plots the DTFS of the sampled sequence in Fig. 2.18(a) with solid dots. Note that the fundamental frequency is denoted by **D** in the program. We see that the four solid dots are quite different from the corresponding hollow dots. In other words, frequency aliasing is appreciable.

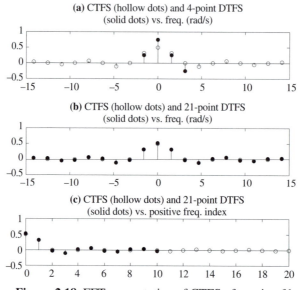

Figure 2.18 FFT computation of CTFS of a train of boxcars.

Next we modify Program 2.3 for a general N. If we take N samples in $[0, 4)$, then the sampling period is $T = P/N = 4/N$. The value of the boxcar is 1 for t in $[0, 1]$ and $[3, 4)$, and 0 for t in $(1, 3)$. Let q=floor(1/T), where floor rounds a number downward to an integer. Then there are $q + 1$ samples in $[0, 1]$, q samples in $[3, 4)$ and $N - 2q - 1$ samples in $(1, 3)$. If we replace the first two lines of Program 2.3 by

```
N= ;P=4;T=P/N;D=2*pi/P;q=floor(1/T);
x=[ones(1,q+1) zeros(1,N-2*q-1) ones(1,q)]
```

then it is applicable for any integer N. For $N = 21$, it generates the solid dots in Fig. 2.18(b). We see that they are quite close to the hollow dots. This verifies (2.66).

Because CTFS coefficients of real-valued signals have even magnitudes and odd phases, we may plot only the positive frequency part or $m \geq 0$. In this case, replacing the last two lines of Program 2.3 by

```
mp=0:ceil((N-1)/2);
stem(mp,X(mp+1),'fill')
```

will yield the solid dots in Fig. 2.18(c). The hollow dots are exact values and are nonzero for all $m \geq 0$, whereas the computed CTFS are implicitly assumed to be zero for $m > \text{ceil}(N - 1)/2$.

Note that the horizontal axes in Figs. 2.18(a) and 2.18(b) are frequency in rad/s, whereas the horizontal axis in Fig. 2.18(c) is frequency index m.

It is possible to compute the largest magnitude percentage error between the N-point FFT computed CTFS coefficients and the exact coefficients. The exact CTFS of the boxcar was computed in Example 2.5, for $P = 4$ and $a = 1$, as $c_0 = 1/2$ and $c_m = 2(\sin \pi m/2)/2\pi m$ for all nonzero m. Consider the program that follows

Program 2.4

```
N= ;P=4;T=P/N;q=floor(1/T);
x=[ones(1,q+1) zeros(1,N-2*q-1) ones(1,q)];
X=fft(x/N);
e1=abs(X(1)-1/2);
k=1:ceil((N-1)/2);
e2=max(abs(abs(X(k+1))-abs(2.*sin(k*pi/2)./(2*pi.*k))));
k1=ceil((N-1)/2)+1:2*N;
e3=max(abs(2.*sin(k1*pi/2)./(2*pi.*k1)));
mp=0:ceil((N-1)/2);
mm=max(abs(X(mp+1)));
pe=max([e1 e2 e3])/mm*100
```

We compare only magnitudes for $m \geq 0$ because of evenness of magnitudes. The index of X starts from 1 instead of 0. Thus the number e1 in the program is the error between the computed and exact magnitudes at $m = 0$; e2 is the largest magnitude error for m=1:ceil((N-1)/2). Because the computed coefficients are automatically assumed to be zero for m>ceil((N-1)/2), the largest error between the exact and computed magnitudes for all m>ceil((N-1)/2) is simply the largest magnitude of the exact CTFS for all m>ceil((N-1)/2). Instead of computing its largest magnitude with m from ceil((N-1)/2)+1 to ∞, we compute it with m from ceil((N-1)/2)+1 to $2N$. The result is listed as e3. Instead of using the exact peak magnitude 0.5, we use the computed peak magnitude, which is computed as mm in the program. Thus pe as defined in the program is the percentage error. The percentage error is 50 for $N = 4$, 6.5 for $N = 21$, and 0.78 for $N = 256$. Thus if we take 256 time samples in one period, then the computed coefficients differ from the exact ones by less than 1% of the peak magnitude.

This example showed that even for CT periodic signals that are not bandlimited, their CTFS coefficients can still be computed from their sampled sequences. The more samples in one period we take, the more accurate the result.

2.7.2 Selecting N to Have Negligible Frequency Aliasing

In the preceding section, we discussed the selection of N with prior knowledge of the CTFS coefficients c_m. If c_m are known, there is no more need to compute them using FFT. Now we discuss how to select N without any knowledge of c_m. Before proceeding, we mention that the larger N is, the more accurate the computed result. However, a large N requires more computation. Thus it is desirable to select a smallest possible N, within an acceptable accuracy, to compute the CTFS coefficients of a CT signal from its sampled sequence.

Consider a CT periodic signal $x(t)$ with period P and with unknown c_m. If we take N samples in one period, then the sampling period is $T = P/N$. We compute the DTFS c_{md} of the sampled sequence $x(nT)$. The question is: Are the computed c_{md} close to c_m in the Nyquist frequency range $(-\pi/T, \ \pi/T]$? Or, equivalently, is the selected N large enough or T is small enough so that the effect of frequency aliasing is negligible? If the computed c_{md} have large magnitudes in the neighborhood of $\pm\pi/T$, then the effect of frequency aliasing is probably significant. On the other hand, if they are identically zero as in Fig. 2.17(c) or practically zero as in Fig. 2.18(b) in the neighborhood of $\pm\pi/T$, then it is highly possible that the effect of frequency aliasing is negligible and c_m roughly equals c_{md} inside the Nyquist frequency range and zero outside the range. Using this property, we will discuss ways of selecting N.

We now use the periodic signal in Example 2.13 to illustrate a procedure of selecting N. Let us select a very large N, say, $N = 4096$. Then the program that follows

Program 2.5
```
P=4;N=4096;T=P/N;q=floor(1/T);D=2*pi/P;
x=[ones(1,q+1) zeros(1,N-2*q-1) ones(1,q)];
X=shift(fft(x/N));
m=-N/2+1:N/2;
stem(m*D,abs(X))
```

will yield the magnitude plot in the Nyquist frequency range $(-\pi/T, \ \pi/T]$ shown in Fig. 2.19(a). Because the magnitudes are practically zero in the neighborhood of $\pm\pi/T = \pm3215.36$ in rad/s, the effect of frequency aliasing due to time sampling is negligible, and the result should be close to the exact one, and we may stop the computation.

Although $N = 4096$ yields a good result, because most significant nonzero magnitudes are limited roughly to $(-200, 200]$ in rad/s, the selected N is unnecessarily large. We compute $400/D = 254.8$, where D is the fundamental frequency. Now we select N in Program 2.5 as 256, then the program will yield the plot in Fig. 2.19(b). This result is not as accurate as the one using $N = 4096$ but is probably acceptable in practice. This is one way of selecting N.

We now discuss a different way. Let us select N_1 and compute its DTFS X_1. If the effect of frequency aliasing is small, then the computed X_1 should be close to the actual CTFS in the Nyquist frequency range, that is, $(-\pi/T_1, \ \pi/T_1]$ with $T_1 = P/N_1$. However, because the actual CTFS are not available, we compare them with those computed using, for example, $N_2 = 2N_1$. Let X_2 be the DTFS computed using N_2. If X_2 is very close to X_1 inside $(-\pi/T_1, \ \pi/T_1]$ and is practically zero outside the range, we stop the computation and accept X_2 as the

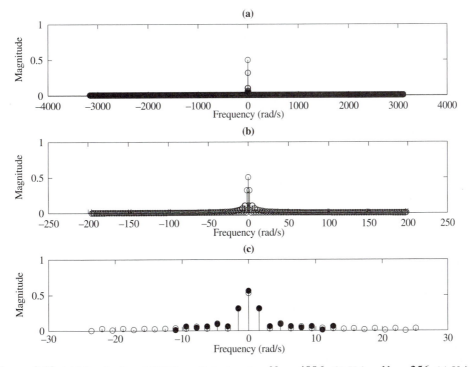

Figure 2.19 (a) Magnitudes of CTFS coefficients using $N = 4096$. (b) Using $N = 256$. (c) Using $= 16$ (solid dots) and $N = 32$ (hollow dots).

CTFS of the CT signal. If not, we repeat the computation. Before proceeding, we mention that the condition $X_1 \approx X_2$ inside the range $(-\pi/T_1, \ \pi/T_1]$ probably implies the condition $X_2 \approx 0$ outside the range and vice versa. Thus we may check only whether $X_1 \approx X_2$ inside the range.

We again use the signal in Example 2.13 to illustrate the procedure. First we select $N_1 = 16$ and compute and plot its result in Fig. 2.19(c) with solid dots. We then repeat the computation with $N_2 = 2N_1$ and plot the result in Fig. 2.19(c) with hollow dots. We see that the solid dots are slightly different from the hollow dots. Thus the effect of frequency aliasing due to time sampling for $N = 16$ is not small and $N = 16$ is not large enough. Whether $N = 32$ is large enough is to be determined by comparing its result with the result using $N = 64$.

The comparison of the solid and hollow dots in Fig. 2.19(c) was done visually. Now we compare them mathematically. We compare them only for $m \geq 0$ because of evenness of magnitudes. Let X1 be the result using $N = 16$. Then the solid dots from $m = 0$ on in Fig. 2.19(c) are the magnitudes of X1(m+1) for m=0:8. Let X2 be the result using $N = 32$. Then the hollow dots from $m = 0$ on in Fig. 2.19(c) are the magnitudes of X2(m+1) for m=0:16. The program that follows

```
k=0:8;
d=max(abs(abs(X1(k+1))-abs(X2(k+1))));
mm=max(abs(X1(k+1)));
pd=d/mm*100
```

computes the largest magnitude difference d between the solid and hollow dots in Fig. 2.19(c). The number mm is the peak magnitude. If d is small comparing to mm, or the percentage difference pd as defined in the program is small, frequency aliasing due to time sampling is small. For the two computations in Fig. 2.19(c), the percentage difference is 5.56%. If this is not acceptable, we may increase N until the percentage magnitude difference is less than a specified value.

Let us summarize the preceding discussion. Consider a CT periodic signal with unknown CTFS coefficients. There are two possible ways to compute its CTFS coefficients using FFT. We may start with a very large N to find the range of nonzero CTFS coefficients and then select a smaller N. Alternatively, we may start with a small N_1 and compute X_1. We then repeat the computation with $N_2 = 2N_1$. If the computed X_2 is practically the same as X_1 for $m = -N_1/2+1 : N_1/2$, we may stop the computation and accept X_2 as the CTFS of the periodic signal. If not, we repeat the computation until the results of two consecutive computations are close. We mention that in this computation, we may select N to be even or a power of 2. If we select N as a power of 2, then FFT computation will be very efficient as we will discuss in Chapter 4.

The preceding procedure can be formulated as follows. Find the smallest integer a in $N = 2^a$ such that the largest magnitude percentage difference between the coefficients computed using N and $N/2$ is less than β. The program that follows will automatically carry out this computation for the CT signal in Example 2.13.

Program 2.6

```
a=1;b=100;P=4;D=2*pi/P;beta=1;
while b>beta
        N1=2^ a;T1=P/N1;q1=floor(1/T1);
        x1=[ones(1,q1+1) zeros(1,N1-2*q1-1) ones(1,q1)];
        X1=fft(x1/N1);
        N2=2*N1;T2=P/N2;q2=floor(1/T2);
        x2=[ones(1,q2+1) zeros(1,N2-2*q2-1) ones(1,q2)];
        X2=fft(x2/N2);
        m1p=0:N1/2;
        d=max(abs(abs(X1(m1p+1))-abs(X2(m1p+1))));
        mm=max(abs(X1(m1p+1)));
        b=d/mm*100;
        a=a+1;
```

```
end
N2,b
m=-N2/2+1:N2/2;
stem(m*D,abs(shift(X2)))
```

If we select $\beta = 1$, then the program will yield $N2 = 256$, $b = 0.77$, and the hollow dots in Fig. 2.19(b). Note that in the program, the initial b can be selected arbitrarily so long as it is larger than beta; the initial a can be selected as 2, 3, or 4, but should not be very large.

We give one example to demonstrate the usefulness of of the FFT method over the analytical method.

◆ **Example 2.14**

Consider the CT periodic signal $x(t)$ with period $P = 2$ and $x(t) = t^2 - 1.5\cos 20t$, for $0 \leq t < 2$, shown in Fig. 2.20(a). Compute its CTFS coefficients.

Analytical computation of its CTFS coefficients requires to carry out the integration that follows

$$c_m = \frac{1}{2} \int_0^2 (t^2 - 1.5\cos 20t)e^{-jm\omega_0 t}\,dt$$

with $\omega_0 = 2\pi/2 = \pi$. It is complicated. We now modify Program 2.6 as

```
Program 2.7
a=1;b=100;P=2;D=2*pi/P;beta= ;
while b>beta
        N1=2^a;T1=P/N1;n1=0:N1-1;
        x1=(n1*T1).^2-1.5*cos(20*n1*T1);
        X1=fft(x1/N1);
        N2=2*N1;T2=P/N2;n2=0:N2-1;
        x2=(n2*T2).^2-1.5*cos(20*n2*T2);
        X2=fft(x2/N2);
        m1p=0:N1/2;
        d=max(abs(abs(X1(m1p+1))-abs(X2(m1p+1))));
        mm=max(abs(X1(m1p+1)));
        b=d/mm*100;
        a=a+1;
    end
    N2,b
```

```
m=-N2/2+1:N2/2;
stem(m∗D,abs(shift(X2)))
```

This differs from Program 2.5 only in the input sequences **x1** and **x2**; the rest are identical. If we select $\beta = 2$, then the program will yield $N2 = 256$, $b = 1.38$, and the plot in Fig. 2.20(b). This result should be close to the exact CTFS coefficients of the periodic signal. Note that the left-most frequency is $(-N/2+1)D = -127\pi = -398.98$ in rad/s and the right-most frequency is $(N/2)D = 128\pi = 402.12$ in rad/s. Thus MATLAB automatically selects the frequency range from -400 to 500 rad/s.

Because the magnitude plot is even, we may plot only the positive-frequency part. Replacing the last two lines of the preceding program by

```
mp=0:N2/2;
stem(mp∗D,abs(X2(mp+1)))
```

yields the plot in Fig. 2.20(c). We see that both Figs. 2.20(b) and (c) can be easily obtained.

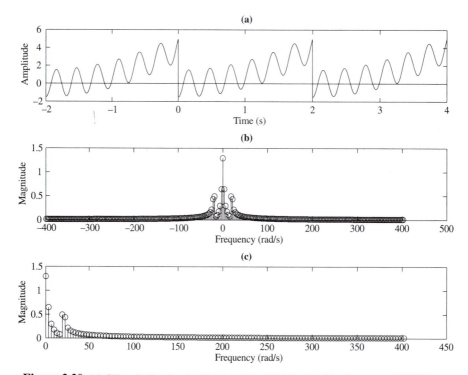

Figure 2.20 (a) CT periodic signal with period 2 s. (b) Magnitudes of computed CTFS c_m. (c) Magnitude plot for $m \geq 0$.

From these examples, we see that CTFS coefficients of CT periodic signals can indeed be computed using FFT.

2.8 Average Power and Its Computation[9]

Consider a CT periodic signal $x(t)$ with period P. Its average power is, as derived in (2.39)

$$P_{av} = \frac{1}{P} \int_0^P |x(t)|^2 dt = \sum_{m=-\infty}^{\infty} |c_m|^2 \tag{2.71}$$

where c_m is its CTFS coefficients. Before discussing its computer computation, we discuss its DT counterpart. Let $x[n] = x(nT)$ be a periodic sequence with period N. Then we have

$$\frac{1}{N} \sum_{n=0}^{N-1} |x[n]|^2 = \sum_{m=<N>} |c_{md}|^2 \tag{2.72}$$

Its derivation is similar to the CT case and will not be repeated.

One way to compute P_{av} in (2.71) on a computer is to approximate it by

$$P_{av} \approx \frac{1}{P} \sum_{n=0}^{N-1} |x(nT)|^2 T = \frac{T}{P} \sum_{n=0}^{N-1} |x(nT)|^2 \tag{2.73}$$

where $T := P/N$. In this equation, $x(t)$ is approximated by a stepwise signal as shown in Fig. 2.21. It is clear that the larger N is or the smaller T is, the better the approximation.

Equation (2.73) can be written as, using $P = NT$,

$$P_{av} \approx \frac{1}{N} \sum_{n=0}^{N-1} |x(nT)|^2 \tag{2.74}$$

which becomes, using (2.72),

$$P_{av} \approx \sum_{m=0}^{N-1} |c_{md}|^2 \tag{2.75}$$

where c_{md}, $m = 0, 1, \ldots, N-1$, is the DT Fourier series of $x(nT)$, $n = 0, 1, \ldots, N-1$, and can be computed using FFT. Thus we can use FFT to compute the average power of CT periodic signals.

---

[9] This section may be skipped without loss of continuity.

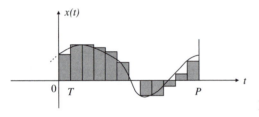

Figure 2.21 Stepwise approximation of $x(t)$.

◆ **Example 2.15**

Consider the CT periodic signal in (2.28) or

$$x(t) = -1.2 + 0.8 \sin 0.6t - 1.6 \cos 1.5t$$

Its fundamental frequency is $\omega_0 = 0.3$, and its fundamental period is $P = 2\pi/\omega_0 = 20.94$. The following program

Program 2.8
```
N=11;P=2*pi/0.3;T=P/N;
n=0:N-1;
x=-1.2+0.8*sin(0.6*n*T)-1.6*cos(1.5*n*T);
P1=sum(x.^2)*T/P
X=fft(x/N);
P2=sum(abs(X).^2)
```

computes the average power using (2.73) and (2.75) for $N = 11$. The average power computed by using (2.73) is denoted by P1 and the one using FFT in (2.75) is denoted by P2. In computing the average power, there is no need to shift the output of FFT. The result is **P1=P2=3.04**. We repeat the computation for $N = 20$, and the result is again **P1=P2=3.04**. In fact, for any N larger than 10, the computed average power will be the exact value. This can be easily explained from the fact that the CT signal $x(t)$ is bandlimited and we have $c_m = c_{md}$ for $m = \text{ceil}(-(N-1)/2) : \text{ceil}((N-1)/2)$ if $N \geq 11$.

Now we discuss the distribution of power in frequencies. In this case, we must use (2.75) or an FFT. Suppose we want to compute the average power in the frequency range $[\omega_l, \ \omega_u]$ with $0 \leq \omega_l < \omega_u \leq \pi/T$. Here we assume implicitly that the power in $[\omega_l, \ \omega_u]$ includes the power in the negative frequency range $[-\omega_u, \ -\omega_l]$. Let $\omega_0 = 2\pi/P$ be the fundamental frequency of a CT periodic signal with period P. Define

$$m_l := \overline{\omega_l/\omega_0} \quad \text{and} \quad m_u := \omega_u/\omega_0 \tag{2.76}$$

where m_l is the least integer larger than or equal to ω_l/ω_0 and can be obtained in MATLAB as $m_l = \text{ceil}(\omega_l/\omega_0)$. The integer m_u is the largest integer smaller than or equal to ω_u/ω_0 and can be obtained in MATLAB as $m_u = \text{floor}(\omega_u/\omega_0)$. Then the power in the frequency range $[\omega_l, \omega_u]$ roughly equals

$$P \approx 2 \left(\sum_{m=m_l}^{m_u} |c_{md}|^2 \right) \tag{2.77}$$

if $\omega_l > 0$ or

$$P \approx |c_{0d}|^2 + 2 \left(\sum_{m=1}^{m_u} |c_{md}|^2 \right) \tag{2.78}$$

if $\omega_l = 0$. The approximation becomes an equality if $x(t)$ is bandlimited and if N is sufficiently large. The two equations in (2.77) and (2.78) can easily be programmed in MATLAB, as the next example illustrates.

◆ **Example 2.16**

Consider the problem in Example 2.7, that is, to compute the power of the periodic signal in Fig. 2.10(a) in the frequency range $[-5, 5]$ in rad/s. The period of the signal is $P = 4$, and its fundamental frequency is $D := \omega_0 = 2\pi/P = 2\pi/NT$. Let us take N=1024 samples in $[0, 4)$. Then the following program will compute the average power in the frequency range $[-5, 5]$.

Program 2.9
```
N=1024;D=2*pi/4;
x=[ones(1,257) zeros(1,511) ones(1,256)];
X=fft(x/N);
mu=floor(5/D);
m=2:mu+1;
p=abs(X(1))^2+2*sum(abs(X(m)).^2)
```

We explain first the program. The train of boxcars is 1 in $0 \le t \le 1$ and $3 \le t < 4$ and is zero in $1 < t < 3$. Because the sampling period is $T = 4/1024 = 1/256$, there are 257 samples of one in $[0, 1]$ and 256 samples of one in $[3, 4)$. Excluding these ones, there are 511 samples of zero in $(1, 3)$. Thus the sampled sequence can be expressed as in x. Its DT Fourier series can be computed as X=fft(x/N). The index of X(m) starts from 1 rather than 0; thus the summation from 1 to m_u in (2.78) becomes the summation from 2 to $mu + 1$ in MATLAB. Note the use of ".^" inside the function sum. The program generates p=0.4761, which differs from the exact value 0.4752 computed in Example 2.7 only by

$$\frac{0.4761 - 0.4752}{0.4752} = \frac{0.0009}{0.4752} = 0.0019$$

or 0.19 percent.

2.9 Concluding Remarks

We recapitulate what we have discussed in this chapter. Consider a DT periodic sequence $x[n] = x(nT)$ with period N and sampling period T. Such a sequence has a different but equivalent representation called the DTFS coefficients c_{md}. Such a representation can reveal more clearly the frequency content of the sequence and can be easily obtained by using FFT as fft(x/N), where x=[x[0] x[1] . . . x[N-1]]. Its outputs are the coefficients c_{md}, for $m = 0, 1, \ldots, N - 1$, and are to be located at discrete "frequencies" $\omega = m(2\pi)/NT$, for $m = 0, 1, \ldots, N - 1$. The output of X=shift(fft(x/N)) are the DTFS coefficients of the periodic sequence located at discrete frequencies $m(2\pi/NT)$ lying inside the Nyquist frequency range $(-\pi/T, \pi/T]$ or with m in (2.50). Program 2.2 is generic and can be used to compute the DTFS coefficients of any periodic sequence. Note that the MATLAB function shift can be replaced by fftshift. In this case, the range of frequency index m must be modified as in Table 2.2.

The same procedure can be applied to compute the CTFS coefficients of any CT periodic signal $x(t)$ with period P from its sampled sequence $x(nT)$. Because the frequency range of $x(t)$ is $(-\infty, \infty)$ and the frequency range of $x(nT)$ is $(-\pi/T, \pi/T]$, we must select T so that all significant frequency components of $x(t)$ are contained in $(-\pi/T, \pi/T]$. In order for $x(nT)$ to be periodic, we select first an integer N, and compute $T = P/N$. Then $x(nT)$ is periodic with period N samples or $NT = P$ s, and has the same fundamental frequency as $x(t)$. One way of selecting N is to compare the results of two consecutive computations as discussed in Programs 2.6 and 2.7. If $x(t)$ is bandlimited, the procedure will yield exact CTFS coefficients; otherwise, it will yield only approximate coefficients due to frequency aliasing. If frequency aliasing occurs, the computed phases are generally useless.

PROBLEMS

2.1   Which of the following are periodic? If any are, find their periods in samples.

(a) $\sin 1.2n$
(b) $\sin 9.7\pi n$
(c) $e^{j1.6\pi n}$
(d) $\cos(3\pi n/7)$

2.2   Find the frequencies of the signals in Problem 2.1 if the sampling period is 2 s. Repeat the problem if the sampling frequency is 2 Hz.

2.3   What is the frequency of the CT signal $\sin(\pi t + \pi/4)$? What is the frequency of its sampled sequence $\sin(\pi n T + \pi/4)$ if the sampling period T is 3 s? Repeat the question for $T = 2$, 1, and 0.5. Under what condition on T will the frequency of the sampled sequence of $\sin(\pi t + \pi/4)$ equal π rad/s?

2.4   Consider (2.10) with unknown A, ω, and θ. Suppose we have

$$x(-T) = 4.4560, \quad x(0) = 4.9875, \quad x(T) = 4.7315$$

with $T = 0.1$. What is its CT signal? [*Answer:* $5\sin(4t + 1.5)$]

2.5   Consider the CT signal

$$x(t) = -0.8 + \cos 0.7t - 1.6\sin 2.1t$$

What are its fundamental frequency and fundamental period? Find its CT Fourier series. Plot its magnitude and phase versus frequency in rad/s. Is the magnitude plot even? Is the phase plot odd?

2.6   Consider a CT periodic signal $x(t)$ with period $P = 2$ and $x(t) = 2t$ for $0 \le t < 2$. Find its CT Fourier series. Plot its magnitude and phase versus frequency in rad/s. Is the magnitude plot even? Is the phase plot odd? Do the magnitude and phase approach 0 as $|m| \to \infty$?

2.7   What is the total energy of the signal in Problem 2.5? What is its average power? Repeat the questions for the signal in Problem 2.6.

2.8   For the signal in Problem 2.5, find the percentage of the average power lying inside the frequency range $[-1, \ 1]$ in rad/s.

2.9   For the signal in Problem 2.6, find the percentage of the average power lying inside the frequency ranges $[1, \ 12]$ and $[-12, \ -1]$.

2.10 Verify the CTFS coefficients in Fig. 2.12.

2.11 Find N distinct harmonically related periodic sequences of (2.40) for $N = 3, 5$, and 6.

2.12 Find the DT Fourier series of a DT periodic sequence with period $N = 3$, sampling period $T = 0.5$, $x[0] = 2$, $x[1] = 1$, and $x[2] = 1$. Plot its coefficients versus frequency in rad/s. Repeat the problem using FFT.

2.13 Find the DT Fourier series of a DT periodic sequence with period $N = 4$, sampling period $T = 1$, $x[-1] = -1$, $x[0] = 2$, $x[1] = 1$, and $x[2] = 1$. Plot its magnitude and phase versus frequency in rad/s.

2.14 Repeat Problem 2.13 using FFT. In using FFT, will you obtain the same result if the inputs of FFT are x=[2 1 1 -1] and x=[-1 2 1 1]? Can you give reasons for their differences?

2.15 Consider the signal

$$x(t) = -0.5 + \cos 2t + 2\sin(2t + \pi/4)$$

What is its CTFS? Is it bandlimited? If yes, what is the minimum number of time samples of $x(t)$ in one period needed to compute the CTFS from its samples? Use N-point FFT to compute its CTFS coefficients from $x(nT)$ with $T = \pi/N$ for $N = 2, 3, \ldots, 6$. Are the results consistent with Theorem 2.1?

2.16 Is the CT signal in Problem 2.5 bandlimited? What is the minimum number of samples in one period in order to compute its CT Fourier series from its sampled sequence? Use an 8-point FFT to compute and plot its CT Fourier series from its sampled sequence. Does the phase plot equal the one obtained in Problem 2.5? If not, how to modify the program so that the resulting phase plot will be the same as the one obtained in Problem 2.5?

2.17 Repeat Problem 2.16 using a 13-point FFT.

2.18 Repeat Problems 2.16 and 2.17 using fftshift instead of shift. If you do not modify the frequency index, will you obtain correct results?

2.19 Is the CT signal in Problem 2.6 bandlimited? Can you compute its CT Fourier series exactly from its sampled sequence? If we compute its CTFS coefficients using $N = 8$ samples in one period, what is the largest error, as a percentage of the peak magnitude, between the computed magnitudes and the exact magnitudes for all m? What is the largest error if $N = 128$?

2.20 Develop for Problem 2.6 a program similar to Program 2.7 to find the smallest a in $N = 2^a$ so that the percentage difference between the magnitudes of the results computed using N and $N/2$ is less than 1%. Is the magnitude plot using $N = 2^a$ close to the one in Problem 2.6? How about the phase plot?

2.21 Repeat Problems 2.8 and 2.9 using FFT.

2.22 Consider the CT periodic signal $x(t) = 1/(1 - t)$, for $0 \le t < 1$, with period $P = 1$ as shown in Fig. 2.22(a). Is it absolutely integrable in one period? Consider the CT

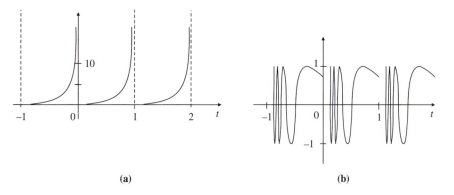

Figure 2.22 Two CT periodic signals with period 1. (a) $x(t) = 1/(1-t)$, $0 \le t < 1$. (b) $x(t) = \sin(1/t)$, $0 < t \le 1$.

periodic signal $x(t) = \sin(1/t)$, for $0 < t \le 1$, with period 1 as shown in Fig. 2.22(b). Is it absolutely integrable? Does it have a finite number of maxima and minima in one period?

2.23   Can you compute analytically the CT Fourier series of the signal in Fig. 2.22(a)? Use $N = 16$- and $2N$-point FFT to compute the DTFS coefficients of its sampled sequence and compare their results. Repeat the computation for $N = 32$. Will they converge? (*Warning*: Do not use a while loop as in Program 2.6 or 2.7; it may not stop.)

2.24   Suppose $x[k] = \cos(2\pi k/N)$, for $k = 0, 1, \ldots, N-1$, are stored in read-only memory (ROM). Let us pick up every sample circularly every second (that is, $T = 1$). By this, we mean that after the sample at $k = N-1$, we go back to $k = 0$. What is the frequency of the generated DT sinusoidal sequence? Now if we pick up every Mth sample, that is, $k = Mn$, for $n = 0, 1, 2, \ldots$, and for $M = 2, 3, \ldots$, what is the frequency of the generated sinusoidal sequence? How many different sinusoidal sequences can you generate?

CHAPTER 3

CT and DT Fourier Transforms— Frequency Spectra

3.1 Introduction

We introduced in Chapter 2 Fourier series to reveal the frequency contents of periodic signals. In this chapter, we shall modify it so that it can also be applied to aperiodic signals. The modified transform is called the CT or DT Fourier transform. If the Fourier transform of a signal is defined, the transform is called the frequency spectrum of the signal, that is

$$\text{Fourier transform} \Leftrightarrow \text{frequency spectrum}$$

In this chapter, we will introduce formally the concept of frequency spectrum and discuss its properties and other related issues.

3.2 CT Fourier Transform (CTFT)

The CT Fourier series of periodic $x(t)$ with period P and fundamental frequency $\omega_0 = 2\pi/P$ is defined in (2.22) and (2.23) as

$$x(t) = \sum_{m=-\infty}^{\infty} c_m e^{jm\omega_0 t} \tag{3.1}$$

with

$$c_m = \frac{1}{P} \int_{t=-P/2}^{P/2} x(t)e^{-jm\omega_0 t}\,dt \tag{3.2}$$

Now we will modify it so that it can also be applied to aperiodic CT signals. First we multiply (3.2) by P to yield

$$X(m\omega_0) := c_m P = \int_{t=-P/2}^{P/2} x(t)e^{-jm\omega_0 t}\,dt \tag{3.3}$$

where we have defined $X(m\omega_0) := c_m P$. This definition is justified because the coefficient c_m is associated with frequency $m\omega_0$. We then use $X(m\omega_0)$ and $\omega_0 = 2\pi/P$ or $1/P = \omega_0/2\pi$ to rewrite (3.1) as

$$x(t) = \frac{1}{P} \sum_{m=-\infty}^{\infty} c_m P e^{jm\omega_0 t} = \frac{1}{2\pi} \sum_{m=-\infty}^{\infty} X(m\omega_0)e^{jm\omega_0 t}\omega_0 \tag{3.4}$$

A periodic signal with period P becomes aperiodic if P approaches infinity. Define $\omega := m\omega_0$. As $P \to \infty$, we have $\omega_0 = 2\pi/P \to 0$. In this case, ω becomes a continuum and ω_0 can be written as $d\omega$. Furthermore, the summation in (3.4) becomes an integration. Thus the modified Fourier series pair in (3.3) and (3.4) becomes, as $P \to \infty$,[1]

$$X(\omega) = \mathcal{F}[x(t)] := \int_{t=-\infty}^{\infty} x(t)e^{-j\omega t}\,dt \qquad \text{(analysis eq.)} \tag{3.5}$$

$$x(t) = \mathcal{F}^{-1}[X(\omega)] := \frac{1}{2\pi} \int_{\omega=-\infty}^{\infty} X(\omega)e^{j\omega t}\,d\omega \qquad \text{(synthesis eq.)} \tag{3.6}$$

This is the *CT Fourier transform* pair. $X(\omega)$ is called the Fourier transform of $x(t)$ and $x(t)$ the inverse Fourier transform of $X(\omega)$. They are also called the analysis and synthesis equations.

The Fourier transform is defined as an integration over $(-\infty, \infty)$, and for some x(t), the integration may diverge and its Fourier transform is not defined. Sufficient conditions for $x(t)$ to have a Fourier transform are

1. $x(t)$ is *absolutely integrable* in the sense

$$\int_{-\infty}^{\infty} |x(t)|\,dt < \infty$$

   and

2. $x(t)$ has a finite number of discontinuities and a finite number of maxima and minima in every finite time interval.

---

[1] We use A:=B to denote that A, by definition, equals B. We use A=:B to denote that B, by definition, equals A.

They are, as in the CT Fourier series, called the *Dirichlet conditions.* If a signal meets the conditions, the infinite integration in (3.6) will converge to a value at every t. The value equals $x(t)$ if $x(t)$ is continuous and the midpoint of $x(t)$ if $x(t)$ is not continuous at t; that is,

$$\frac{x(t+) + x(t-)}{2} = \frac{1}{2\pi} \int_{\omega=-\infty}^{\infty} X(\omega)e^{j\omega t} d\omega$$

Except for some mathematically contrived functions, most signals meet the second condition of the Dirichlet conditions. Therefore, we discuss only the first condition. The signals $x(t) = 1$, $\sin 2t$, and $e^{0.1t}$ for all t in $(-\infty, \infty)$ are not absolutely integrable. The signal $x(t) = e^{-0.1t}$, for $t \geq 0$, and $x(t) = 0$, for $t < 0$, is absolutely integrable, because

$$\int_{-\infty}^{\infty} |x(t)| dt = \int_{0}^{\infty} e^{-0.1t} dt = \frac{1}{-0.1} e^{-0.1t} \Big|_{0}^{\infty}$$
$$= -10(0 - 1) = 10$$

If a signal is *bounded* in the sense

$$|x(t)| \leq q \quad \text{for all } t \text{ in } (-\infty, \infty)$$

for some constant q, and of *finite duration* in the sense

$$x(t) = 0 \quad \text{for } t < t_1 \text{ and } t > t_2$$

for some finite time instants t_1 and t_2 with $t_1 < t_2$, then we have

$$\int_{-\infty}^{\infty} |x(t)| dt \leq \int_{t_1}^{t_2} q dt = q(t_2 - t_1) < \infty$$

Thus if a signal is bounded and of finite duration, it is absolutely integrable and its Fourier transform is always defined. If the Fourier transform of a CT signal is defined, the transform is called the *frequency spectrum* or, simply, the *spectrum* of the signal. From now on, the Fourier transform, frequency spectrum, and spectrum, will be used interchangeably. The frequency spectrum is generally complex valued. Its magnitude $|X(\omega)|$ is called the *magnitude spectrum* and its phase $\not{\hspace{-2pt}\times} X(\omega)$, the *phase spectrum*. Here are some examples.

◆ **Example 3.1**

Consider the function

$$x(t) = \begin{cases} e^{at} & \text{for } t \geq 0 \\ 0 & \text{for } t < 0 \end{cases}$$

where a can be real or complex. If Re $a > 0$, where Re stands for the real part, the function grows unbounded either exponentially or oscillatorily, and its CT Fourier transform is not defined. If Re $a < 0$, the function is absolutely integrable and its frequency spectrum is[2]

$$X(\omega) = \int_0^\infty e^{at} e^{-j\omega t} dt = \frac{1}{a - j\omega} e^{(a-j\omega)t} \Big|_{t=0}^{\infty}$$

$$= 0 - \frac{1}{(a - j\omega)} = \frac{1}{j\omega - a} \tag{3.7}$$

This spectrum is a complex-valued function of ω. If we express $X(\omega)$, for $a = -0.1$, as

$$X(\omega) = \frac{1}{\sqrt{\omega^2 + 0.1^2} e^{j \tan^{-1}(\omega/0.1)}} = \frac{1}{\sqrt{\omega^2 + 0.01}} e^{-j \tan^{-1} 10\omega}$$

then its magnitude and phase spectra are

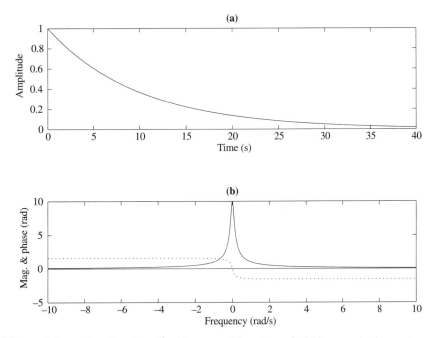

Figure 3.1 (a) Time function e^{at} with $a = -0.1$ and $t \geq 0$. (b) Its magnitude spectrum (solid line) and phase spectrum (dotted line).

---

[2] If Re $a = 0$, its CT Fourier transform is still defined but is fairly complex. See Fig. 3.4(d).

$$|X(\omega)| = \frac{1}{\sqrt{\omega^2 + 0.01}} \qquad \sphericalangle X(\omega) = -\tan^{-1} 10\omega$$

Figure 3.1(a) shows the time signal for $a = -0.1$, and Fig. 3.1(b) shows its magnitude spectrum (solid line) and phase spectrum (dotted line). Note that the phase approaches $\pm\pi/2 = \pm1.57$ rad/s as $|\omega| \to \infty$. The magnitude and phase spectra are obtained in MATLAB by typing

Program 3.1

```
w=-10:0.02:10;
X=1.0./(j*w+0.1);
plot(w,abs(X),w,angle(X),':')
```

The first line generates ω from -10 to 10 with increment 0.02. The second line is (3.7) with $a = -0.1$. Note the use of dot division (./) for element-by-element division in MATLAB. If we use division without a dot (/), then no graph will be generated. We see that using MATLAB to plot frequency spectra is very simple. From the plot, we see that the magnitude spectrum is an even function of ω, and the phase spectrum is an odd function of ω. This is a general property and will be discussed shortly. We also mention that if we use (3.6) to compute $x(t)$, we will obtain $x(0) = 0.5$, the average of $x(0-) = 0$ and $x(0+) = 1$. Other than this point, (3.6) will equal 0 for $t < 0$ and $e^{-0.1t}$ for $t > 0$.

Before proceeding, we mention that if a spectrum is complex-valued, we generally plot its magnitude and phase spectra. If a spectrum is real-valued, we can plot it directly. Of course we can also plot its magnitude and phase spectra.

◆ Example 3.2 CT Rectangular Windows

Consider the function shown in Fig. 3.2(a) or

$$w_a(t) = \begin{cases} 1 & \text{for } |t| \le a \\ 0 & \text{for } |t| > a \end{cases} \tag{3.8}$$

It is called a CT *rectangular window* or *boxcar* with width $L := 2a$. Its Fourier transform is

$$
\begin{aligned}
W_a(\omega) &:= \int_{-\infty}^{\infty} w_a(t)e^{-j\omega t}\,dt = \int_{-a}^{a} e^{-j\omega t}\,dt = \left. \frac{1}{-j\omega}e^{-j\omega t} \right|_{t=-a}^{a} \\
&= \frac{e^{-j\omega a} - e^{j\omega a}}{-j\omega} = \frac{2(e^{j\omega a} - e^{-j\omega a})}{j2\omega} = \frac{2\sin a\omega}{\omega}
\end{aligned}
\tag{3.9}
$$

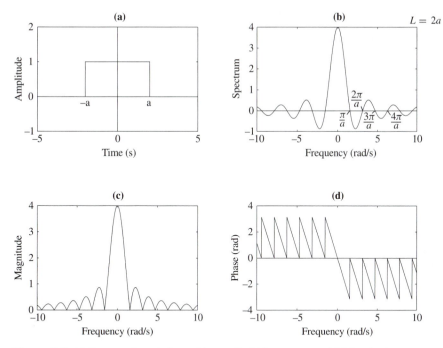

Figure 3.2 (a) Rectangular window with $a = 2$. (b) Its spectrum. (c) Magnitude spectrum of shifted window in (3.10) with $L = 2a = 4$. (d) Phase spectrum of (3.10).

This is the frequency spectrum of the window. Clearly it is a real-valued function of ω and can be plotted directly in Fig. 3.2(b). Because

$$W_a(-\omega) = \frac{2\sin(-a\omega)}{-\omega} = \frac{2\sin a\omega}{\omega} = W_a(\omega)$$

it is even. Because $\sin a\omega = 0$ at $\omega = k\pi/a$, for all integer k, we have $W_a(k\pi/a) = 0$ for all nonzero integer k. The value of $W_a(0)$, however, is different from zero and can be computed, using l'Hôpital's rule, as

$$W_a(0) = \left.\frac{d(2\sin a\omega)/d\omega}{d\omega/d\omega}\right|_{\omega=0} = \left.\frac{2a\cos a\omega}{1}\right|_{\omega=0} = 2a = L$$

The spectrum consists of one main lobe with base width $2\pi/a$ and height $L = 2a$, and infinitely many side lobes with base width π/a and decreasing amplitudes.

In addition to the window defined in (3.8), we will also encounter the window defined by

$$w_L(t) = \begin{cases} 1 & \text{for } 0 \le t \le L \\ 0 & \text{for } t < 0 \text{ and } t > L \end{cases} \tag{3.10}$$

It is the shifting of the window in (3.8) by a to the right and can be called a CT *shifted rectangular window* with width L. Its Fourier transform is

$$W_L(\omega) := \int_{-\infty}^{\infty} w_a(t)e^{-j\omega t}\,dt = \int_0^L e^{-j\omega t}\,dt = \left.\frac{1}{-j\omega}e^{-j\omega t}\right|_{t=0}^L$$

$$= \frac{e^{-j0.5\omega L}\left(e^{j0.5\omega L} - e^{-j0.5\omega L}\right)}{j\omega}$$

$$= e^{-j0.5\omega L}\cdot\frac{2\sin 0.5\omega L}{\omega} = e^{-ja\omega}\frac{2\sin a\omega}{\omega} \tag{3.11}$$

It is not a real-valued function as in (3.9). Its magnitude and phase spectra are plotted in Figs. 3.2(c) and (d). Its magnitude spectrum is identical to the one in (3.9). The phase in Fig. 3.2(b) is either 0 or π depending on whether $W_a(\omega)$ is positive or negative. The phase in Fig. 3.2(d) differs from the one in (3.9) by a linear phase, as we will discuss shortly.

Let us define

$$\text{sinc } \theta := \frac{\sin \theta}{\theta} \tag{3.12}$$

It is called the *sinc function*. Using this function, we can write (3.9) as $W_a(\omega) = 2a \text{ sinc } a\omega$. Thus the frequency spectrum of a rectangular window is a sinc function. As plotted in Fig. 3.2(b), we have

$$|W_a(\omega)| = |2a \text{ sinc } a\omega| \leq W_a(0) = 2a$$

for all ω. Thus we conclude $|\text{sinc } \theta| \leq 1$, for all θ.

◆ **Example 3.3 Analog Ideal Low-Pass Filter**

In this example, we will search for a time function that has the spectrum shown in Fig. 3.3(a) or

$$H(\omega) = \begin{cases} 1 & \text{for } |\omega| \leq \omega_c \\ 0 & \text{for } \omega_c < |\omega| < \infty \end{cases}$$

This function, as we will discuss in Chapter 6, characterizes an analog ideal low-pass filter. Using (3.6), we have

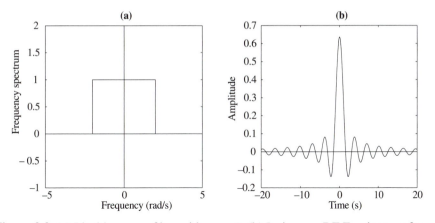

Figure 3.3 (a) Ideal low-pass filter with $\omega_c = 2$. (b) Its inverse DT Fourier transform.

$$h(t) = \frac{1}{2\pi} \int_{-\infty}^{\infty} H(\omega)e^{j\omega t}d\omega = \frac{1}{2\pi} \int_{-\omega_c}^{\omega_c} e^{j\omega t}d\omega$$

$$= \frac{1}{2\pi jt} e^{j\omega t}\Big|_{\omega=-\omega_c}^{\omega_c} = \frac{e^{j\omega_c t} - e^{-j\omega_c t}}{2j\pi t}$$

$$= \frac{\sin \omega_c t}{\pi t} = \frac{\omega_c}{\pi} \text{sinc } \omega_c t \qquad (3.13)$$

It is also a sinc function and is plotted in Fig. 3.3(b). The value of $h(t)$ at $t = 0$ can be computed, using l'Hôpital rule's rule, as ω_c/π.

We mention that $h(t)$ can also be obtained from Example 3.2 by using the duality property of the CT Fourier transform and its inverse. See Problem 3.8.

3.2.1 Frequency Spectrum of CT Periodic Signals

This subsection will develop the frequency spectrum of periodic signals. Periodic signals are not absolutely integrable, but their Fourier transforms are still defined. Note that absolute integrability is a sufficient but not a necessary condition for the existence of Fourier transforms.

Instead of computing directly the frequency spectrum of periodic signals, we compute first the inverse Fourier transform of $\delta(\omega - \omega_0)$, an impulse at $\omega = \omega_0$. See the appendix. Using the sifting property in (A.6), we compute

$$\mathcal{F}^{-1}[\delta(\omega - \omega_0)] = \frac{1}{2\pi} \int_{-\infty}^{\infty} \delta(\omega - \omega_0) e^{j\omega t} d\omega = \frac{1}{2\pi} e^{j\omega t} \Big|_{\omega = \omega_0} = \frac{1}{2\pi} e^{j\omega_0 t}$$

which implies

$$\mathcal{F}[e^{j\omega_0 t}] = 2\pi \delta(\omega - \omega_0) \tag{3.14}$$

Thus the frequency spectrum of the complex exponential $e^{j\omega_0 t}$ is an impulse at ω_0 with weight 2π. If $\omega_0 = 0$, (3.14) becomes

$$\mathcal{F}[1] = 2\pi \delta(\omega)$$

where 1 is the function $x(t) = 1$ for all t in $(-\infty, \infty)$ as shown in Fig. 3.4(a). Its frequency spectrum is zero everywhere except at $\omega = 0$. Because $x(t) = 1$ contains no nonzero frequency component, it is called a dc signal.

The CT Fourier transform is a linear operator in the sense that

$$\mathcal{F}[a_1 x_1(t) + a_2 x_2(t)] = a_1 \mathcal{F}[x_1(t)] + a_2 \mathcal{F}[x_2(t)]$$

for any constants a_1 and a_2. Using this property and (3.14), we have

$$\mathcal{F}[\sin \omega_0 t] = \mathcal{F}\left[\frac{e^{j\omega_0 t} - e^{-j\omega_0 t}}{2j}\right] = \frac{\pi}{j}[\delta(\omega - \omega_0) - \delta(\omega + \omega_0)]$$

$$= -j\pi \delta(\omega - \omega_0) + j\pi \delta(\omega + \omega_0) \tag{3.15}$$

and

$$\mathcal{F}[\cos \omega_0 t] = \mathcal{F}\left[\frac{e^{j\omega_0 t} + e^{-j\omega_0 t}}{2}\right] = \pi \delta(\omega - \omega_0) + \pi \delta(\omega + \omega_0) \tag{3.16}$$

They are plotted in Figs. 3.4(b) and (c). The magnitude spectra of $\sin \omega_0 t$ and $\cos \omega_0 t$ are the same, but their phase spectra are different. Their frequency spectra are zero everywhere except at $\pm \omega_0$. Thus the frequency spectrum as defined is consistent with our perception of the frequency of $\sin \omega_0 t$ and $\cos \omega_0 t$.

The frequency spectrum of a dc signal is zero everywhere except at $\omega = 0$. This is so only if $x(t) = 1$ for all t in $(-\infty, \infty)$. If $x(t)$ is a step function defined as $x(t) = 1$ for $t \geq 0$ and $x(t) = 0$ for $t < 0$, as shown in Fig. 3.4(d), then its spectrum can be computed as

$$X(\omega) = \pi \delta(\omega) + \frac{1}{j\omega}$$

(See Ref. 6, p. 347) and is plotted in Fig. 3.4(d). It has nonzero frequency components for all ω. Even though the step function is also considered as a dc signal, its spectrum contains nonzero frequency components. Likewise, if $\sin \omega_0 t$ and $\cos \omega_0 t$ are defined only for $t \geq 0$, then their

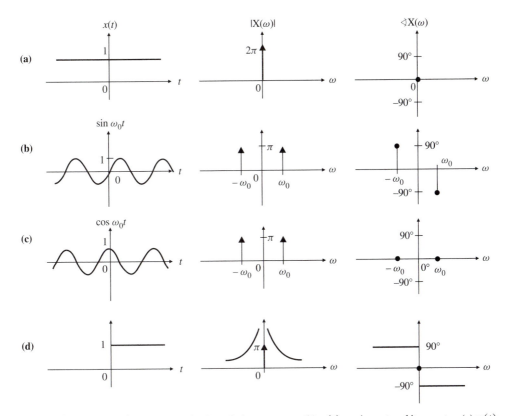

Figure 3.4 (a) $x(t) = 1$ and its magnitude and phase spectra. (b) $x(t) = \sin \omega_0 t$ and its spectra. (c) $x(t) = \cos w_0 t$ and its spectra. (d) $x(t) = 1$ for $t \geq 0$ and its spectra.

spectra contain nonzero frequency components for all ω other than the impulses at $\pm \omega_0$ and their analyses will be complicated. For this reason, in discussing frequency spectra, periodic signals are always defined for all t in $(-\infty, \infty)$.

Most CT periodic signal with period P can be expressed in the CT Fourier series as

$$x(t) = \sum_{m=-\infty}^{\infty} c_m e^{jm\omega_0 t}$$

with $\omega_0 = 2\pi/P$. Thus the frequency spectrum of $x(t)$ is, using (3.14),

$$X(\omega) = \mathcal{F}[x(t)] = \sum_{m=-\infty}^{\infty} c_m \mathcal{F}\left[e^{jm\omega_0 t}\right] = \sum_{m=-\infty}^{\infty} 2\pi c_m \delta(\omega - m\omega_0) \qquad (3.17)$$

It is a sequence of impulses in the frequency range $(-\infty, \infty)$.

◆ **Example 3.4 Sampling Function**

Consider the sampling function discussed in Example 2.6 or

$$r(t) = \sum_{k=-\infty}^{\infty} \delta(t - kT) \tag{3.18}$$

Its CT Fourier series was computed in (2.35) as

$$r(t) = \frac{1}{T} \sum_{m=-\infty}^{\infty} e^{jm\omega_0 t}$$

with $\omega_0 = 2\pi/T$. Thus the frequency spectrum of $r(t)$ is

$$R(\omega) = \mathcal{F}[r(t)] = \frac{2\pi}{T} \sum_{m=-\infty}^{\infty} \delta(\omega - m\omega_0) \tag{3.19}$$

It consists of a sequence of impulses.

3.3 Properties of Frequency Spectra

In this section, we discuss some properties of the frequency spectrum.

3.3.1 Boundedness and Continuity

We first show that if a signal is absolutely integrable, then its frequency spectrum is bounded and continuous. Because $|e^{-j\omega t}| = 1$, for all ω and t, we have

$$|X(\omega)| = \left| \int_{-\infty}^{\infty} x(t) e^{-j\omega t} dt \right| \leq \int_{-\infty}^{\infty} |x(t)| dt$$

Thus if $x(t)$ is absolutely integrable, then its spectrum is bounded. Let us consider, for any $\mu > 0$,

$$|X(\omega + \mu) - X(\omega)| = \left| \int_{-\infty}^{\infty} x(t) \left(e^{-j(\omega+\mu)t} - e^{-j\omega t} \right) dt \right|$$

$$= \left| \int_{-\infty}^{\infty} x(t) \left[e^{-j(\omega+0.5\mu)t} \left(e^{-j0.5\mu t} - e^{j0.5\mu t} \right) \right] \right|$$

$$\leq \int_{-\infty}^{\infty} |x(t)| |2 \sin 0.5\mu t| dt \tag{3.20}$$

where we have used $|e^{-j(\omega+0.5\mu)t}| = 1$ and $e^{-j0.5\mu t} - e^{j0.5\mu t} = -2j\sin 0.5\mu t$. Because $|2\sin 0.5\mu t| \leq 2$, $|\text{sinc }\theta| \leq 1$, and

$$|2\sin 0.5\mu t| = \left|\mu t\left(\frac{\sin 0.5\mu t}{0.5\mu t}\right)\right| = |\mu t||\text{sinc }(0.5\mu t)| \leq |\mu t|$$

we can write (3.20) as

$$|X(\omega + \mu) - X(\omega)| \leq 2\left(\int_{-\infty}^{-L} + \int_{L}^{\infty}\right)|x(t)|dt + \int_{-L}^{L}|x(t)|\left|\mu t\frac{\sin 0.5\mu t}{0.5\mu t}\right|dt$$

$$\leq 2\left(\int_{-\infty}^{-L} + \int_{L}^{\infty}\right)|x(t)|dt + \mu L\int_{-L}^{L}|x(t)|dt$$

$$\leq 2\left(\int_{-\infty}^{-L} + \int_{L}^{\infty}\right)|x(t)|dt + \mu L M$$

where $M = \int_{-\infty}^{\infty}|x(t)|dt$. Now for any $\epsilon > 0$, we can choose an L so large such that the first term after the last inequality is smaller than $\epsilon/2$. We then select $\mu = \epsilon/2LM$. Thus we conclude that for any $\epsilon > 0$, if $x(t)$ is absolutely integrable, there exists a μ such that

$$|X(\omega + \mu) - X(\omega)| < \epsilon$$

This shows that $X(\omega)$ is continuous. This establishes the assertion that if $x(t)$ is absolutely integrable, then its frequency spectrum is bounded and continuous.

If a CT signal is bounded and of finite duration, then it is automatically absolutely integrable. Thus the spectrum of such a signal is bounded and continuous. Periodic CT signals, which are defined from $-\infty$ to ∞, are not absolutely integrable and their frequency spectra consist of impulses that are not bounded nor continuous. The time function in (3.13) can be shown to be not absolutely integrable, and its spectrum is not continuous, as shown in Fig. 3.3(a). Note that the function in (3.13) is square integrable and its spectrum is still defined. Absolute integrability and periodicity are two sufficient conditions for the existence of a spectrum; they are not necessary conditions.

3.3.2 Even and Odd

In the preceding chapter, we showed that if a periodic signal is real, then its Fourier series has even magnitude and odd phase. If a periodic signal is real and even, then its Fourier series is real and even. The same procedure can be used to show that the frequency spectrum has the same properties. To avoid repetition, we will use a different procedure to establish some of the properties listed in Table 3.1.[3] The reader may skip, without loss of continuity, the proof of Table

---

[3] In discussing the oddness of the phase of CTFS, we must exclude the phase at $\omega = 0$ as in Example 2.4. It is, however, not necessary to do so here, because CTFT $X(\omega)$ is defined for all ω and the phase at an isolated ω is not important.

Table 3.1 Properties of Spectrum

$x(t)$	$X(\omega)$	$|X(\omega)|$	$\not< X(\omega)$
Real	Congugate symmetric	Even	Odd
Even	Even	Even	Even
Odd	Odd	Odd	Odd
Real & even	Real & even	Even	0° or ±180°, even or odd
Real & odd	Imaginary & odd	Even	±90° odd

3.1, and go directly to the topic of *time shifting*.

A real- or complex-valued function $f(p)$ is even or symmetric if $f(p) = f(-p)$; odd or antisymmetric if $f(p) = -f(-p)$. It is *conjugate symmetric* if $f(p) = f^*(-p)$ or $f^*(p) = f(-p)$. If $f(p)$ is written as

$$f(p) = f_r(p) + jf_i(p)$$

where $f_r(p)$ and $f_i(p)$ denote, respectively, the real and imaginary parts of $f(p)$, then we have

- $f(p)$ is real $\Leftrightarrow$ $f(p) = f^*(p)$
- $f(p)$ is imaginary $\Leftrightarrow$ $f(p) = -f^*(p)$
- $f(p)$ is even $\Leftrightarrow$ $f_r(p)$ and $f_i(p)$ are even
- $f(p)$ is odd $\Leftrightarrow$ $f_r(p)$ and $f_i(p)$ are odd
- $f(p)$ is conjugate symmetric $\Leftrightarrow$ $f_r(p)$ is even and $f_i(p)$ is odd

The preceding discussion can be applied to $x(t)$, $X(\omega)$, c_m, and c_{md}.

First we show that if a CT signal $x(t)$ is real, then its spectrum is conjugate symmetric. Taking the complex conjugate of $X(\omega)$ in (3.5), we obtain

$$X^*(\omega) = \left(\int_{-\infty}^{\infty} x(t)e^{-j\omega t}dt \right)^* = \int_{-\infty}^{\infty} x^*(t)e^{j\omega t}dt$$

which becomes, if $x(t)$ is real or $x^*(t) = x(t)$,

$$X^*(\omega) = \int_{-\infty}^{\infty} x(t)e^{j\omega t}dt = X(-\omega)$$

where we have replaced ω by $-\omega$ in (3.5). Thus we conclude that if $x(t)$ is real, then $X(-\omega) = X^*(\omega)$ and

$$\text{Re } X(\omega) = \text{Re } X(-\omega) \quad \text{and} \quad \text{Im } X(\omega) = -\text{Im } X(-\omega)$$

From these, we can readily show

$$|X(\omega)| = |X(-\omega)| \quad \text{and} \quad \sphericalangle X(\omega) = - \sphericalangle X(-\omega)$$

Thus the spectrum of any real signal has even amplitude and odd phase.

Next we show that if $x(t)$ is even $[x(t) = x(-t)]$, then its frequency spectrum is even $[X(\omega) = X(-\omega)]$. Replacing ω by $-\omega$ in (3.5) yields

$$X(-\omega) = \int_{t=-\infty}^{\infty} x(t) e^{j\omega t} dt = \int_{\tau=\infty}^{-\infty} x(-\tau) e^{-j\omega \tau}(-d\tau)$$

$$= \int_{\tau=-\infty}^{\infty} x(-\tau) e^{-j\omega \tau} d\tau = \int_{t=-\infty}^{\infty} x(-t) e^{-j\omega t} dt \qquad (3.21)$$

where we have introduced the new variable $\tau = -t$ and then renamed it back to t after the last equality. If $x(t)$ is even or $x(t) = x(-t)$, then (3.21) becomes

$$X(-\omega) = \int_{-\infty}^{\infty} x(t) e^{-j\omega t} dt = X(\omega)$$

Thus $X(\omega)$ is even. If $X(\omega)$ is even, then both its real and imaginary parts are even. Therefore, its magnitude and phase are also even.

If $x(t)$ is real, then its spectrum is conjugate symmetric or $X(-\omega) = X^*(\omega)$. If $x(t)$ is even, then its spectrum is even or $X(-\omega) = X(\omega)$. Thus, if $x(t)$ is real and even, we have

$$X(-\omega) = X^*(\omega) = X(\omega)$$

The second equality implies that $X(\omega)$ is real. Thus, if $x(t)$ is real and even, so is $X(\omega)$. If $X(\omega)$ is real, then its phase is either $0°$ or $\pm 180°$. If all negative $X(\omega)$ is selected to have $180°$ or $-180°$, then the phase will be even. If negative $X(\omega)$ is selected to have $180°$ for $\omega \geq 0$ and $-180°$ for $\omega < 0$ or vice versa, then the phase will be odd. If $X(\omega)$ is real, its values are called amplitudes.

If $x(t)$ is odd or $x(t) = -x(-t)$, then (3.21) becomes

$$X(-\omega) = - \int_{t=-\infty}^{\infty} x(t) e^{-j\omega t} dt = -X(\omega)$$

which implies that $X(\omega)$ is odd. If $x(t)$ is real, then $X(-\omega) = X^*(\omega)$. Thus if $x(t)$ is real and odd, we have

$$X(-\omega) = X^*(\omega) = -X(\omega)$$

The second equality implies that $X(\omega)$ is pure imaginary. Thus, if $x(t)$ is real and odd, then $X(\omega)$ is pure imaginary and odd. If $X(\omega)$ is pure imaginary, then its phase is either $90°$ or $-90°$.

3.3.3 Time Shifting

Consider a CT signal $x(t)$ with frequency spectrum $X(\omega)$. Let t_0 be a constant. Then the signal $x(t - t_0)$ is the shifting of $x(t)$ to t_0 as shown in Fig. 3.5. Let $X_0(\omega)$ be the spectrum of $x(t - t_0)$. Then we have

$$X_0(\omega) := \mathcal{F}[x(t - t_0)] = \int_{-\infty}^{\infty} x(t - t_0)e^{-j\omega t}\,dt$$

$$= e^{-j\omega t_0} \int_{t=-\infty}^{\infty} x(t - t_0)e^{-j\omega(t-t_0)}\,dt$$

which becomes, after introducing the new variable $\tau := t - t_0$,

$$X_0(\omega) = e^{-j\omega t_0} \int_{\tau=-\infty}^{\infty} x(\tau)e^{-j\omega\tau}\,d\tau = e^{-j\omega t_0} X(\omega) \tag{3.22}$$

Because $|e^{-j\omega t_0}| = 1$ and $\not{\angle} e^{-j\omega t_0} = -\omega t_0$, we have

$$|X_0(\omega)| = |e^{-j\omega t_0}||X(\omega)| = |X(\omega)|$$

and

$$\not{\angle} X_0(\omega) = \not{\angle} e^{-j\omega t_0} + \not{\angle} X(\omega) = \not{\angle} X(\omega) - \omega t_0 \tag{3.23}$$

Thus time shifting will not affect the amplitude spectrum but will introduce a linear phase into the phase spectrum.

The window defined in (3.10) is the shifting of the window in (3.8) by a; therefore, their phase spectra differ by $-a\omega$, as implied by (3.23). However, MATLAB plots phases only between $\pm 180°$ or $\pm\pi$ rad/s by adding or subtracting repeatly 360°. Thus, instead of one single straight line, we see sections of straight lines in Fig. 3.2(d).

3.3.4 Frequency Shifting

Let $x(t)$ be a CT signal with spectrum $X(\omega)$. Then we have

$$\mathcal{F}[e^{j\omega_0 t}x(t)] = \int_{-\infty}^{\infty} e^{j\omega_0 t}x(t)e^{-j\omega t}\,dt$$

$$= \int_{-\infty}^{\infty} x(t)e^{-j(\omega-\omega_0)t}\,dt$$

$$= X(\omega - \omega_0) \tag{3.24}$$

This is called frequency shifting because the frequency spectrum of $x(t)$ is being shifted to ω_0 after $x(t)$ is multiplied by $e^{j\omega_0 t}$. Using (3.24) and the linearity of the Fourier transform, we have

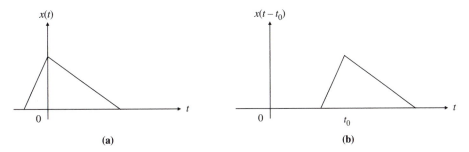

Figure 3.5 (a) A signal. (b) The signal shifted to t_0.

$$\mathcal{F}[x(t)\cos\omega_0 t] = \mathcal{F}\left[x(t)\frac{e^{j\omega_0 t} + e^{-j\omega_0 t}}{2}\right]$$

$$= 0.5[X(\omega - \omega_0) + X(\omega + \omega_0)] \tag{3.25}$$

If $x(t)$ has a spectrum as shown in Fig. 3.6(a), then the spectrum of $x(t)\cos\omega_0 t$ is as shown in Fig. 3.6(b). It consists of the spectra of $x(t)$ shifted to $\pm\omega_0$ and the magnitude is cut in half. The multiplication of $x(t)$ by $\cos\omega_0 t$ is called *frequency modulation* and is important in communication.

3.3.5 Time Compression and Expansion

Consider the signal $x(t)$ shown in Fig. 3.7(a). We plot in Figs. 3.7(b) and (c) $x(2t)$ and $x(0.5t)$. We see that the time duration of $x(2t)$ is cut in half and the duration of $x(0.5t)$ is doubled. In general, we have

- $x(at)$ with $a > 1$: time compression
- $x(at)$ with $0 < a < 1$: time expansion

Now we study their effects on frequency spectra. By definition, we have

$$\mathcal{F}[x(at)] = \int_{-\infty}^{\infty} x(at)e^{j\omega t}\,dt = \frac{1}{a}\int_{-\infty}^{\infty} x(at)e^{j(\omega/a)at}\,d(at)$$

$$= \frac{1}{a}\int_{-\infty}^{\infty} x(\tau)e^{j(\omega/a)\tau}\,d\tau$$

where we have used $\tau = at$. Thus we conclude

$$\mathcal{F}[x(at)] = \frac{1}{a}X(\omega/a)$$

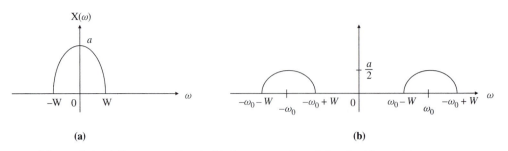

Figure 3.6 (a) Spectrum of $x(t)$. (b) Spectrum of modulated $x(t)$ or $x(t) \cos \omega_0 t$.

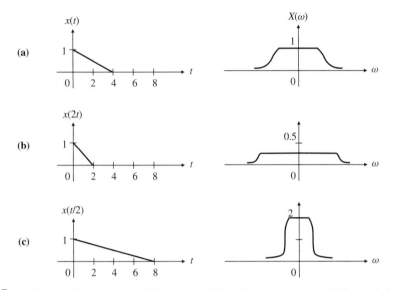

Figure 3.7 (a) Signal and its spectrum. (b) Compressed signal and its spectrum. (c) Expanded signal and its spectrum.

Note that this holds only for a positive. If the spectrum of $x(t)$ is as shown on the right-hand side of Fig. 3.7(a), then the spectra of $x(2t)$ and $x(0.5t)$ are as shown in Figs. 3.7(b) and (c). We see that

- Time compression $\Leftrightarrow$ frequency expansion
- Time expansion $\Leftrightarrow$ frequency compression

This fact can be used to explain the change of pitch of an audio tape when its speed is changed. When we increase the speed of the tape (time compression), its frequency spectrum is expanded and, consequently, contains higher-frequency components. Thus the pitch is higher. When we decrease the speed (time expansion), its spectrum is compressed and contains lower frequency components. Thus its pitch is lower.

The preceding discussion can be stated more generally as: The duration of time signals and bandwidth of their frequency spectra are inversely related. See Problem 3.9.

3.4 Distribution of Energy in Frequencies

We now discuss the physical meaning of the frequency spectrum. Every nonzero *periodic* CT signal has infinite total energy; therefore, we discussed in Section 2.4.1 only its average power. If an *aperiodic* CT signal has a finite total energy, then its average power as defined in (2.37) is always zero. Therefore, we do not discuss average power for aperiodic signals. Instead, we discuss their total energy.

The total energy of a CT signal $x(t)$ was defined in (2.36) as

$$E := \int_{t=-\infty}^{\infty} x(t)x^*(t)dt = \int_{-\infty}^{\infty} |x(t)|^2 dt \tag{3.26}$$

where $x^*(t)$ is the complex conjugate of x(t). An absolutely integrable CT signal may not have a finite total energy. See Problem 3.10. However, if it is also bounded, then it has a finite total energy. Indeed, if $x(t)$ is bounded [$|x(t)| < q$ for all t and for some constant q] and absolutely integrable ($\int_{-\infty}^{\infty} |x(t)|dt < \infty$), then we have

$$E = \int_{-\infty}^{\infty} |x(t)|^2 dt < q \int_{-\infty}^{\infty} |x(t)|dt < \infty$$

This establishes the assertion. A bounded and finite-duration CT signal is always absolute integrable and, therefore, has a finite total energy.

The total energy of a CT signal can be computed directly in the time domain or indirectly from its frequency spectrum. Let $X(\omega)$ be the CT Fourier transform of x(t), or

$$x(t) = \frac{1}{2\pi} \int_{\omega=-\infty}^{\infty} X(\omega)e^{j\omega t} d\omega$$

Substituting this into (3.26) yields

$$E = \int_{t=-\infty}^{\infty} x(t) \left(\frac{1}{2\pi} \int_{\omega=-\infty}^{\infty} X(\omega)e^{j\omega t} d\omega \right)^* dt$$

$$= \frac{1}{2\pi} \int_{\omega=-\infty}^{\infty} X^*(\omega) \left(\int_{t=-\infty}^{\infty} x(t)e^{-j\omega t} dt \right) d\omega$$

where we have changed the order of integrations. The term inside the large parentheses equals $X(\omega)$; thus we have

$$E = \int_{-\infty}^{\infty} |x(t)|^2 dt = \frac{1}{2\pi} \int_{-\infty}^{\infty} X^*(\omega)X(\omega)d\omega = \frac{1}{2\pi} \int_{-\infty}^{\infty} |X(\omega)|^2 d\omega \tag{3.27}$$

This is called *Parseval's formula*.

Using Parseval's formula, the total energy of $x(t)$ can also be computed from its magnitude spectrum. More important, the magnitude spectrum reveals the distribution of energy in frequencies. For example, the energy contained in the frequency range $[\omega_1, \omega_2]$, with $\omega_1 < \omega_2$, is given by

$$\frac{1}{2\pi} \int_{\omega_1}^{\omega_2} |X(\omega)|^2 d\omega$$

We mention that if $x(t)$ has finite energy, then its spectrum contains no impulses and

$$\int_{\omega_0-}^{\omega_0+} |X(\omega)|^2 d\omega = 0$$

Thus it is meaningless to talk about energy at a discrete or isolated frequency. For this reason, $X(\omega)$ is also called the *spectral density* of $x(t)$.

The integration in (3.27) is carried out over positive and negative frequencies. If $x(t)$ is real, then $|X(-\omega)| = |X(\omega)|$, and (3.27) can also be written as

$$E = \frac{1}{2\pi} \int_{-\infty}^{\infty} |X(\omega)|^2 d\omega = \frac{1}{\pi} \int_{0}^{\infty} |X(\omega)|^2 d\omega \qquad (3.28)$$

Thus the total energy of a real-valued signal can be computed over positive frequencies. The quantity $|X(\omega)|^2/\pi$ may be called the *energy spectral density*. Note that the energy of a signal is independent of its phase spectrum. Furthermore, time shifting of a signal will not affect its total energy.

To conclude this section, we compare periodic and aperiodic signals, and their CT Fourier series c_m and CT Fourier transform $X(\omega)$:

- The total energy of every nonzero CT periodic signal is infinite, and its average power equals

$$P_{av} = \sum_{m=-\infty}^{\infty} |c_m|^2$$

  Nonzero power appears only at discrete frequency $m\omega_0$.
- The total energy of a bounded and absolutely integrable signal equals

$$E = \frac{1}{2\pi} \int_{-\infty}^{\infty} |X(\omega)|^2 d\omega$$

  The energy at any isolated frequency is zero; therefore, we discuss its energy distribution only over nonzero frequency intervals. Its average power, defined by

$$P_{av} = \lim_{L \to \infty} \frac{1}{2L} \int_{t=-L}^{L} |x(t)|^2 dt \qquad (3.29)$$

  is always zero.

- The frequency spectrum of a CT periodic signal with Fourier series c_m is

$$X(\omega) = \sum_{m=-\infty}^{\infty} 2\pi c_m \delta(\omega - m\omega_0)$$

It is nonzero only at discrete frequencies and, therefore, can be called the discrete frequency spectrum. If we also call c_m the discrete frequency spectrum, then confusion may occur. Thus in this text, we call c_m CTFS coefficients or frequency components. Note that if $X(\omega)$ contains impulses, then

$$E = \frac{1}{2\pi} \int_{-\infty}^{\infty} |X(\omega)|^2 d\omega = \infty$$

3.5 Effects of Truncation

Consider a CT signal of infinite duration. If we want to compute its frequency spectrum using a computer, we must truncate and then discretize it or discretize and then truncate it. Thus truncation and discretization are two important issues in spectral computation. In this section, we study the effect of truncation on the spectrum of CT signals.

Consider two CT signals $x_i(t)$ with spectra $X_i(\omega)$, for $i = 1, 2$. Clearly $x_i(t)$ can be computed from $X_i(\omega)$ using the inverse CT Fourier transform as

$$x_i(t) = \frac{1}{2\pi} \int_{\omega=-\infty}^{\infty} X_i(\omega)e^{j\omega t} d\omega = \frac{1}{2\pi} \int_{\bar{\omega}=-\infty}^{\infty} X_i(\bar{\omega})e^{j\bar{\omega}t} d\bar{\omega}$$

Note that the integration variable ω can be replaced by any other variable as shown. Let us define a new signal $x(t) := x_1(t)x_2(t)$. It is obtained by multiplying $x_1(t)$ and $x_2(t)$ point by point. The CT Fourier transform of $x(t)$ is

$$X(\omega) = \int_{t=-\infty}^{\infty} x_1(t)x_2(t)e^{-j\omega t} dt$$

$$= \int_{t=-\infty}^{\infty} \left(\frac{1}{2\pi} \int_{\bar{\omega}=-\infty}^{\infty} X_1(\bar{\omega})e^{j\bar{\omega}t} d\bar{\omega} \right) x_2(t)e^{-j\omega t} dt$$

$$= \frac{1}{2\pi} \int_{\bar{\omega}=-\infty}^{\infty} X_1(\bar{\omega}) \left(\int_{t=-\infty}^{\infty} x_2(t)e^{-j(\omega-\bar{\omega})t} dt \right) d\bar{\omega}$$

where we have used $\bar{\omega}$ in the inverse Fourier transform of x_1 to avoid confusion with the ω in $X(\omega)$ and have changed the order of integrations. The term inside the large parentheses equals $X_2(\omega - \bar{\omega})$; thus we have

$$\mathcal{F}[x_1(t)x_2(t)] = \frac{1}{2\pi} \int_{\bar{\omega}=-\infty}^{\infty} X_1(\bar{\omega})X_2(\omega - \bar{\omega})d\bar{\omega} \tag{3.30}$$

This equation has an alternative form. Define a new variable $\hat{\omega} := \omega - \bar{\omega}$, where ω is fixed. Then we have $\bar{\omega} = \omega - \hat{\omega}$, $d\hat{\omega} = -d\bar{\omega}$ and

$$\int_{\bar{\omega}=-\infty}^{\infty} X_1(\bar{\omega})X_2(\omega - \bar{\omega})d\bar{\omega} = \int_{\hat{\omega}=\infty}^{-\infty} X_1(\omega - \hat{\omega})X_2(\hat{\omega})(-d\hat{\omega})$$

$$= \int_{\hat{\omega}=-\infty}^{\infty} X_1(\omega - \hat{\omega})X_2(\hat{\omega})d\hat{\omega} = \int_{\bar{\omega}=-\infty}^{\infty} X_1(\omega - \bar{\omega})X_2(\bar{\omega})d\bar{\omega}$$

where we have renamed $\hat{\omega}$ as $\bar{\omega}$ after the last equality. Thus (3.30) can be written as

$$\mathcal{F}[x_1(t)x_2(t)] = \frac{1}{2\pi} \int_{-\infty}^{\infty} X_1(\bar{\omega})X_2(\omega - \bar{\omega})d\bar{\omega}$$

$$= \frac{1}{2\pi} \int_{-\infty}^{\infty} X_2(\bar{\omega})X_1(\omega - \bar{\omega})d\bar{\omega} \tag{3.31}$$

The integration in (3.31) is called a (continuous-frequency) *convolution*. It has the *commutative* property or the property that the roles of X_1 and X_2 can be interchanged. Thus multiplications in the time domain becomes convolutions in the transform domain denoted as

time multiplication $\Leftrightarrow$ frequency convolution

With the preceding discussion, we are ready to study the effects of truncation. Consider a CT signal $x(t)$ with spectrum $X(\omega)$. Let us truncate the signal before $t = -a$ and after a. This truncation is the same as multiplying $x(t)$ by the CT rectangular window $w_a(t)$ defined in (3.8) and plotted in Fig. 3.2(a). The spectrum of $w_a(t)$ was computed in (3.9) as

$$W_a(\omega) = \frac{2\sin a\omega}{\omega} \tag{3.32}$$

Thus the spectrum of the truncated $x(t)$ or the signal $x(t)w_a(t)$ is, using (3.31),

$$\mathcal{F}[x(t)w_a(t)] = \frac{1}{2\pi} \int_{-\infty}^{\infty} X(\bar{\omega})W_a(\omega - \bar{\omega})d\bar{\omega} \tag{3.33}$$

$$= \frac{1}{2\pi} \int_{-\infty}^{\infty} W_a(\bar{\omega})X(\omega - \bar{\omega})d\bar{\omega} \tag{3.34}$$

which implies that the spectrum of a truncated signal equals the convolution of $W_a(\omega)$ and the spectrum of the untruncated signal. Thus the waveform of $W_a(\omega)$ plays a crucial role in determining the effects of truncation. As shown in Fig. 3.2(b), for $L = 2a$, the spectrum has a main lobe with base width $4\pi/L$ and height L. All side lobes have base width $2\pi/L$, half of the base width of the main lobe. The largest magnitude of the side lobe on either side is about 20% of the height of the main lobe.

Before proceeding, we mention that the preceding discussion still holds if the rectangular window $w_a(t)$ is replaced by the shifted rectangular window $w_L(t)$ defined in (3.10). Both $w_L(t)$ and $w_a(t)$ will be used in the examples that follow.

Now let us use examples to study the effects of truncation. Consider $\cos \omega_0 t$, for all t. Its spectrum is, as computed in (3.16),

$$X(\omega) = \pi \delta(\omega - \omega_0) + \pi \delta(\omega + \omega_0) \tag{3.35}$$

and is plotted in Fig. 3.8(a) for $\omega_0 = 0.5$ and in Fig. 3.8(d) for $\omega_0 = 3$. If we truncate $\cos \omega_0 t$ before $t = 0$ and after L or, equivalently, consider $(\cos \omega_0 t) w_L(t)$, then its spectrum can be computed as, using the sifting property of impulses discussed in (A.6),

$$\mathcal{F}[\cos \omega_0 t \cdot w_L(t)] = \frac{1}{2\pi} \int_{-\infty}^{\infty} [\pi \delta(\bar{\omega} - \omega_0) + \pi \delta(\bar{\omega} + \omega_0)] W_L(\omega - \bar{\omega}) d\bar{\omega}$$

$$= 0.5[W_L(\omega - \omega_0) + W_L(\omega + \omega_0)] \tag{3.36}$$

where $W_L(\omega)$ is defined in (3.11). Thus the spectrum of the truncated $\cos \omega_0 t$ is the sum of $0.5 W_L(\omega)$ shifted to $\pm \omega_0$. Note that the height of $0.5 W_L(\omega)$ is $0.5L$ at $\omega = 0$. Because the spectrum in (3.36) is complex, and because its phase spectrum is less useful, we plot only its magnitude spectrum. Figures 3.8(b) and (e) show the magnitude spectra of truncated $\cos 0.5t$ and $\cos 3t$ with $L = 8$. For $L = 8$, the width of the main lobe is $4\pi/L = 1.57$ and the two main lobes shifted to ± 0.5 overlap with each other. Thus the resulting plot shows only one lobe in Fig. 3.8(b). The two main lobes shifted to ± 3 do not overlap with each other; but they still overlap with side lobes. Therefore, even though the spectrum of the truncated $\cos 3t$ shows two lobes at ± 3 as shown in Fig. 3.8(e), the height of the spikes is not exactly $L/2 = 4$. Thus we conclude that truncation may introduce *leakage* or *smearing* into a spectrum. The wider the main lobe, the larger the leakage.

The spectrum of the window consists of infinitely many side lobes, all with the same base width but decreasing magnitudes. When these side lobes convolve with a spectrum, they will introduce ripples into the spectrum as shown in Figs. 3.8(b) and (e). Thus truncation will also introduce *ripples* into a spectrum.

Figures 3.8(c) and (f) show the spectra of truncated $\cos 0.5$ and $\cos 3t$ for $L = 102.4$. The effects of truncation (leakage and ripples) are clearly visible but not as large as the ones in Figs. 3.8(b) and (e). As L increases, the two spikes become higher and narrower. Eventually, the two spikes become the two impulses shown in Figs. 3.8(a) and (d).

3.5.1 Gibbs Phenomenon

In Fig. 3.8, we see that truncation introduces leakage and ripples. As the window length L increases, one may expect that these effects will decrease. This is indeed the case if the original spectrum contains no discontinuity. If the spectrum of an untruncated signal contains discontinuity as in Fig. 3.8, then as L increases, the leakage becomes less and the ripples become narrow and move closer to $\pm \omega_0$. However, there is an unusual phenomenon: The largest magnitude of the ripples, relative to the peak magnitude of the spectrum, will not decrease as L increases. It remains roughly constant as shown in Fig. 3.8. This phenomenon is called the *Gibbs phenomenon*.

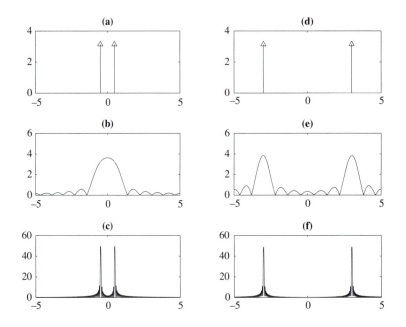

Figure 3.8 (a) Spectrum of $\cos 0.5t$, for all t. (b) Magnitude spectrum of $\cos 0.5t$ with $0 \leq t \leq L = 8$. (c) Magnitude spectra of $\cos 0.5t$ with $0 \leq t \leq L = 102.4$. (d, e, f) Corresponding spectra for $\cos 3t$. All horizontal coordinates are frequency in rad/s.

We demonstrate again the Gibbs phenomenon for the analog ideal lowpass filter studied in Example 3.3 or

$$h(t) = \frac{\sin \omega_c t}{\pi t} \tag{3.37}$$

Its spectrum is

$$H(\omega) = \begin{cases} 1 & \text{for } |\omega| \leq \omega_c \\ 0 & \text{for } |\omega| > \omega_c \end{cases}$$

In this example, we will use the rectangular window in (3.8). Using (3.33) and (3.32), we have

$$\mathcal{F}[h(t)w_a(t)] = \frac{1}{2\pi} \int_{-\infty}^{\infty} H(\bar{\omega}) W_a(\omega - \bar{\omega}) d\bar{\omega} = \int_{-\omega_c}^{\omega_c} \frac{\sin a(\omega - \bar{\omega})}{\pi(\omega - \bar{\omega})} d\bar{\omega} \tag{3.38}$$

This is the spectrum of the truncated ideal low-pass filter in (3.37). If $\omega_c = 2$, then the spectra of the truncated CT ideal low-pass filter for $a = 5$ and 50 are as shown in Figs. 3.9(a) and (b). We see that the sharp edges at $\omega_c = \pm 2$ are smeared and ripples are introduced into

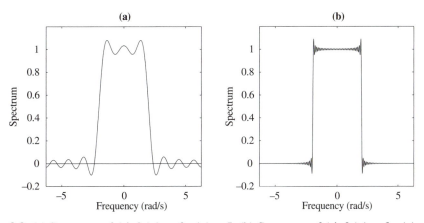

Figure 3.9 (a) Spectrum of $(\sin 2t)/\pi t$, for $|t| \leq 5$. (b) Spectrum of $(\sin 2t)/nt$, for $|t| \leq 50$.

the spectra. However, the magnitude of the largest ripple remains roughly the same, about 9% of the amount of the discontinuity. This is the Gibbs phenomenon. This phenomenon can be eliminated using some windows other than rectangular. This will be discussed in Chapter 7.

We now discuss the plotting of the spectra in Fig. 3.9. The integration in (3.38) can be expressed, using an integration table, as an infinite power series. Its computation involves the convergence problem. Alternatively, we can approximate (3.38) directly by a summation for each selected ω. If we select 50 different ω, then we must compute the summation 50 times. The plot in Fig. 3.9 is not obtained by either method. It is obtained by using an FFT, as we will discuss in Example 4.8.

◆ **Example 3.5**

Consider the CT signal $x(t) = e^{-0.1t}$ shown in Fig. 3.1(a). It is a signal with infinite duration, and its spectrum was computed in Example 3.1 as $X(\omega) = 1/(j\omega + 0.1)$ and plotted in Fig. 3.1(b). Let us truncate the signal after L or consider $e^{-0.1t}w_L(t)$. Although the spectrum can be computed using a frequency convolution as in (3.33), it is simpler to compute it directly from the truncated signal. As in Example 3.1, we have

$$X_L(\omega) = \mathcal{F}[e^{-0.1t}w_L(t)] = \int_0^L e^{-0.1t}e^{-j\omega t}dt$$

$$= \frac{1}{-0.1 - j\omega}e^{-(0.1+j\omega)t}\Big|_{t=0}^L = \frac{-1}{j\omega + 0.1}\left(e^{-(j\omega+0.1)L} - 1\right)$$

$$= \frac{1}{j\omega + 0.1}\left(1 - e^{-j\omega L}e^{-0.1L}\right)$$

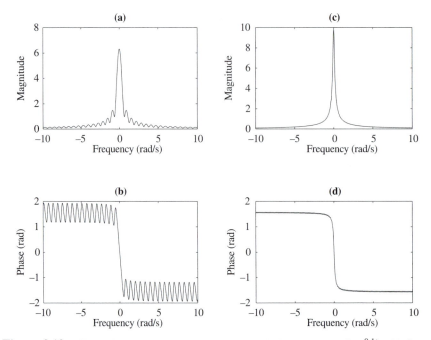

Figure 3.10 Effects of truncation: (a,b) Magnitude and phase spectra of $e^{-0.1t}$ with $0 \leq t \leq L = 10$. (c,d) Magnitude and phase spectra of $e^{-0.1t}$ with $0 \leq t \leq L = 40$.

Figures 3.10(a) and (b) show, respectively, the magnitude and phase spectra of $X_{10}(\omega)$; they are obtained using a program similar to Program 3.1. Comparing with the plots in Fig. 3.1(b), which is replotted in Figs. 3.10(c) and (d) with solid lines, we see that truncation introduces ripples but no appreciable leakage. For $L = 40$, the magnitude and phase spectra, as shown in Figs. 3.10(c) and (d) with dotted lines, are indistinguishable from those of $X(\omega) = 1/(j\omega + 1)$; they have no ripples and no leakage. This example does not show the Gibbs phenomenon because its original spectrum does not have any discontinuity.

3.6 DT Fourier Transform (DTFT)

In this and the following sections, we discuss the discrete-time counterparts of what has been discussed for CT signals. Consider a DT periodic signal $x(nT)$ with period N. Then it can be expressed in the DT Fourier series as

$$x(nT) = \sum_{m=0}^{N-1} c_{md} e^{jm\omega_0 nT} \tag{3.39}$$

with $\omega_0 = 2\pi/NT$ and

$$c_{md} = \frac{1}{N} \sum_{n=0}^{N-1} x(nT)e^{-jm\omega_0 nT}$$

$$= \frac{1}{N} \sum_{n=-\gamma}^{\overline{\gamma}} x(nT)e^{-jm\omega_0 nT} \tag{3.40}$$

where $\gamma = (N-1)/2$ and the overline rounds a number upward to an integer. Now we shall modify the DT Fourier series so that it can also be applied to aperiodic DT signals. First we multiply (3.40) by N to yield

$$X_d(m\omega_0) := c_{md}N = \sum_{n=-\gamma}^{\overline{\gamma}} x(nT)e^{-jm\omega_0 nT} \tag{3.41}$$

We then use $X_d(m\omega_0)$ and $\omega_0 = 2\pi/NT$ or $1/N = (T/2\pi)\omega_0$ to write (3.39) as

$$x(nT) = \frac{1}{N} \sum_{m=0}^{N-1} N c_{md} e^{jm\omega_0 nT} = \frac{T}{2\pi} \sum_{m=0}^{N-1} X_d(m\omega_0)e^{jm\omega_0 nT}\omega_0 \tag{3.42}$$

A periodic sequence with period N becomes aperiodic if N approaches infinity. Define $\omega := m\omega_0$. As $N \to \infty$, we have $\omega_0 = 2\pi/NT \to 0$. In this case, ω becomes a continuum, ω_0 can be written as $d\omega$, and the summation in (3.42) becomes an integration. Thus the modified discrete-time Fourier series pair in (3.41) and (3.42) becomes, as $N \to \infty$,

$$X_d(\omega) = \sum_{n=-\infty}^{\infty} x(nT)e^{-j\omega nT} =: \mathcal{F}_d[x(nT)] \tag{3.43}$$

$$x[n] := x(nT) = \frac{T}{2\pi} \int_{\omega=0}^{2\pi/T} X_d(\omega)e^{j\omega nT}d\omega =: \mathcal{F}_d^{-1}X_d(\omega) \tag{3.44}$$

Note that the integration in (3.44) is from 0 to $2\pi/T$ because as m in $m\omega_0 = m(2\pi/NT)$ ranges from 0 to $N-1$, the frequency $\omega = m\omega_0$ ranges from 0 to $(N-1)2\pi/NT$, which approaches $2\pi/T$ as $N \to \infty$. We mention that it is possible to assume $T = 1$ in (3.43) and (3.44) as in most DSP texts, and the subsequent discussion can be simplified. However, it will be more complex to relate them to the CT Fourier transform.

Next we discuss an important property of $X_d(\omega)$. Using

$$e^{j(\omega+2\pi/T)nT} = e^{j\omega nT}e^{j2\pi n} = e^{j\omega nT}$$

for all integer n, we can readily show

$$X_d(\omega + 2\pi/T) = X_d(\omega)$$

for all ω. Thus $X_d(\omega)$ is periodic with period $2\pi/T$. Because $e^{j\omega nT}$ is also periodic with period $2\pi/T$, we can write (3.44) more generally as

$$x(nT) = \frac{T}{2\pi} \int_{<2\pi/T>} X_d(\omega)e^{j\omega nT} d\omega =: \mathcal{F}_d^{-1} X_d(\omega) \qquad (3.45)$$

The pair in (3.43) and (3.44) or (3.45) is called the *discrete-time Fourier transform pair* and is the DT counterpart of (3.5) and (3.6). As in (3.5) and (3.6), (3.43) may be called the analysis equation and (3.44) the synthesis equation. The integration in (3.45) is over an interval of $2\pi/T$. If the integration is over $(-\infty, \infty)$ as in (3.6), then the integration will diverge and the equation is meaningless. This is an important difference between the CT and DT Fourier transforms.

The DT Fourier transform is defined as a summation over all integers from $-\infty$ to ∞. For some $x(nT)$, the summation in (3.43) may diverge and its transform is not defined. If $x(nT)$ is *absolutely summable* in the sense

$$\sum_{n=-\infty}^{\infty} |x(nT)| \leq q < \infty$$

then its discrete-time Fourier transform is defined. In such a case, the transform is called the *frequency spectrum* or, simply, *spectrum* of the DT signal. As in the CT case, the DT Fourier transform, frequency spectrum, and spectrum will be used interchangeably.

The spectrum $X_d(\omega)$ is periodic with period $2\pi/T$; therefore, we consider $X_d(\omega)$ only over a frequency interval of $2\pi/T$. As in Section 2.5, the frequency range will be chosen as

$$\frac{-\pi}{T} < \omega \leq \frac{\pi}{T} \text{ (in rad/s)} \qquad -0.5f_s < f \leq 0.5f_s \text{ (in Hz)}$$

where $f_s = 1/T$ and $f = \omega/2\pi$. This range is consistent with our perception of frequency.

In general, the spectrum $X_d(\omega)$ is complex valued. Its magnitude $|X_d(\omega)|$ is called the *magnitude spectrum*, and its phase $\not{\angle} X_d(\omega)$, the *phase spectrum*. As in the CT case, if $x(nT)$ is real, then its magnitude spectrum is even and its phase spectrum is odd. If $x(nT)$ is real and even, so is $X_d(\omega)$. If $X_d(\omega)$ is real, we can plot it directly; otherwise, we plot its magnitude and phase. Here are some examples.

◆ Example 3.6

Consider the sequence

$$x[n] = x(nT) = \begin{cases} a^n & \text{for } n \geq 0 \\ 0 & \text{for } n < 0 \end{cases}$$

where a is a real or complex constant. Because the sequence is identically zero for $n < 0$, it is called a positive-time sequence. Using

$$\sum_{n=0}^{\infty} r^n = \frac{1}{1-r} \quad \text{if } |r| < 1$$

we can readily show that the sequence is absolutely summable if $|a| < 1$. Under the assumption of $|a| < 1$, the frequency spectrum of the sequence is

$$X_d(\omega) = \sum_{-\infty}^{\infty} x(nT)e^{-jn\omega T} = \sum_{n=0}^{\infty} a^n e^{-jn\omega T} = \sum_{n=0}^{\infty} \left(ae^{-j\omega T}\right)^n = \frac{1}{1 - ae^{-j\omega T}}$$

We plot in Figs. 3.11(a) and (c), with respect to time, the time sequences for $T = 1$ and 0.5 and with $a = 0.6$. The two plots will be identical if they are plotted with respect to time index. We plot in Figs. 3.11(b) and (d) their magnitude spectra (solid line) and phase spectra (dotted line). Clearly, the magnitude spectrum is even and the phase spectrum is odd. They are periodic with period $2\pi/T$ and need to be plotted over the Nyquist frequency interval $(-\pi/T, \pi/T]$. From the plots we can also see the effect of T. A smaller T will compress the time duration but expand the bandwidth of the frequency spectrum. Thus, as in the CT case, time duration and frequency bandwidth are inversely related.

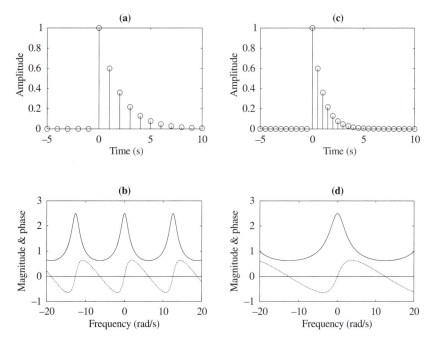

Figure 3.11 (a) Time sequence $x(nT)$ with $T = 1$. (b) The magnitude (solid line) and phase (dotted line) spectra of (a). Its Nyquist frequency range is $(-\pi, \pi]$. (c) Time sequence $x(nT)$ with $T = 0.5$. (d) The magnitude (solid line) and phase (dotted line) spectra of (c). Its Nyquist frequency range is $(-2\pi, 2\pi]$.

If $|a| < 1$, the frequency spectrum of the sequence is defined as discussed above. If $|a| > 1$, the sequence grows unbounded, its DT Fourier transform diverges, and its frequency spectrum is not defined. If $|a| = 1$, the sequence is not absolutely summable but its spectrum is still defined. See Problems 5.24 and 5.25.

A beginner often thinks that the spectrum of a DT sequence is defined only at discrete frequencies. This is certainly incorrect. In fact, if a DT sequence is absolutely summable, then its spectrum is, as in the CT case, a bounded and continuous function of ω. Every sequence of finite length is absolutely summable; therefore, its spectrum is bounded and continuous.

◆ **Example 3.7 DT Rectangular Window**

Consider the sequence shown in Fig. 3.12(a) or

$$w_d(nT) = \begin{cases} 1 & \text{for } |n| \leq M \\ 0 & \text{for } |n| > M \end{cases} \tag{3.46}$$

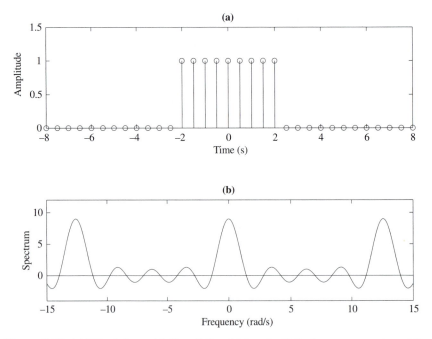

Figure 3.12 (a) DT window with $T = 0.5$ and $M = 4$. (b) Its frequency spectrum. The Nyquist frequency range is $(-2\pi, \ 2\pi]$.

with $T = 0.5$. It is a sequence of length $N := 2M + 1$ and is called a *DT rectangular window* with length N. Its DT Fourier transform is

$$W_d(\omega) := \sum_{n=-\infty}^{\infty} w(nT)e^{-j\omega nT} = \sum_{n=-M}^{M} e^{-j\omega nT}$$

which can be written, using the new index $\bar{n} = M + n$, as

$$W_d(\omega) = \sum_{\bar{n}=0}^{2M} e^{-j\omega(\bar{n}-M)T} = e^{j\omega MT} \sum_{\bar{n}=0}^{2M} \left(e^{-j\omega T}\right)^{\bar{n}}$$

This becomes, using (2.43),

$$\begin{aligned}
W_d(\omega) &= e^{j\omega MT} \frac{1 - e^{-j\omega(2M+1)T}}{1 - e^{-j\omega T}} = \frac{e^{j\omega MT} - e^{-j\omega(M+1)T}}{1 - e^{-j\omega T}} \\
&= \frac{e^{-j0.5\omega T}(e^{j\omega(M+0.5)T} - e^{-j\omega(M+0.5)T})}{e^{-j0.5\omega T}(e^{j0.5\omega T} - e^{-j0.5\omega T})} \\
&= \frac{2j \sin\left[(M+0.5)\omega T\right]}{2j \sin(0.5\omega T)} = \frac{\sin(0.5N\omega T)}{\sin(0.5\omega T)}
\end{aligned} \tag{3.47}$$

where we have used $N = 2M + 1$. The spectrum is real-valued and we plot it in Fig. 3.12(b) for $M = 4$, $N = 9$, and $T = 0.5$. The rectangular window is real and even; therefore so is its spectrum. The spectrum is periodic with period $2\pi/T = 4\pi = 12.56$. In practice, we need to plot it only in the Nyquist frequency range $(-6.28, 6.28]$. In the range, it has one main lobe and a number of side lobes. The number of side lobes equals $N - 2$, as we will show shortly.

We discuss some general properties of (3.47). Its value at $\omega = 0$ can be computed, using l'Hôpital's rule, as

$$\begin{aligned}
W_d(0) &= \frac{d[\sin(0.5N\omega T)]/d\omega}{d\sin(0.5\omega T)/d\omega}\bigg|_{\omega=0} \\
&= \frac{0.5NT\cos(0.5N\omega T)}{0.5T\cos(0.5\omega T)}\bigg|_{\omega=0} = N
\end{aligned}$$

Because $\sin(0.5N\omega T) = 0$ for

$$\omega = m\frac{2\pi}{NT} \quad m = 0, \pm 1, \pm 2, \ldots$$

we can readily show that the main lobe has base width $4\pi/NT$ and every side lobe has base width $2\pi/NT$. If the number of side lobes is $N - 2$, then we have

$$\frac{4\pi}{NT} + (N-2)\frac{2\pi}{NT} = \frac{2\pi(2+N-2)}{NT} = \frac{2\pi}{T}$$

This equals the length of the Nyquist frequency range. Thus, unlike the CT case where the spectrum of the rectangular window has infinitely many side lobes, the spectrum of the DT rectangular window has only $(N - 2)$ number of side lobes.

◆ **Example 3.8 Digital Ideal Low-Pass Filter**

Consider the spectrum shown in Fig. 3.13(a) or

$$H_d(\omega) = \begin{cases} 1 & \text{for } |\omega| \leq \omega_c \\ 0 & \text{for } \omega_c < |\omega| \leq \pi \end{cases}$$

with $\omega_c = 1$. Here we have assumed implicitly that the sampling period T is 1. This function, as we will discussed in Chapter 6, characterizes digital ideal low-pass filters. Using (3.45), we compute its time sequence as

$$h_d[n] = h_d(n) = \frac{1}{2\pi} \int_{-\pi}^{\pi} H_d(\omega) e^{j\omega n} d\omega = \frac{1}{2\pi} \int_{-\omega_c}^{\omega_c} e^{j\omega n} d\omega$$

$$= \frac{e^{j\omega_c n} - e^{-j\omega_c n}}{2j\pi n} = \frac{\sin \omega_c n}{\pi n} \tag{3.48}$$

for $n = 0, \pm 1, \pm 2, \ldots$. It is plotted in Fig. 3.13(b); its value at $n = 0$ is ω_c/π. This sequence is not absolutely summable, and its spectrum shows discontinuity. It is the DT counterpart of the CT function in (3.13).

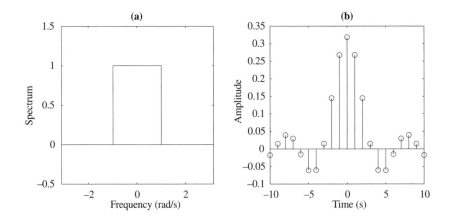

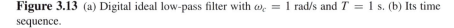

Figure 3.13 (a) Digital ideal low-pass filter with $\omega_c = 1$ rad/s and $T = 1$ s. (b) Its time sequence.

3.6.1 Frequency Spectrum of DT Periodic Signals

Now we consider the frequency spectrum of periodic sequences. We compute first the inverse DT Fourier transform of the impulse $\delta(\omega - \omega_0)$, where ω_0 is assumed to lie inside the Nyquist frequency range. Using the sifting property of impulses discussed in the appendix, we have

$$\mathcal{F}_d^{-1}[\delta(\omega - \omega_0)] = \frac{T}{2\pi} \int_{<2\pi/T>} \delta(\omega - \omega_0)e^{jn\omega T} d\omega$$

$$= \frac{T}{2\pi} e^{jn\omega T}\Big|_{\omega=\omega_0} = \frac{T}{2\pi} e^{jn\omega_0 T}$$

which implies

$$\mathcal{F}_d[e^{jn\omega_0 T}] = \frac{2\pi}{T}\delta(\omega - \omega_0) \tag{3.49}$$

Thus the frequency spectrum of the complex exponential sequence $e^{jn\omega_0 T}$ is an impulse at ω_0 with weight $2\pi/T$. If $\omega_0 = 0$, (3.49) becomes

$$\mathcal{F}_d[1] = \frac{2\pi}{T}\delta(\omega)$$

where 1 is the sequence $x(nT) = 1$ for all integers n in $(-\infty, \infty)$. Thus the spectrum of $x(nT) = 1$, for all n in $(-\infty, \infty)$, is an impulse with weight $2\pi/T$. The spectrum of $x(nT) = 1$, for $n \geq 0$, and $x(nT) = 0$, for $n < 0$, can be computed as $X_d(\omega) = \frac{\pi}{T}\delta(\omega) + \frac{1}{1-e^{-j\omega t}}$. See Problem 5.25. This is similar to the CT case as shown in Figs. 3.4 (a) and (d).

As in the CT Fourier transform, the DT Fourier transform is a linear operator. Thus we have

$$\mathcal{F}_d[\sin \omega_0 nT] = \mathcal{F}_d\left[\frac{e^{j\omega_0 nT} - e^{-j\omega_0 nT}}{2j}\right] = \frac{\pi}{jT}[\delta(\omega - \omega_0) - \delta(\omega + \omega_0)]$$

$$= \frac{-j\pi}{T}\delta(\omega - \omega_0) + \frac{j\pi}{T}\delta(\omega + \omega_0)$$

and

$$\mathcal{F}_d[\cos \omega_0 nT] = \mathcal{F}_d\left[\frac{e^{j\omega_0 nT} + e^{-j\omega_0 nT}}{2}\right] = \frac{\pi}{T}\delta(\omega - \omega_0) + \frac{\pi}{T}\delta(\omega + \omega_0)$$

They are the DT counterparts of (3.15) and (3.16). Note that $\sin \omega_0 nT$ and $\cos \omega_0 nT$ are not periodic if $\omega_0 T$ is not a rational number multiple of π. However, their frequencies are still defined. Thus we consider them, from now on, to be periodic for all ω_0 and T.

Every periodic sequence $x(nT)$ with period N can be expressed in the discrete-time Fourier series as

$$x(nT) = \sum_{m=<N>} c_{md}e^{jm\omega_0 nT}$$

with $\omega_0 = 2\pi/NT$. Thus the frequency spectrum of $x(nT)$ is

$$X_d(\omega) = \mathcal{F}_d[x(nT)] = \sum_{m=<N>} \frac{2\pi}{T} c_{md} \delta(\omega - m\omega_0) \tag{3.50}$$

It consists of a sequence of impulses at discrete frequencies; therefore, it can be called the discrete frequency spectrum. The spectrum is periodic with period $2\pi/T$. In practice, we need to plot the spectrum with m in (2.50) or with ω in the Nyquist frequency range.

From the preceding development, we see that most discussion for CT signals are applicable to DT signals if the frequency range is changed from $(-\infty, \infty)$ to $(-\pi/T, \pi/T]$. In fact, all properties discussed in Section 3.3 are also applicable to the DT case. For example, if a DT signal $x[n] = x(nT)$ is shifted to $x[n - n_0]$, then the shifting will not affect its magnitude spectrum but will introduce a linear phase into the phase spectrum. To be more specific, if $X_d(\omega) = \mathcal{F}_d[x(nT)]$, then

$$X_{d0}(\omega) := \mathcal{F}_d[x((n - n_0)T)] = e^{-jn_0\omega T} X_d(\omega) \tag{3.51}$$

and

$$|X_{d0}(\omega)| = |X_d(\omega)| \quad \text{and} \quad \measuredangle X_{d0}(\omega) = \measuredangle X_d(\omega) - n_0\omega T \tag{3.52}$$

They are the DT counterparts of (3.22) and (3.23). The DT counterpart of (3.27) is

$$E_d := \sum_{n=-\infty}^{\infty} |x(nT)|^2 = \frac{T}{2\pi} \int_{-\pi/T}^{\pi/T} X_d^*(\omega) X_d(\omega) d\omega \tag{3.53}$$

To conclude this section, we mention that if a spectrum does not contain impulses, there is no difference in using $(-\pi/T, \pi/T]$, $[-\pi/T, \pi/T]$, or $(-\pi/T, \pi/T)$ as its frequency range. If a frequency spectrum contains an impulse at π/T, then the spectrum will also contain, due to periodicity, an impulse at $-\pi/T$, which is actually a different representation of the impulse at π/T. In this case, to avoid redundancy, we must use $(-\pi/T, \pi/T]$ or $[-\pi/T, \pi/T)$ as the frequency range.

3.7 Effects of Truncation

Before discussing the effects of truncation, we need some preliminary results. Let $X_{di}(\omega)$ be the DT Fourier transforms of $x_i(nT)$, for $i = 1, 2$. Then we have

$$\mathcal{F}_d[x_1(nT)x_2(nT)] = \frac{T}{2\pi} \int_{<2\pi/T>} X_{d1}(\bar{\omega}) X_{d2}(\omega - \bar{\omega}) d\bar{\omega}$$

$$= \frac{T}{2\pi} \int_{<2\pi/T>} X_{d2}(\bar{\omega}) X_{d1}(\omega - \bar{\omega}) d\bar{\omega} \tag{3.54}$$

This is the DT counterpart of (3.31) and can be established using the same procedure. Therefore, its derivation will not be repeated.

Now consider a discrete-time sequence $x(nT)$ with spectrum $X_d(\omega)$. Suppose we truncate the sequence before $-M$ and after M. This is the same as multiplying the sequence $x(nT)$ by the rectangular window $w_d(nT)$ defined in (3.46). The spectrum of $w_d(nT)$ was computed in (3.47) as

$$W_d(\omega) = \frac{\sin 0.5N\omega T}{\sin 0.5\omega T} \tag{3.55}$$

where $N = 2M + 1$. The spectrum of the truncated $x(nT)$ or $x(nT)w_d(nT)$ is, using (3.54),

$$\mathcal{F}_d[x(nT)w_d(nT)] = \frac{T}{2\pi} \int_{-\pi/T}^{\pi/T} X_d(\bar{\omega}) W_d(\omega - \bar{\omega}) d\bar{\omega}$$

$$= \frac{T}{2\pi} \int_{-\pi/T}^{\pi/T} W_d(\bar{\omega}) X_d(\omega - \bar{\omega}) d\bar{\omega} \tag{3.56}$$

which implies that the spectrum of a truncated sequence equals the convolution of $W_d(\omega)$ and the spectrum of the untruncated sequence. Thus the waveform of $W_d(\omega)$ plays a crucial role in determining the effects of truncation. Because the waveform of $W_M(\omega)$ is similar to the one of the CT case, all discussion for the CT case applies here. That is, truncation will introduce leakage and ripples, as will be demonstrated by examples.

Consider the sequence $\cos \omega_0 nT$. Its spectrum was computed in Section 3.6.1 as

$$X_d(\omega) = \frac{\pi}{T}[\delta(\omega - \omega_0) + \delta(\omega - \omega_0)]$$

and is plotted in Figs. 3.14(a) and (d) for $\omega_0 = 0.5$, $\omega_0 = 3$, and $T = 1$. Note that the spectrum is periodic with period $2\pi/T = 2\pi$; thus we have two impulses at $\omega = \pm 3.28$ in Fig. 3.14(d), which are the shiftings of the ones at $\omega = \mp 3$ by $\pm 2\pi = \pm 6.28$. Although the two impulses at $\omega = \pm 3.28$ are outside the Nyquist frequency range $(-3.14, \ 3.14]$ and are redundant, they nevertheless will affect the truncated spectrum. If we truncate $\cos \omega_0 nT$ for $|n| > M$, then the spectrum of the truncated $\cos \omega_0 nT$ is

$$\mathcal{F}_d[(\cos \omega_0 nT)w_d(nT)] = \frac{T}{2\pi} \int_{-\pi/T}^{\pi/T} \frac{\pi}{T}[\delta(\bar{\omega} - \omega_0) + \delta(\bar{\omega} + \omega_0)] W_d(\omega - \bar{\omega}) d\bar{\omega}$$

$$= 0.5[W_d(\omega - \omega_0) + W_d(\omega + \omega_0)] \tag{3.57}$$

It is real valued. Figures 3.14(b) and (e) show the frequency spectra of truncated $\cos 0.5nT$ and $\cos 3nT$ for $M = 4$, and Figs. 3.14(c) and (f) show those for $M = 30$. They are similar to those in Fig. 3.8, where we plotted only the magnitude spectra. Thus, as in the CT case, truncation also introduces leakage and ripples in the DT case. The effects, however, will be more severe because the impulses outside the Nyquist frequency range will also affect those inside the range.

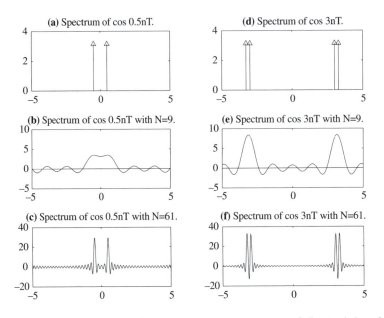

Figure 3.14 (a) Spectrum of cos $0.5n$ for all n. (b) Spectrum of cos $0.5n$, for $|n| \leq 4$. (c) Spectrum of cos $0.5n$, for $|n| \leq 30$. (d, e, f) Corresponding spectra of cos $3nT$ with $T = 1$. All horizontal coordinates are frequency in rad/s. The Nyquist frequency range is $(-\pi, \pi]$.

Figure 3.14 also show the Gibbs phenomenon. We will show this again for the digital ideal low-pass filter, with $T = 1$, discussed in Example 3.8. Consider

$$h_d[n] = \frac{\sin \omega_c n}{\pi n}$$

for all integer n. Its spectrum is

$$H_d(\omega) = \begin{cases} 1 & \text{for } |\omega| \leq \omega_c \\ 0 & \text{for } \omega_c < |\omega| \leq \pi \end{cases}$$

Using (3.56) with $T = 1$, we can readily show

$$\mathcal{F}_d[h_d[n]w_d[n]] = \frac{1}{2\pi} \int_{-\pi}^{\pi} H_d(\bar{\omega}) W_d(\omega - \bar{\omega}) d\bar{\omega}$$

$$= \frac{1}{2\pi} \int_{-\omega_c}^{\omega_c} \frac{\sin [0.5N(\omega - \bar{\omega})]}{\sin [0.5(\omega - \bar{\omega})]} d\bar{\omega} \qquad (3.58)$$

This is the DT counterpart of (3.38). If $\omega_c = 1$, then the spectra of the truncated DT ideal low-pass filter for $M = 5$ and 50 are as shown in Figs. 3.15(a) and (b). This is obtained not from

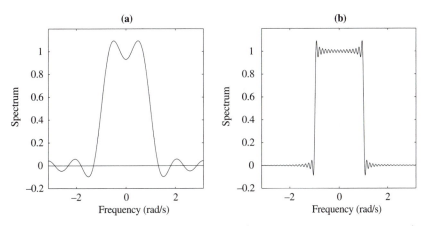

Figure 3.15 (a) Spectrum of $(\sin n)/\pi n$, for $|n| \leq 5$ in the Nyquist frequency range $(-\pi, \pi]$. (b) Spectrum of $(\sin n)/\pi n$, for $|n| \leq 50$.

(3.58) but from the DTFT of $(2M + 1)$ terms of $h_d[n]$; it can also be obtained using FFT. See Problem 4.13. The frequency spectra of truncated sequences show ripples, leakage, and Gibbs phenomena. The situation is identical to the CT case.

3.8 Nyquist Sampling Theorem

Consider a CT signal $x(t)$ and its sampled sequence $x(nT)$. The spectrum of $x(t)$ is defined as its CT Fourier transform, and the spectrum of $x(nT)$ is defined as its DT Fourier transform; that is,

$$X(\omega) = \mathcal{F}[x(t)] \quad \text{and} \quad X_d(\omega) = \mathcal{F}_d[x(nT)]$$

This section will develop the relationship between $X(\omega)$ and $X_d(\omega)$, and then establish the Nyquist sampling theorem.

If we apply the CT Fourier transform to $x(nT)$ directly, the result will be identically zero; that is, $\mathcal{F}[x(nT)] \equiv 0$. Now we shall modify the sequence as

$$x_s(t) := \sum_{n=-\infty}^{\infty} x(nT)\delta(t - nT) \tag{3.59}$$

It is a sequence of impulses at $t = nT$ with weight $x(nT)$. It is defined for all t but is zero everywhere except at sampling instants. Thus $x_s(t)$ can be considered as a continuous-time representation of the discrete-time sequence $x(nT)$. The application of the CT Fourier transform to (3.59) yields

$$\mathcal{F}[x_s(t)] = \mathcal{F}\left[\sum_{n=-\infty}^{\infty} x(nT)\delta(t - nT)\right] = \sum_{n=-\infty}^{\infty} x(nT)\mathcal{F}[\delta(t - nT)]$$

Using the sifting property of impulses, we have

$$\mathcal{F}[\delta(t - nT)] = \int_{-\infty}^{\infty} \delta(t - nT)e^{-j\omega t}dt = e^{-j\omega t}\Big|_{t=nT} = e^{-j\omega nT}$$

Thus the CT Fourier transform of $x_s(t)$ is

$$\mathcal{F}[x_s(t)] = \sum_{n=-\infty}^{\infty} x(nT)e^{-j\omega nT}$$

which is the DT Fourier transform in (3.43). Thus we have established

$$X_d(\omega) = \mathcal{F}_d[x(nT)] = \mathcal{F}[x_s(t)] \tag{3.60}$$

This shows the close relationship between the CT and DT Fourier transforms.

Because of (3.60), finding the relationship between $X(\omega)$ and $X_d(\omega)$ is the same as finding the relationship between the CT Fourier transforms of $x(t)$ and $x_s(t)$. Consider

$$r(t) := \sum_{n=-\infty}^{\infty} \delta(t - nT) \tag{3.61}$$

It is the sampling function defined in (2.33) and plotted in Fig. 2.11. Because

$$x(nT)\delta(t - nT) = x(t)\delta(t - nT)$$

[see (A.5)], we have

$$x_s(t) = \sum_{n=-\infty}^{\infty} x(nT)\delta(t - nT) = x(t)\sum_{n=-\infty}^{\infty} \delta(t - nT) = x(t)r(t) \tag{3.62}$$

Thus $x_s(t)$ is the product of $x(t)$ and the sampling function $r(t)$. The process of multiplying $x(t)$ by $r(t)$ is called a *modulation*. Thus the sampling process in Fig. 3.16(a) can be interpreted as the modulation shown in Fig. 3.16(b). Because $r(t)$ is periodic with period T, we can express it in the CT Fourier series as

$$r(t) = \sum_{n=-\infty}^{\infty} \delta(t - nT) = \frac{1}{T}\sum_{m=-\infty}^{\infty} e^{jm\omega_0 t} \tag{3.63}$$

with $\omega_0 = 2\pi/T$. See (2.35). Thus (3.62) implies

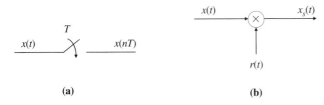

(a)   **(b)**   **Figure 3.16** (a) Sampling. (b) Modulation.

$$x_s(t) = x(t)\left(\frac{1}{T}\sum_{m=-\infty}^{\infty} e^{jm\omega_0 t}\right) = \frac{1}{T}\sum_{m=-\infty}^{\infty} x(t)e^{jm\omega_0 t} \tag{3.64}$$

Because of (3.60), we have

$$X_d(\omega) = \mathcal{F}[x_s(t)] = \frac{1}{T}\sum_{m=-\infty}^{\infty} \mathcal{F}[x(t)e^{jm\omega_0 t}] \tag{3.65}$$

If $X(\omega) = \mathcal{F}[x(t)]$, then we have

$$\mathcal{F}[x(t)e^{jm\omega_0 t}] = X(\omega - m\omega_0) \tag{3.66}$$

as shown in (3.24). Substituting (3.66) into (3.65) yields

$$X_d(\omega) = \frac{1}{T}\sum_{m=-\infty}^{\infty} X(\omega - m\omega_0) = \frac{1}{T}\sum_{m=-\infty}^{\infty} X\left(\omega - \frac{2\pi m}{T}\right) \tag{3.67}$$

This relates the frequency spectra of $x(t)$ and its sampled sequence $x(nT)$.

Let us discuss the implication of (3.67). For easy plotting, we assume $X(\omega)$ to be real valued and of the form shown in Fig. 3.17(a). It is *frequency bandlimited* to W in the sense that

$$X(\omega) = 0 \qquad \text{for } |\omega| > W$$

or $X(\omega)$ has the highest nonzero frequency component at $\omega = W$. The function $X(\omega - m\omega_0)$ has the same wave form as $X(\omega)$ except that it is shifted to $\omega = m\omega_0$. Thus $X_d(\omega)$ is the sum of all repetitive shiftings of $X(\omega)/T$ to $m2\pi/T$, $m = 0, \pm1, \pm2, \ldots$. Figure 3.17 plots the frequency spectra $X_d(\omega)$ for three different T values. Note that the vertical coordinates of Figs. 3.17(b)–(d) are $TX_d(\omega)$, not $X_d(\omega)$. We see that if $\pi/T > W$, then the repetitions will not overlap as in Fig. 3.17(b) and the resulting $TX_d(\omega)$ in the frequency range $(-\pi/T, \pi/T]$ is identical to $X(\omega)$. Note that they are different outside the range because $X_d(\omega)$ can be extended periodically but not $X(\omega)$. In conclusion, if $x(t)$ is bandlimited to W rad/s and if the sampling period T is chosen to be $\pi/T > W$ or $T < \pi/W$, then

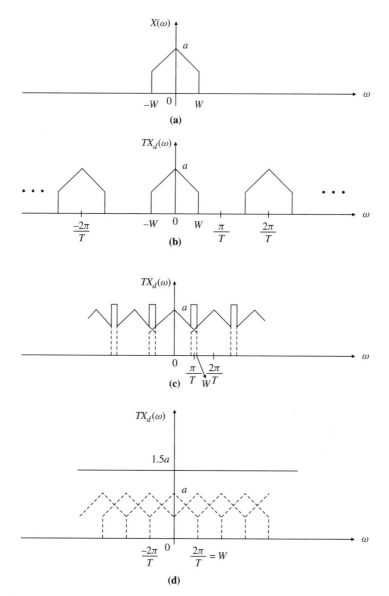

Figure 3.17 (a) Spectrum, bandlimited to W, of CT $x(t)$. (b) Spectrum of $Tx(nT)$ with $T < \pi/W$. (c) With $\pi/W < T < 2\pi/W$. (d) With $T = 2\pi/W$.

$$\text{Spec. of } x(t) = \begin{cases} T \times [\text{Spec. of } x(nT)] & \text{for } |\omega| \le \pi/T \\ 0 & \text{for } |\omega| > \pi/T \end{cases} \qquad (3.68)$$

where Spec. stands for frequency spectrum.

Let $f_{max} := W/2\pi$ and $f_s := 1/T$ (in Hz). Then the condition $T < \pi/W$ is the same as the condition $f_s > 2f_{max}$. Thus the preceding statement can also be stated as: If $x(t)$ has the highest nonzero frequency f_{max}, then its frequency spectrum can be computed from its sampled sequence if the sampling frequency is larger than $2f_{max}$.

Consider a CT signal $x(t)$. Suppose only its values at sampling instants are known, can we determine all values of $x(t)$? Or can $x(t)$ be recovered from $x(nT)$? This is answered by the theorem that follows.

> **Nyquist Sampling Theorem** Let $x(t)$ be a CT signal bandlimited to W (in rad/s) or have the highest frequency $f_{max} = W/2\pi$ (in Hz); that is,
>
> $$X(\omega) = 0 \quad \text{for } |\omega| > W$$
>
> where $X(\omega)$ is the frequency spectrum of x(t). Then $x(t)$ can be recovered from its sampled sequence $x(nT)$ if the sampling period T is less than π/W, or the sampling frequency $f_s = 1/T$ is larger than $2f_{max}$.

If $x(t)$ has the highest frequency f_{max} and if the sampling frequency is larger than $2f_{max}$, then the frequency spectrum of $x(t)$ can be computed from $x(nT)$. We can then compute $x(t)$ from its frequency spectrum by taking the inverse CT Fourier transform. To be more specific, we have

$$x(t) = \frac{1}{2\pi} \int_{\omega=-\infty}^{\infty} X(\omega) e^{j\omega t} d\omega$$

$$= \frac{1}{2\pi} \int_{\omega=-\pi/T}^{\pi/T} X(\omega) e^{j\omega t} d\omega$$

where $\pi/T > W$. This equation becomes, after substituting (3.68) and then using the integration in (3.13),

$$x(t) = \frac{1}{2\pi} \int_{\omega=-\pi/T}^{\pi/T} \left(T \sum_{n=-\infty}^{\infty} x(nT) e^{-j\omega nT} \right) e^{j\omega t} d\omega$$

$$= \frac{T}{2\pi} \sum_{n=-\infty}^{\infty} x(nT) \int_{\omega=-\pi/T}^{\pi/T} e^{j\omega(t-nT)} d\omega$$

$$= \frac{T}{2\pi} \sum_{n=-\infty}^{\infty} x(nT) \frac{e^{j(\pi/T)(t-nT)} - e^{-j(\pi/T)(t-nT)}}{j(t-nT)}$$

$$= \sum_{n=-\infty}^{\infty} x(nT) \frac{\sin[\pi(t-nT)/T]}{\pi(t-nT)/T}$$

$$= \sum_{n=-\infty}^{\infty} x(nT) \operatorname{sinc}[\pi(t-nT)/T] \tag{3.69}$$

where sinc $c = \sin c / c$ is the sinc function. Thus $x(t)$ is a weighted sum of sinc functions. Note that because sinc $0 = 1$ and sinc $k\pi = 0$ for every nonzero integer k, the infinite summation in (3.69) reduces to $x(mT)$ if $t = mT$. If $x(nT)$ for all n are known, the CT signal $x(t)$ for all t can be computed from (3.69). Thus we conclude that if $x(t)$ is bandlimited to W and if $T < \pi/W$, then $x(t)$ can be recovered from $x(nT)$ by using (3.69). We call (3.69) the *ideal interpolation formula*. This establishes the Nyquist sampling theorem. The sampling theorem discussed in Section 2.6 is a special case of this theorem. The frequency $2 f_{max}$ is often called the *Nyquist rate*.

3.8.1 Frequency Aliasing due to Time Sampling

Consider again (3.67) and the CT signal with frequency spectrum in Fig. 3.17(a). If $\pi/T \leq W$, then the repetitions of $X(\omega - m\omega_0)$ will overlap as shown in Figs. 3.17(c) and (d). This type of overlapping is called *frequency aliasing*. In this case, the resulting $T X_d(\omega)$ in the frequency range $(-\pi/T, \pi/T]$ will be different from $X(\omega)$, and we cannot compute the spectrum of $x(t)$ from its sampled sequence.

  If a CT signal is not bandlimited, then frequency aliasing will always occur no matter how small the sampling period chosen. However, if a CT signal, such as the one in Example 3.1 or 3.2, has a finite total energy, then its frequency spectrum decreases to zero as $|\omega| \to \infty$. In this case, if T is sufficiently small, then we have

$$X\left(\omega - \frac{2\pi m}{T}\right) \approx 0 \tag{3.70}$$

or

$$\text{Re } X\left(\omega - \frac{2\pi m}{T}\right) \approx 0 \ \text{ and } \ \text{Im } X\left(\omega - \frac{2\pi m}{T}\right) \approx 0$$

for $m \neq 0$ and for $|\omega| \leq \pi/T$. Let us write (3.67) as

$$T X_d(\omega) = \cdots + X\left(\omega + \frac{2\pi}{T}\right) + X(\omega) + X\left(\omega - \frac{2\pi}{T}\right) + \cdots$$

Then (3.70) implies

$$X(\omega) \approx \begin{cases} T X_d(\omega) & \text{for } |\omega| \leq \pi/T \\ 0 & \text{for } |\omega| > \pi/T \end{cases} \tag{3.71}$$

for T sufficiently small. This equation implies

$$|X(\omega)| \approx \begin{cases} T|X_d(\omega)| & \text{for } |\omega| \leq \pi/T \\ 0 & \text{for } |\omega| > \pi/T \end{cases} \tag{3.72}$$

$$\text{Re}[X(\omega)] \approx \begin{cases} \text{Re}[T X_d(\omega)] & \text{for } |\omega| \leq \pi/T \\ 0 & \text{for } |\omega| > \pi/T \end{cases} \tag{3.73}$$

and

$$\text{Im}[X(\omega)] \approx \begin{cases} \text{Im}[TX_d(\omega)] & \text{for } |\omega| \leq \pi/T \\ 0 & \text{for } |\omega| > \pi/T \end{cases} \tag{3.74}$$

However, the phase of a number can be very large even if its real and imaginary parts are very small, as discussed in Example 2.11. Thus in general we have

$$\angle X(\omega) \neq \angle (TX_d(\omega)) \quad \text{for all } \omega \tag{3.75}$$

In conclusion, if a CT signal has a finite total energy, then its magnitude spectrum can be computed from the DT Fourier transform of its sampled sequence if the sampling period T is chosen to be sufficiently small. This is illustrated by examples that follow.

◆ Example 3.9

Consider the CT and DT rectangular windows in (3.8) and (3.46). The frequency spectrum $W_a(\omega)$ of the CT window was computed in (3.9) as $2(\sin a\omega)/\omega$, and the frequency spectrum $W_d(\omega)$ of the DT window was computed in (3.47) as $\sin(0.5N\omega T)/\sin(0.5\omega T)$. Because the two spectra differ by a factor of T, and because they are real, we will compare $W_a(\omega)$ and $TW_d(\omega)$ for $a = 2$. If we sample the CT window with sampling period T, then M in (3.46) equals $\underline{a/T}$, where the underline rounds a number downward to an integer, and can be obtained in MATLAB as M=floor(a/T).

First we choose $T = 0.2$. Then the frequency range of the DT window is $(-\pi/T, \pi/T] = (-15.7, 15.7]$, and the spectrum will be periodic with period 31.4 rad/s. The program that follows

Program 3.4

```
a=2;T=0.2;w=-pi/T:0.02:pi/T;
Wa=2*sin(a*w)./w;
M=floor(a/T);N=2*M+1;
Wd=T*sin(0.5*N*w*T)./sin(0.5*w*T);
plot(w,Wa,w,Wd,':')
axis([-2*pi 2*pi -1.5 5])
```

yields $W_a(\omega)$ (solid line) and $TW_d(\omega)$ (dotted line) in Fig. 3.18(a), where we plot only the frequency range $[-2\pi, 2\pi]$ (rad/s) for easier comparison. We see that, because of appreciable frequency aliasing, the two spectra differ considerably. Next we select $T = 0.05$. Figure 3.18(b) shows the two spectra for ω in $(-62.8, 62.8]$. In order to make a better comparison, we zoom in Fig. 3.18(c) the frequency ranges $[-5, 5]$ of Fig. 3.18(b), and in Fig. 3.18(d) the frequency range $[50, 62.8]$ of Fig. 3.18(b). Figure 3.18(c) is obtained from Program 3.4 by changing T to 0.05 and its

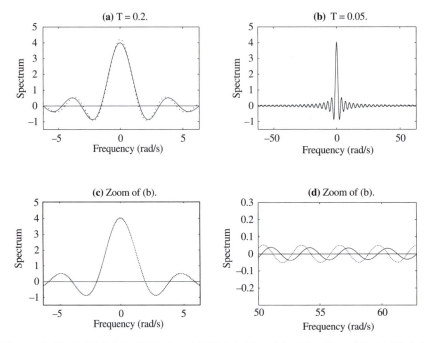

Figure 3.18 (a) $W_a(\omega)$ (solid line) and $TW_d(\omega)$ (dotted line) for $T = 0.2$. (b) $W_a(\omega)$ (solid line) and $TW_d(\omega)$ (dotted line) for $T = 0.05$. (c) Plot of (b) in [-5, 5]. (d) Plot of (b) in [50, 62.8].

last line to **axis([-5 5 -1.5 5])**. Figure 3.18(d) can be similarly obtained. We see that the two spectra are almost indistinguishable in the frequency range $[-5, 5]$. Their largest difference in $[50, 2\pi]$ is about 0.05 or $0.05/4 = 0.0125 = 1.25\%$ of their peak magnitude 4. Thus the effect of frequency aliasing is negligible. In conclusion, even though the CT window $w_a(t)$ is not bandlimited, its frequency spectrum can be computed approximately from its sampled sequence if the sampling period is sufficiently small.

◆ **Example 3.10**

Consider the exponential function

$$x(t) = \begin{cases} e^{-0.1t} & \text{for } t \geq 0 \\ 0 & \text{for } t < 0 \end{cases}$$

Its spectrum was computed in Example 3.1 as $X(\omega) = 1/(0.1 + j\omega)$ and is not bandlimited. The spectrum is complex valued.

The sampled sequence of $x(t)$ is $x(nT) = e^{-0.1nT}$ for $n \geq 0$. Using

$$\sum_{n=0}^{\infty} r^n = 1 + r + r^2 + r^3 + \cdots = \frac{1}{1-r}$$

for $|r| < 1$, we can compute the DT Fourier transform of $x(nT)$ as

$$X_d(\omega) = \sum_{n=0}^{\infty} e^{-0.1nT} e^{-j\omega nT} = \frac{1}{1 - e^{-0.1T} e^{-j\omega T}}$$

Figure 3.19(a) shows the magnitude spectra of $X(\omega)$ (solid line) and $TX_d(\omega)$ (dotted line) for $T = 0.1$ and $|\omega| < \pi/T = 31.4$. We zoom in Figs. 3.19(c) and (d) the plot of Fig. 3.19(a) in $[-2, \ 2]$ and $[29, \ 31.4]$. We see that the largest error between the two magnitude spectra is about $0.02/10 = 0.002 = 0.2\%$ of the peak magnitude. Thus the magnitude spectrum of the exponential function can be computed from its sampled sequence. The situation in the phase spectra is different. Figure 3.19(b) shows the phase spectra of $X(\omega)$ (solid line) and $TX_d(\omega)$ (dotted line). They are the same only in the neighborhood of $\omega = 0$.

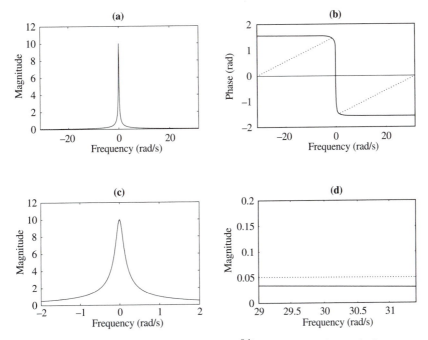

Figure 3.19 (a) Exact magnitude spectrum of $e^{-0.1t}$ (solid line) and magnitude spectrum computed from its samples (dotted line). (b) Corresponding phase spectra. (c) Plot of (a) in $[-2, \ 2]$. (d) Plot of (a) in $[29, \ 10\pi]$.

From the preceding two examples, we see that even if $x(t)$ is not bandlimited, its spectrum can still be computed from its sampled sequence. The smaller T is, the lesser the frequency aliasing and, consequently, the more accurate the result. Thus it is possible to compute the spectrum of $x(t)$, with any degree of accuracy, from its sampled sequence.

In conclusion, if frequency aliasing due to time sampling is negligible, the spectrum of $x(t)$ can be computed from its sampled sequence. One way to check frequency aliasing is to compare the spectra of $x(t)$ and $x[nT]$ for all ω in $(-\pi/T, \pi/T]$. If the spectrum of $x(t)$ is not known, the way to check frequency aliasing is to check the spectrum of $x[nT]$ in the neighborhood of $\pm\pi/T$. If the spectrum in the neighborhood of $\pm\pi/T$ is significantly different from 0, then frequency aliasing is large and we must select a smaller T. If the spectrum in the neighborhood of $\pm\pi/T$ is practically zero, then frequency aliasing is negligible and (3.71) holds. This is the case in Fig. 1.19(b). Thus the magnitude spectrum of the sentence in Fig. 1.19(a) can be computed from its samples with sampling period $T = 1/8000$.

3.9 Time-limited Bandlimited Theorem

To conclude this chapter, we discuss a fundamental relationship between a time signal and its frequency spectrum. A CT signal $x(t)$ is bandlimited to W if its frequency spectrum $X(\omega)$ is 0 for $|\omega| > W$. It is time-limited to b if

$$x(t) = 0 \quad \text{for } |t| > b$$

It turns out that if a CT signal is time-limited, then it cannot be bandlimited and vice versa. The only exception is the trivial case $x(t) = 0$ for all t. To establish this assertion, we show that if $x(t)$ is bandlimited to W and time limited to b, then it must be identically zero. Indeed, if $x(t)$ is bandlimited to W, then (3.6) implies

$$x(t) = \frac{1}{2\pi} \int_{-W}^{W} X(\omega) e^{j\omega t} d\omega \tag{3.76}$$

Its differentiation repeatedly with respect to t yields

$$x^{(k)}(t) = \frac{1}{2\pi} \int_{-W}^{W} X(\omega)(j\omega)^k e^{j\omega t} d\omega \tag{3.77}$$

for $k = 0, 1, 2, \ldots$, where $x^{(k)}(t) := d^k x(t)/dt^k$. Because $x(t)$ is time-limited to b, its derivatives are identically zero for all $|t| > b$. Thus (3.77) implies

$$\int_{-W}^{W} X(\omega)(\omega)^k e^{j\omega a} d\omega = 0 \tag{3.78}$$

for any a with $a > b$. Next we use

$$e^c = 1 + \frac{c}{1!} + \frac{c^2}{2!} + \cdots = \sum_{k=0}^{\infty} \frac{c^k}{k!}$$

to rewrite (3.76) as

$$x(t) = \frac{1}{2\pi} \int_{-W}^{W} X(\omega) e^{j\omega(t-a)} e^{j\omega a} d\omega$$

$$= \frac{1}{2\pi} \int_{-W}^{W} X(\omega) \left[\sum_{k=0}^{\infty} \frac{(j\omega(t-a))^k}{k!} \right] e^{j\omega a} d\omega$$

$$= \sum_{k=0}^{\infty} \frac{(j(t-a))^k}{2\pi k!} \int_{-W}^{W} X(\omega)(\omega)^k e^{j\omega a} d\omega$$

which is identically zero following (3.78). Thus a bandlimited and time-limited CT signal must be identically zero. In conclusion, no nontrivial CT signal can be both time limited and bandlimited.

The preceding fact has an important implication in digital signal processing of CT signals. It states that a CT signal must be infinitely long or its spectrum must be infinitely wide or both. This imposes a fundamental limitation in digital processing of CT signals. If a CT signal is bandlimited, then its *exact* frequency spectrum can be computed from its sampled sequence by choosing a sufficiently small sampling period. However, the sequence will be infinitely long because the signal cannot be time limited. Thus the sequence must be truncated in computer computation of its spectrum. Thus the computed spectrum cannot be an exact one. Conversely, if a CT signal is time limited, then its spectrum cannot be bandlimited. Thus frequency aliasing will occur in using its sampled sequence to compute its frequency spectrum. In conclusion, theoretically speaking, error will always occur in using digital techniques to process CT signals.

Although no CT signal can be both time limited and bandlimited in theory, most CT signals can be considered to be so in practice. For example, consider the exponential signal in Fig. 3.1(a). It is not time limited. However, the signal decays to zero rapidly; its value is less than 1% of its peak value for $t \geq 50$ and less than 0.01% for $t \geq 100$. Therefore, the exponential signal can be considered to be time limited in practice. Likewise, its frequency spectrum approaches 0, with rate $1/|\omega|$, as $|\omega|$ approaches infinity. Thus its frequency spectrum can also be considered to be bandlimited. In conclusion, most practical signals can be considered to be both time limited and bandlimited and can be processed digitally with negligible errors.

3.9.1 Practical Reconstruction of $x(t)$ from $x(nT)$

Because of the many advantages of digital techniques, many CT signals such as audio signals are now transmitted or stored digitally. In order to do so, CT signals must be sampled into discrete-time signals. Clearly, the sampling frequency must equal the Nyquist rate or be larger. After transmission or storage, the CT signal must be recovered from its sampled sequence. The CT signal can be computed from its sampled sequence by using the ideal interpolation formula in (3.69). However, (3.69) is not used in practice for two reasons. First, it requires infinitely many numbers of additions and multiplications. This is not feasible in practice. More seriously, we can start to compute $x(t)$ only after $x(nT)$, for all n, become available. Thus it cannot be carried out in real time. In practice, $x(t)$ can be recovered from $x(nT)$ by using a zero-order hold as shown in Fig. 3.20(a), or a first-order hold as shown in Fig. 3.20(b). The zero-order hold holds the current sample constant until the next sample arrives; the first-order hold connects the current and past samples by a straight line and then extends it until the next sample arrives. We see that there are discontinuities in the reconstructed CT signals. These discontinuities can be removed by passing the reconstructed CT signals through a CT low-pass filter, which we will discuss in a later chapter. Clearly, it is possible to develop higher-order holds by using three or more current and past samples. However, most practical D/A converters are based on the zero-order hold.

When we use a zero-order hold to recover a CT signal, error always occurs. Clearly, the smaller the sampling period, the smaller the error. However, a smaller sampling period requires more memory location, more computation, and faster hardware speed. Depending on applications, a very small sampling period may become unnecessary. Therefore, the selection of a sampling period depends on many factors; the Nyquist sampling theorem is merely one of them.

Speech and music are clearly time-limited CT signals. Therefore, their spectra are not bandlimited. However, the energy above some high frequency may become extremely small. Furthermore, human hearing capability is limited. Therefore, speech and music can be considered to be bandlimited. What is its highest frequency, however, is not so clear cut.

The frequency spectrum of human voices is generally limited to 4 kHz, as shown in Fig. 1.19(b). This is the reason that the sampling period for the sentence recorded in Fig. 1.19(a) was chosen as $T = 1/8000$. The computed spectrum in Fig. 1.19(b) is practically zero for $|f| > 3800$ Hz. Thus there is no frequency aliasing in Fig. 1.19(b) due to time sampling. This confirms the assertion that the spectrum of human voices is limited to 4 kHz.

A violin can produce a sound with frequency as high as 15 kHz. As mentioned earlier, methods of recovering audio signals from their sampled sequences should also play a role in determining the sampling period. Cost and performance requirements are also important factors. For example, on a telephone line, the sampling frequency has been selected as 8 kHz. Broadcasting companies are now using digital audio recording on their networks with a sampling frequency of 32 kHz. Music compact discs that require higher fidelity contain signals sampled at 44.1 kHz. In conclusion, the sampling frequency is not dictated by the Nyquist rate alone.

PROBLEMS

3.1   What is the frequency spectrum of the time function $\delta(t)$? Is the time function absolutely integrable? Is its spectrum bounded and continuous?

3.2   Find the spectrum of

$$x(t) = \begin{cases} e^{-2t} & \text{for } t \geq 0 \\ 0 & \text{for } t < 0 \end{cases}$$

and plot its magnitude and phase spectra.

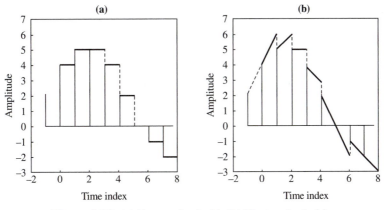

Figure 3.20 (a) Zero-order hold. (b) First-order hold.

3.3 Is the spectrum of e^{-2t}, for all t, defined? Is the spectrum of $e^{-2|t|}$, for all t, defined? Is so, find its spectrum. Is the spectrum real and even?

3.4 Find the frequency spectrum of

$$x(t) = \begin{cases} e^{-2t} + 0.5 \sin 1.5t & \text{for } t \geq 0 \\ 0.5 \sin 1.5t & \text{for } t < 0 \end{cases}$$

3.5 Consider the signal

$$x(t) = 1 + \sin 2t + \cos \pi t$$

Is it periodic? Is it absolutely integrable? Find its frequency spectrum.

3.6 Find the spectrum of

$$x(t) = \begin{cases} 0.5 \sin 1.5t & \text{for } |t| \leq \pi/1.5 \\ 0 & \text{for } |t| > \pi/1.5 \end{cases}$$

and plot its magnitude and phase spectra. Is the spectrum odd and pure imaginary?

3.7 Find the spectrum of

$$\bar{x}(t) = \begin{cases} 0.5 \sin 1.5t & \text{for } 0 \leq t \leq 2\pi/1.5 \\ 0 & \text{for } t < 0 \text{ and } t > 2\pi/1.5 \end{cases}$$

by using the result of Problem 3.6.

3.8   Show that if $X(\omega) = \mathcal{F}[x(t)]$, then

$$x(-\omega) = \frac{1}{2\pi} \mathcal{F}[X(t)]$$

This is called the *duality property* of the CT Fourier transform.

3.9   Consider a real signal $x(t)$ with spectrum $X(\omega)$ that has the property $X(0) \geq |X(\omega)|$, for all ω. The time duration of $x(t)$ can be defined as

$$D = \frac{\left(\int_{-\infty}^{\infty} |x(t)| dt\right)^2}{\int_{-\infty}^{\infty} |x(t)|^2 dt}$$

and the frequency bandwidth of of $X(\omega)$ can be defined as

$$B = \frac{\int_{-\infty}^{\infty} |X(\omega)|^2 d\omega}{2|X(0)|^2}$$

Show $DB \geq \pi$. Thus, roughly, time duration is inversely proportional to frequency bandwidth.

3.10   Is it true that if a CT signal is absolutely integrable, then it is magnitude-squared integrable or has a finite total energy? If not, find a counter example. Is it true that if a DT signal is absolutely summable, then it is magnitude-squared summable?

3.11   Compute the total energy of the CT rectangular window in (3.8) with $a = 2$. Compute also the energy in the main lobe. What is its percentage of the total energy?

3.12   An impulse sequence is defined as

$$\delta[n] = \delta(nT) = \begin{cases} 1 & n = 0 \\ 0 & n \neq 0 \end{cases}$$

What is its frequency spectrum? Is it bounded and continuous?

3.13   Find the spectra of the following sequences, all with $T = 0.5$: (a) $x[-1] = 1, x[0] = 0, x[1] = -1, x[n] = 0$ otherwise; (b) $x[0] = 1, x[1] = -2, x[2] = 1, x[n] = 0$ otherwise; (c) $x[0] = -1, x[1] = 2, x[2] = 2, x[3] = -1, x[n] = 0$ otherwise; (d) $x[n] = 0.9^n$ for all n. Sketch their magnitude and phase spectra by hand. What is the effect of changing T? Can the result in Problem 3.9 be used to explain the effect?

3.14   Plot the magnitude and phase spectra of $x[n] = 0.9^n$, for $n \geq 0$, and $x[n] = 0$, for $n < 0$, and $T = 1$. Is it a high- or low-frequency signal?

3.15 Repeat Problem 3.14 for $x[n] = (-0.9)^n$, for $n \geq 0$, and $x[n] = 0$, for $n < 0$, and $T = 1$.

3.16 Compute the spectrum $X(\omega)$ of $e^{-0.2t}$, for $t \geq 0$ and the spectrum $X_d(\omega)$ of $e^{-0.2nT}$, for $n \geq 0$ and $T = 0.5$. Compare $X(\omega)$ and $TX_d(\omega)$. Is frequency aliasing due to time sampling significant?

3.17 Repeat Problem 3.16 for $T = 0.05$.

3.18 Let $X_d(\omega)$ be the frequency spectrum of $x[n] = x(nT)$ and let $X_{0d}(\omega)$ be the frequency spectrum of $x[n - n_0]$. Show

$$|X_{0d}(\omega)| = |X_d(\omega)|$$

$$\measuredangle X_{0d}(\omega) = \measuredangle X_d(\omega) - \omega n_o T$$

3.19 Use the integration formula

$$\int e^{at} \cos pt \, dt = \frac{e^{at}(a \cos pt + p \sin pt)}{a^2 + p^2}$$

and

$$\int t^2 e^{at} \, dt = \frac{e^{at}}{a}\left(t^2 - \frac{2}{a^2}\left(e^{at} - 1\right)\right)$$

to find the CTFT (frequency spectrum) of

$$x(t) = \begin{cases} t^2 - 1.5\cos 20t & \text{for } 0 \leq t \leq 2 \\ 0 & \text{otherwise} \end{cases}$$

Plot its magnitude and phase spectra. This problem will be recomputed using FFT in Problem 4.19.

3.20 Suppose $x[n] = \cos(2\pi n/N)$, for $n = 0, 1, \ldots, N-1$, are stored in read-only memory (ROM). Let us pick the sequence circularly every T seconds. By this, we mean that after $n = N - 1$, we go back to $n = 0$. If the sequence is converted to a CT signal using the ideal interpolation formula in (3.69), what is the frequency of the resulting CT sinusoidal signal? This is one way of generating a CT sinusoidal signal.

3.21 In Problem 3.20, suppose $N = 100$ and the sampling period must be $T \geq 0.01$. Is it possible to generate a sinusoid with frequency 3π rad/s from the stored data?

CHAPTER 4

DFT and FFT—Spectral Computation

4.1 Introduction

We introduced in the preceding chapter frequency spectra for CT signals and DT signals. The signals in the preceding chapter are all given in closed form and their spectra can be computed analytically. Analytical spectral computation is not possible if a signal is obtained by measurement or cannot be expressed in a simple closed form. In this case, the only way to compute the spectrum is by using a numerical method. This chapter introduces one such method. In fact, even for the signals discussed in the preceding chapter, it is simpler to compute their spectra using the method to be introduced.

We start the chapter by showing that computer computation of the discrete-time Fourier transform (DTFT) leads naturally to the discrete Fourier transform (DFT). We then discuss one version of the fast Fourier transform (FFT) to compute DFT efficiently. Finally, we use FFT to compute frequency spectra of DT and CT signals and use inverse FFT to compute DT and CT signals from their frequency spectra.

4.2 Discrete Fourier Transform (DFT)

This section introduces the discrete Fourier transform, or DFT for short. Unlike the discrete-time Fourier transform, which is defined for finite or infinite sequences, the DFT is defined only for sequences of finite length. Before giving a formal definition, we shall develop the DFT from the DTFT with computer computation in mind.

Consider the sequence $x(nT)$ of length N

$$x[n] := x(nT) \qquad \text{for } n = 0, 1, 2, \dots, N - 1 \tag{4.1}$$

It is assumed that $x[n] = 0$ for $n < 0$ and $n \geq N$. Its DT Fourier transform is

$$X_d(\omega) = \sum_{n=-\infty}^{\infty} x(nT)e^{-j\omega nT} = \sum_{n=0}^{N-1} x(nT)e^{-j\omega nT} \tag{4.2}$$

for all ω. Recall that $X_d(\omega)$ is periodic with period $2\pi/T$, and we are interested in $X_d(\omega)$ with ω in the Nyquist frequency range $(-\pi/T, \ \pi/T]$. There are infinitely many ω in the range. If we use a digital computer to compute (4.2), we can compute $X_d(\omega)$ only at a finite number of ω. Suppose we compute $X_d(\omega)$ at N equally spaced ω in $(-\pi/T, \ \pi/T]$. Then these N points should be located at

$$\omega_m = m\frac{2\pi}{NT}$$

with

$$m = \overline{-(N-1)/2} : \overline{(N-1)/2} \tag{4.3}$$

for N even or odd, where the overline rounds a number upward to an integer. This indexing is complicated. Instead, we shall compute N equally spaced ω in $[0, \ 2\pi/T)$ with

$$\omega_m = m\frac{2\pi}{NT} \qquad \text{for } m = 0, 1, 2, \dots, N - 1$$

Once $X_d(\omega_m)$ with ω_m in $[0, \ 2\pi/T)$ are computed, $X_d(\omega_m)$ with ω_m in the Nyquist frequency range $(-\pi/T, \ \pi/T]$ can be obtained by periodic extension or by shifting as discussed in Section 2.6.1.

We define $X_d[m] := X_d(\omega_m)$ and $x[n] := x(nT)$. Then (4.2) implies

$$X_d[m] := X_d\left(m\frac{2\pi}{NT}\right) = \sum_{n=0}^{N-1} x[n]e^{-j(m2\pi/NT)nT}$$

$$= \sum_{n=0}^{N-1} x[n]e^{-j2\pi mn/N} \tag{4.4}$$

for $m = 0, 1, \dots, N - 1$. This is, by definition, the discrete Fourier transform or DFT. Thus *the DFT computes N equally spaced frequency samples of the discrete-time Fourier transform.* Note that there are two indices n and m, both ranging from 0 to $N - 1$. The integer n denotes time instant and is called the *time index*. The integer m denotes discrete frequency and is called the *frequency index*. Note that the sampling period T does not appear in (4.4).

The time sequence $x(nT)$, for $n = 0, 1, \ldots, N - 1$, can be computed from its frequency spectrum $X_d(\omega)$ by using the inverse DT Fourier transform as

$$x[n] = x(nT) = \frac{T}{2\pi} \int_0^{2\pi/T} X_d(\omega) e^{j\omega nT} d\omega \tag{4.5}$$

If the values of $X_d(\omega)$ are available only at $\omega = m(2\pi/NT)$, for $m = 0, 1, \ldots, N - 1$, then we must approximate (4.5) by a summation. The simplest approximation is to assume $X_d(\omega) e^{j\omega nT}$ constant between samples. Then the integration in (4.5) can be approximated by a summation as

$$x[n] \approx \frac{T}{2\pi} \sum_{m=0}^{N-1} X_d \left(m \frac{2\pi}{NT} \right) e^{jm(2\pi/NT)nT} \left(\frac{2\pi}{NT} \right)$$

$$= \frac{1}{N} \sum_{m=0}^{N-1} X_d[m] e^{j2\pi mn/N} \tag{4.6}$$

for $n = 0, 1, \ldots, N - 1$. This is in fact the inverse DFT. We see that computer computation of the DTFT leads naturally to the DFT.

We have developed the DFT and its inverse from the DTFT in (4.4) and (4.6). Now we shall formally establish them as a transform pair. Consider a sequence of N real or complex numbers $x[n]$, $n = 0, 1, \ldots, N - 1$. We define

$$W = e^{-j2\pi/N} \tag{4.7}$$

Then the DFT of $x[n]$ is defined as

$$X_d[m] := \mathcal{D}[x[n]] := \sum_{n=0}^{N-1} x[n] e^{-j2\pi nm/N} = \sum_{n=0}^{N-1} x[n] W^{nm} \tag{4.8}$$

for $m = 0, 1, \ldots, N - 1$, and the inverse DFT of $X_d[m]$ is

$$x[n] = \mathcal{D}^{-1}[X_d[m]] = \frac{1}{N} \sum_{m=0}^{N-1} X_d[m] e^{j2\pi mn/N} = \frac{1}{N} \sum_{m=0}^{N-1} X_d[m] W^{-mn} \tag{4.9}$$

for $n = 0, 1, \ldots, N - 1$. To show that they are indeed a transform pair, we substitute (4.8), after changing the summation index from n to k, into (4.9) to yield

$$\mathcal{D}^{-1}[X_d[m]] = \frac{1}{N} \sum_{m=0}^{N-1} \left(\sum_{k=0}^{N-1} x[k] e^{-j2\pi km/N} \right) e^{j2\pi mn/N} \tag{4.10}$$

Note that if the index n in (4.8) is not changed to k in the substitution, it will be confused with the n in (4.9). Changing the order of summations and using the orthogonality property in (2.44), we obtain

$$\mathcal{D}^{-1}[X_d[m]] = \frac{1}{N} \sum_{k=0}^{N-1} x[k] \sum_{m=0}^{N-1} e^{j2\pi(n-k)m/N} = \frac{1}{N} x[n]N = x[n]$$

This establishes the DFT pair. It is rather surprising that, although (4.6) was obtained by approximation, it turns out to be an equality.

Let us discuss an important property of the DFT. Because $W^{\pm mkN} = e^{\mp j2\pi mk} = 1$, for every integer m and k [see Fig. 2.2(b)], we have

$$W^{-(m\pm kN)n} = W^{-mn}W^{\mp knN} = W^{-mn}$$

and

$$W^{m(n\pm kN)} = W^{mn}W^{\pm mkN} = W^{mn}$$

Thus (4.8) and (4.9) imply

$$X_d[m] = X_d[m \pm N] = X_d[m \pm 2N] = \cdots \tag{4.11}$$

and

$$x[n] = x[n \pm N] = x[n \pm 2N] = \cdots \tag{4.12}$$

Therefore, even though the DFT is defined for a finite time sequence of length N, it can be considered to be defined for an infinite periodic sequence with period N. Define

$$\tilde{x}[n] := \begin{cases} x[n] & \text{for } 0 \leq n \leq N-1 \\ \text{periodic extension of } x[n] \text{ with period } N \end{cases}$$

or

$$\tilde{x}[n] := \sum_{k=-\infty}^{\infty} x[n+kN] \tag{4.13}$$

It is the periodic extension of $x[n]$. Then the DFT pair can be written more generally as

$$X_d[m] = \mathcal{D}[x[n]] = \mathcal{D}[\tilde{x}[n]] = \sum_{n=<N>} \tilde{x}[n]W^{nm} \tag{4.14}$$

for $m = 0, 1, \ldots, N-1$, and

$$\tilde{x}[n] = \mathcal{D}^{-1}[X_d[m]] = \frac{1}{N} \sum_{m=<N>} X_d[m]W^{-mn} \tag{4.15}$$

for $n = 0, 1, \ldots, N-1$, where $<N>$ denotes any summation over N consecutive integers, for example, from $n = 0$ to $N-1$ or from $n = n_0$ to $N+n_0-1$ for any integer n_0. Because of (4.14),

DFT can be considered to be defined for periodic sequences and can be computed from any one period. Thus given a finite sequence $x[n]$ with $n_1 \leq n \leq n_2$ and length $N = n_2 - n_1 + 1$, its DFT is

$$X_d[m] = \mathcal{D}[x[n]] = \sum_{n=n_1}^{n_2} x[n]W^{nm} = \sum_{n=0}^{N-1} \tilde{x}[n]W^{nm} \tag{4.16}$$

with $W = e^{-2\pi/N}$ and $\tilde{x}[n]$ is the periodic extension of $x[n]$. It is important to mention that the sampling period T does not appear explicitly in (4.14) and (4.15). The sampling period T comes into the picture only when we place $X_d[m]$ at frequency $\omega = m(2\pi/NT)$. We now give some examples and compare their DTFT and DFT.

◆ **Example 4.1**

Consider the finite sequence of length 3 shown in Fig. 4.1(a); that is, $x[0] = 1$, $x[1] = 0.5$, $x[2] = -0.5$, and $T = 0.5$. Its frequency spectrum or DTFT is

$$X_d(\omega) = \sum_{n=-\infty}^{\infty} x[n]e^{-j\omega nT} = 1 + 0.5e^{-j\omega T} - 0.5e^{-j2\omega T}$$

$$= 1 + 0.5e^{-1.5j\omega T}(e^{j0.5\omega T} - e^{-j0.5\omega T})$$

$$= 1 + je^{-j1.5\omega T}\sin 0.5\omega T \tag{4.17}$$

Figure 4.1(b) shows its magnitude spectrum (solid line) and phase spectrum (dotted line) in the Nyquist frequency range $(-\pi/T, \pi/T] = (-6.28, 6.28]$ in rad/s. It is obtained in MATLAB by typing

Program 4.1
```
w=-8:0.05:13;
X=1+0.5*exp(-j*0.5*w)-0.5*exp(-j*1.0*w);
plot(w,abs(X),w,angle(X),':')
```

The spectrum is periodic with period $2\pi/T = 4\pi = 12.56$ in rad/s. The spectrum is bounded and continuous.

Next we compute the DFT of the sequence. We define

$$W := e^{-j2\pi/3} = -0.5 - j0.866$$

It is periodic with period $N = 3$. We compute

$$W^2 = W^{-1} = e^{j2\pi/3} = -0.5 + j0.866; \qquad W^3 = W^0 = 1$$

$$W^4 = W = -0.5 - j0.866; \quad W^{-2} = W = -0.5 - j0.866$$

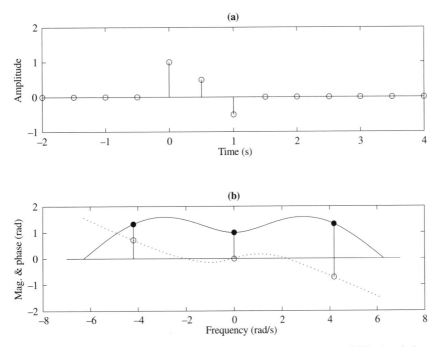

Figure 4.1 (a) Time sequence of length 3. (b) Its DTFT [magnitude (solid line) and phase (dotted line) spectra] and its three-point DFT [magnitude (solid dots) and phase (hollow dots)].

The DFT of the sequence is

$$X_d[0] = \sum_{n=0}^{2} x[n]W^{0 \cdot n} = 1 + 0.5 - 0.5 = 1 = 1e^{j0}$$

$$X_d[1] = \sum_{n=0}^{2} x[n]W^{1 \cdot n} = 1 + 0.5W - 0.5W^2$$

$$= 1 + 0.5(-0.5 - j0.866) - 0.5(-0.5 + j0.866)$$

$$= 1 - j0.866 = 1.3229e^{-j0.7137}$$

$$X_d[2] = \sum_{n=0}^{2} x[n]W^{2 \cdot n} = 1 + 0.5W^2 - 0.5W^4$$

$$= 1 + j0.866 = 1.3229e^{j0.7137}$$

They are located at $m(2\pi/NT) = m(4.19)$ for $m = 0, 1, 2$. Because $X_d[m]$ is periodic with period $N = 3$, we have $X_d[-1] = X_d[2]$. The magnitudes (solid dots) and phases (hollow dots) of $X_d[m]$, for $m = -1, 0, 1$, are plotted in Fig. 4.1(b). They are indeed three samples of $X_d(\omega)$.

◆ Example 4.2

Consider the finite sequence of length 3 shown in Fig. 4.2(a); that is, $x[-1] = -0.5$, $x[0] = 1$, $x[1] = 0.5$, and $T = 0.5$. Its frequency spectrum or DTFT is

$$X_d(\omega) = \sum_{n=-\infty}^{\infty} x[n]e^{-j\omega nT} = -0.5e^{j\omega T} + 1 + 0.5e^{-j\omega T}$$

$$= 1 - 0.5(e^{j\omega T} - e^{-j\omega T}) = 1 - j\sin\omega T \qquad (4.18)$$

Figure 4.2(b) shows its magnitude spectrum (solid line) and phase spectrum (dotted line) in the frequency range $(-6.28, 6.28]$. It is obtained using Program 4.1 with X replaced by (4.19). The spectrum is different from the one in Fig. 4.1(b).

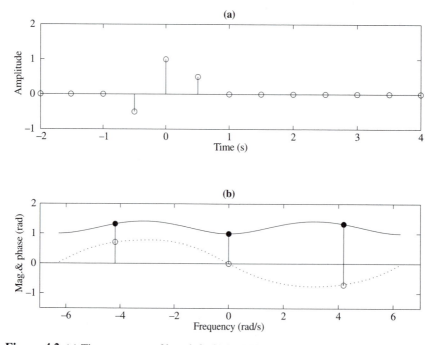

Figure 4.2 (a) Time sequence of length 3. (b) Its DTFT [magnitude (solid line) and phase (dotted line) spectra] and its three-point DFT [magnitude (solid dots) and phase (hollow dots)].

Next we compute the DFT of the sequence. Its DFT should be the same as the one computed in Example 4.1 because the periodic extensions of the two sequences in Figs. 4.1(a) and 4.2(a) are the same. To verify this, we compute its DFT by using the data from $n = -1$ to 1 as

$$X_d[0] = \sum_{n=-1}^{1} x[n]W^{0 \cdot n} = -0.5 + 1 + 0.5 = 1$$

$$X_d[1] = \sum_{n=-1}^{1} x[n]W^{1 \cdot n} = -0.5W^{-1} + 1 - 0.5W$$

$$= -0.5(-0.5 + j0.866) + 1 + 0.5(-0.5 - j0.866)$$

$$= 1 - j0.866 = 1.3229e^{-j0.7137}$$

$$X_d[2] = \sum_{n=-1}^{1} x[n]W^{2 \cdot n} = -0.5W^{-2} + 1 + 0.5W^2$$

$$= 1 + j0.866 = 1.3229e^{j0.7137}$$

They indeed equal those in Example 4.1. They are three samples of the spectrum as shown in Fig. 4.2(b). Note that although the spectra of the sequences in Figs. 4.1(a) and 4.2(a) are different, their DFTs are the same.

If a time sequence is of finite length, its spectrum is bounded and continuous and its Nyquist frequency range can be selected as $(-\pi/T, \pi/T]$, $[-\pi/T, \pi/T]$, $(-\pi/T, \pi/T)$, or $[-\pi/T, \pi/T)$. However, it is still better to use the first one, as the next example illustrates.

◆ **Example 4.3**

Consider the sequence of length 4 in Fig. 4.3(a); that is, $x[0] = 2, x[1] = -1, x[2] = 1, x[3] = 1$, and $T = 0.5$. Its spectrum is $X_d(\omega) = 2 - e^{-j\omega T} + e^{-j2\omega T} + e^{-j3\omega T}$. Figure 4.3(b) shows its magnitude spectrum (solid line) and phase spectrum (dotted line). To compute its DFT, we first compute

$$W = e^{-2\pi/N} = e^{-\pi/2} = -j; \quad W^2 = e^{-j\pi} = -1; \quad W^3 = j; \quad W^4 = W^0 = 1$$

Then we have

$$X_d[0] = 2 - 1 + 1 + 1 = 3 = 3^{j0^\circ}$$
$$X_d[1] = 2 - W + W^2 + W^3 = 2 - (-j) - 1 + j = 1 + 2j = 2.24e^{j1.1}$$

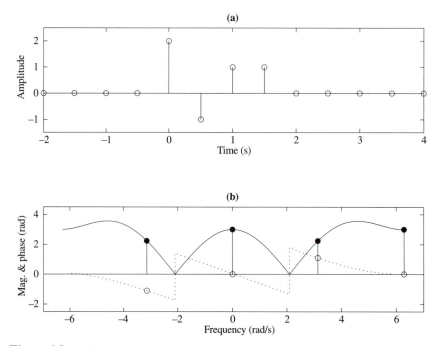

Figure 4.3 (a) Time sequence of length 4. (b) Its DTFT [magnitude (solid line) and phase (dotted line) spectra] and its four-point DFT [magnitude (solid dots) and phase (hollow dots)].

$$X_d[2] = 2 - W^2 + W^4 + W^6 = 2 - (-1) + 1 - 1 = 3$$
$$X_d[3] = 2 - W^3 + W^6 + W^9 = 2 - j - 1 - j = 1 - 2j = 2.24e^{-j1.1}$$

They are located at $\omega = m(2\pi/(4 \times 0.5)) = m(3.14)$ for $m = 0, 1, 2, 3$. Because $X_d[m]$ is periodic with period 4, we have $X_d[-1] = X_d[3]$. Its DFT with m in (4.3) or $m = -1, 0, 1, 2$ is plotted in Fig. 4.3(b) with solid and hollow dots. They are indeed four samples of the spectrum in the Nyquist frequency range $(-6.28, 6.28]$. There is one sample at $\omega = \pi/T = 6.28$ but no sample at $\omega = -6.28$.

4.2.1 Relationship between DFT and DTFS

This section discusses the relationship between the discrete Fourier transform and the discrete-time Fourier series. Consider the periodic sequence $\tilde{x}[n]$ in (4.13) with period N. Its DTFS is, as discussed in Section 2.5,

$$\tilde{x}[n] = \sum_{n=<N>} c_{md} e^{j2\pi mn/N}$$

with

$$c_{md} = \frac{1}{N} \sum_{n=<N>} \tilde{x}[n]e^{-j2\pi mn/N} = \frac{1}{N} \sum_{n=<N>} \tilde{x}[n]W^{nm}$$

[see (2.45) and (2.46)]. Comparing this with (4.16), we obtain

$$X_d[m] = Nc_{dm} \qquad \text{or} \qquad c_{dm} = \frac{X_d[m]}{N} \tag{4.19}$$

Thus *the discrete Fourier transform differs from the discrete-time Fourier series only by a factor.* Consequently, all discussion in Chapter 2 for DTFS is applicable to DFT.

4.2.2 Inverse DFT and Inverse DTFT—Time Aliasing due to Frequency Sampling

The DFT computes samples of the discrete-time Fourier transform. Now we discuss the relationship between the inverse DFT and the inverse DTFT. Consider a sequence $x[n] := x(nT)$ (not necessarily a finite sequence) with frequency spectrum

$$X_d(\omega) = \sum_{-\infty}^{\infty} x(nT)e^{-j\omega nT} \tag{4.20}$$

with ω in $(-\pi/T, \ \pi/T]$. The time sequence $x[n]$ can be computed uniquely from the inverse DTFT of $X_d(\omega)$ as

$$x[n] = \frac{T}{2\pi} \int_{\omega=-\pi/T}^{\pi/T} X_d(\omega)e^{j\omega nT}d\omega$$

$$= \frac{T}{2\pi} \int_{\omega=0}^{2\pi/T} X_d(\omega)e^{j\omega nT}d\omega \tag{4.21}$$

To compute this equation, we need $X_d(\omega)$ for all ω in $(-\pi/T, \ \pi/T]$ or $[0, \ 2\pi/T)$. Now if only N equally spaced samples of $X_d(\omega)$ or

$$X_d[m] := X_d(m(2\pi/NT)) \qquad \text{for } m = 0, 1, \ldots, N-1 \tag{4.22}$$

are available, then we cannot use (4.21) to compute $x[n]$. But we can compute the inverse DFT of (4.22). Let $\hat{x}[n]$, for $n = 0, 1, \ldots, N-1$, be the inverse DFT of $X_d(\omega)$, that is,

$$\hat{x}[n] = \frac{1}{N} \sum_{m=0}^{N-1} X_d[m]e^{j2\pi mn/N} \tag{4.23}$$

which implies

$$X_d[m] = \sum_{n=0}^{N-1} \hat{x}[n]e^{-j2\pi mn/N} \tag{4.24}$$

The question is: What is the relationship between $x[n]$ and $\hat{x}[n]$? To establish the relationship, we substitute $\omega = m(2\pi/NT)$ into (4.20) and then divide the infinite summation into sections, each section consisting of N terms, as

$$X_d[m] := X_d\left(m\frac{2\pi}{NT}\right) = \sum_{n=\infty}^{\infty} x[n]e^{-j2\pi mn/N}$$

$$= \sum_{k=-\infty}^{\infty} \sum_{n=kN}^{kN+N-1} x[n]e^{-j2\pi mn/N} \tag{4.25}$$

which becomes, after introducing the new index $\bar{n} := n - kN$,

$$X_d[m] = \sum_{k=-\infty}^{\infty} \sum_{\bar{n}=0}^{N-1} x[\bar{n} + kN]e^{-j2\pi m\bar{n}/N}e^{-j2\pi mk} \tag{4.26}$$

We use the property $e^{-j2\pi mk} = 1$, for every integers k and m, interchange the order of the summations, and drop the overbar of n to write (4.26) as

$$X_d[m] = \sum_{n=0}^{N-1} \sum_{k=-\infty}^{\infty} x[n + kN]e^{-j2\pi mn/N} \tag{4.27}$$

Comparing (4.24) and (4.27) yields

$$\hat{x}[n] = \sum_{k=-\infty}^{\infty} x[n + kN] \tag{4.28}$$

This relates the inverse DTFT of $X_d(\omega)$ and the inverse DFT of the N frequency samples of $X_d(\omega)$.

Let us discuss the physical implication of (4.28). Let $x[n]$ be the inverse DTFT of $X_d(\omega)$ and of length 4 as shown in Fig. 4.4(a). If $N = 6$, then $x[n + N]$ and $x[n - N]$ are the shiftings of $x[n]$ to the left and right 6 samples, as shown in Fig. 4.4(b) with solid dots. Thus $\hat{x}[n]$ is the sum of $x[n]$ and all its repetitive shiftings to the right and left every N samples. This $\hat{x}[n]$ is the inverse DFT of the $N = 6$ frequency samples of $X_d(\omega)$ in $[0, 2\pi)$. If N is equal to or larger than the length of $x[n]$, then $x[n \pm kN]$, for all integer k, will not overlap with each other, as in Fig. 4.4(b), and $\hat{x}[n]$ is the periodic extension of $x[n]$ with period N. In this case, we have $x[n] = \hat{x}[n]$, for $n = 0, 1, \ldots, N - 1$.

If $N = 3$, less than the length of $x[n]$, then $x[n]$ and its repetitive shiftings will overlap with each other. This is called *time aliasing*, and the resulting sequence is as shown in Fig. 4.4(c). This

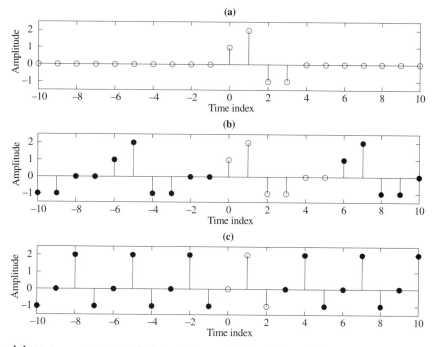

Figure 4.4 (a) Inverse DTFT of $X_d(\omega)$. (b) Inverse DFT of $N = 6$ frequency samples of $X_d(\omega)$. (c) Inverse DFT of $N = 3$ frequency samples of $X_d(\omega)$.

time aliasing is due to *frequency sampling* of the spectrum, and is dual to the frequency aliasing due to *time sampling* discussed in Section 3.8.1. When time aliasing occurs, the resulting $\hat{x}[n]$, as shown in Fig. 4.4(c), will be different from $x[n]$ for $n = 0, 1, \ldots, N - 1$. In conclusion, given a frequency spectrum in $[0, \ 2\pi/T)$, if it is known to be the spectrum of a finite sequence and if the number of frequency samples N is equal to or larger than the length of the sequence, then we have

$$\text{inverse DTFT} = \begin{cases} \text{inverse DFT} & \text{for } n_1 \leq n \leq n_1 + N - 1 \\ 0 & \text{for } n < n_1 \text{ and } n > n_1 + N - 1 \end{cases} \quad (4.29)$$

for some integer n_1. The integer n_1 will be discussed in Section 4.7 when we apply (4.29) in actual computation. If N is less than the length of a sequence (including an infinite sequence), then time aliasing will occur and we have

$$\text{inverse DTFT} \neq \text{inverse DFT}$$

However, if an infinite sequence is absolutely summable, then the sequence will approach 0 as $n \to \pm\infty$. In this case, if N is selected to be sufficiently large, time aliasing will be negligible, and we have

$$\text{inverse DTFT} \approx \begin{cases} \text{inverse DFT} & \text{for } n_1 \le n \le n_1 + N - 1 \\ 0 & \text{for } n < n_1 \text{ and } n > n_1 + N - 1 \end{cases} \qquad (4.30)$$

for some n_1. This will be discussed further in Section 4.7.

4.3 Properties of DFT

This section discusses some properties of DFT. The discussion will be brief because those properties are similar to those of Fourier series and Fourier transforms.

4.3.1 Even and Odd

In discussing the properties of DFT and inverse DFT, the finite time sequence must be extended periodically to all n. Let $x[n]$, for $n = 0, 1, \ldots, N - 1$, be a sequence of length N and let $\tilde{x}[n]$ be its periodic extension with period N or

$$\tilde{x}[n] = \sum_{k=-\infty}^{\infty} x[n + kN]$$

The sequence $x[n]$ is defined to be even if $\tilde{x}[n]$ is even. It is odd if $\tilde{x}[n]$ is odd. If $\tilde{x}[n]$ is even, then

$$\tilde{x}[n] = \tilde{x}[-n] = \tilde{x}[N - n]$$

for all integers n. This implies

$$x[n] = x[N - n]$$

for $n = 0, 1, 2, \ldots, N/2$, where the underline rounds a number downward to an integer. This condition is plotted in Figs. 4.5(a) and (b) for $N = 8$ and 9. We see that no conditions are imposed on $x[0]$ and $x[N/2]$ for N even, and no condition is imposed on $x[0]$ for N odd. If $\tilde{x}[n]$ is odd, then

$$\tilde{x}[n] = -\tilde{x}[-n] = -\tilde{x}[N - n]$$

for all integers n, or

$$x[n] = -x[N - n]$$

for $n = 0, 1, 2, \ldots, N/2$, as shown in Figs. 4.5(c) and (d). We see that the condition requires $x[0] = 0$ and $x[N/2] = 0$ for N even, and $x[0] = 0$ for N odd.

Let $X_d[m]$ be the DFT of $x[n]$. Then we have

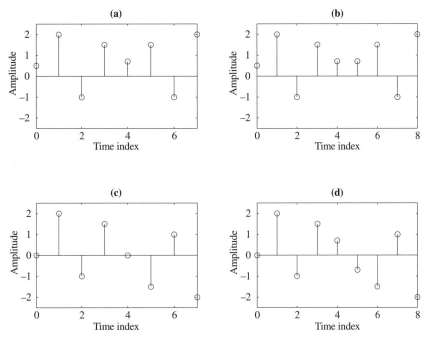

Figure 4.5 (a) Even sequence of length 8. (b) Even sequence of length 9. (c) Odd sequence of length 8. (d) Odd sequence of length 9. All horizontal coordinates are time index.

- $x[n]$ or $\tilde{x}[n]$ is real $\Leftrightarrow X_d[m]$ has even magnitude and odd phase
- $x[n]$ or $\tilde{x}[n]$ is real and even $\Leftrightarrow X_d[m]$ is real and even
- $x[n]$ or $\tilde{x}[n]$ is real and odd $\Leftrightarrow X_d[m]$ is imaginary and odd

where $X_d[m]$ is implicitly assumed to be extended, with period N, to all m. These properties are similar to those of the Fourier series and Fourier transforms.

4.3.2 Periodic Shifting

In CT and DT Fourier transforms, time shifting will not affect the magnitude spectrum but will introduce a linear phase into the phase spectrum. We have a similar situation for DFT. Let $x[n]$ be a sequence of length N and let $\tilde{x}[n]$ be its periodic extension. Let $X_d[m]$ be the DFT of $x[n]$ or $\tilde{x}[n]$ and let n_0 be an integer. Then we have

$$\mathcal{D}[\tilde{x}[n - n_0]] = \sum_{n=<N>} \tilde{x}[n - n_0]W^{mn}$$

$$= W^{mn_0} \sum_{n=<N>} \tilde{x}[n - n_0]W^{m(n-n_0)}$$

$$= W^{mn_0} X_d[m]$$

where $W = e^{-j2\pi/N}$, or

$$\mathcal{D}[\tilde{x}[n - n_0]] = e^{-j2\pi m n_0/N} X_d[m] \tag{4.31}$$

This equation is similar to (3.51) if ω is replaced by $2\pi m/NT$ and is identical to (2.51). Thus time shifting does not affect the magnitude of DFT but introduces a linear phase (in terms of m) into the phase of DFT. Time shifting in $\tilde{x}[n]$ can be interpreted as circular or periodic shifting in $x[n]$. For example, if we plot $x[n]$ for $n = 0, 1, 2, 3, 4, 5$ in the counterclockwise direction on a circle as shown in Fig. 4.6(a), then the linear shifting $\tilde{x}[n - 2]$ is the same as the circular shifting of $x[n]$ in the same direction (counterclockwise) two samples as shown in Fig. 4.6(b). The linear shifting $\tilde{x}[n + 1]$ is the same as the circular shifting of $x[n]$ in the opposite direction (clockwise) one sample as shown in Fig. 4.6(c). These shifting can be tabulated as in Table 4.1.

4.4 Fast Fourier Transform (FFT)

Consider a complex-valued sequence $x[n]$, for $n = 0, 1, \ldots, N - 1$. It is a finite sequence of length N. Its discrete Fourier transform is

$$X_d[m] = \sum_{n=0}^{N-1} x[n] W^{nm} \tag{4.32}$$

with

$$W = e^{-j2\pi/N} \tag{4.33}$$

for $m = 0, 1, \ldots, N - 1$. Direct computation of (4.32) for each m requires N complex multiplications and $(N - 1)$ complex additions. Thus computing $X_d[m]$ for $m = 0, 1, \ldots, N - 1$ requires N^2 complex multiplications and $N(N - 1)$ complex additions.[1] As N increases, these numbers increase rapidly. Therefore, it is desirable to find a more efficient way of computing DFT. Let us discuss one such method.

Suppose N in (4.32) is divisible by 2, then we can decompose $x[n]$ into two subsequences $x_1[n]$ and $x_2[n]$ as

$$x_1[n] := x[2n] \quad x_2[n] := [2n + 1] \quad \text{for } n = 0, 1, \ldots, (N/2) - 1 \tag{4.34}$$

where $x_1[n]$ and $x_2[n]$ consist of, respectively, the even and odd terms of $x[n]$. Clearly, both have length $N_1 := N/2$. Their DFT are, respectively,

---

[1] Note that each complex multiplication requires four real multiplications and two real additions, and each complex addition requires two real additions.

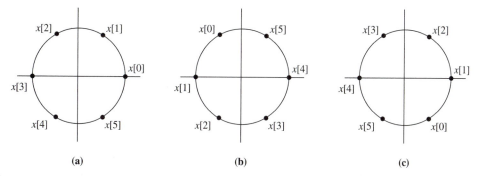

Figure 4.6 Circular or periodic shifting: (a) $x[n]$. (b) $x[n-2] = x[n+4]$. (c) $x[n+1] = x[n-5]$.

Table 4.1 Circular Shifting for $N = 6$

n	0	1	2	3	4	5
$x[n]$	$x[0]$	$x[1]$	$x[2]$	$x[3]$	$x[4]$	$x[5]$
$\tilde{x}[n-2]$	$x[4]$	$x[5]$	$x[0]$	$x[1]$	$x[2]$	$x[3]$
$\tilde{x}[n+1]$	$x[1]$	$x[2]$	$x[3]$	$x[4]$	$x[5]$	$x[0]$

$$X_{d1}[m] = \sum_{n=0}^{N_1-1} x_1[n] W_1^{mn} \tag{4.35}$$

and

$$X_{d2}[m] = \sum_{n=0}^{N_1-1} x_2[n] W_1^{mn} \tag{4.36}$$

with

$$W_1 := e^{-j2\pi/N_1} = e^{-j4\pi/N} = W^2 \tag{4.37}$$

for $m = 0, 1, \ldots, N_1 - 1$. Note that $X_{d1}[m]$ and $X_{d2}[m]$ are periodic with period $N_1 = N/2$ or

$$X_{di}[m] = X_{di}[m + kN_1] \tag{4.38}$$

for any integer k and for $i = 1, 2$. Let us express $X_d[m]$ in terms of $X_{d1}[m]$ and $X_{d2}[m]$. We use $W_1 = W^2$ to write (4.32) explicitly as

$$X_d[m] = x[0]W^0 + x[2]W^{2m} + \cdots + x[N-2]W^{(N-2)m}$$
$$+ x[1]W^m + x[3]W^{3m} + \cdots + x[N-1]W^{(N-1)m}$$

$$= x_1[0]W_1^0 + x_1[1]W_1^m + \cdots + x_1[N_1 - 1]W_1^{(N_1-1)m}$$

$$+ W^m \left[x_2[0]W_1^0 + x_2[1]W_1^m + \cdots + x_2[N_1 - 1]W_1^{(N_1-1)m} \right]$$

which becomes, after substituting (4.35) and (4.36),

$$X_d[m] = X_{d1}[m] + W^m X_{d2}[m] \tag{4.39}$$

for $m = 0, 1, \ldots, N - 1$. Let us compute the number of complex multiplications needed in (4.39). Each $X_{di}[m]$ in (4.35) and (4.36) requires $N_1^2 = (N/2)^2$ complex multiplications. In addition, we need N complex multiplications in (4.39) to generate $X_d[m]$. Thus the total number of complex multiplications for computing $X_d[m]$ from (4.35), (4,36), and (4.39) is

$$\left(\frac{N}{2}\right)^2 + \left(\frac{N}{2}\right)^2 + N = N + \frac{N^2}{2}$$

Similarly, the total number of complex additions needed in (4.35), (4.36), and (4.39) is

$$\frac{N}{2}\left(\frac{N}{2} - 1\right) + \frac{N}{2}\left(\frac{N}{2} - 1\right) + N = \frac{N^2}{2}$$

Computing $X_d[m]$ directly from (4.32) requires N^2 complex multiplications and $N(N - 1)$ complex additions. If we decompose $x[n]$ into two subsequences of length $N/2$ and compute $X_d[m]$ using (4.39), then the amount of computation is reduced by almost a factor of 2. If we decompose $x_i[n]$, for $i = 1, 2$, into two subsequences, then the amount of computation can again be cut in half. This process of computing DFT is one version of the fast Fourier transform.

Let us develop the preceding version of FFT in detail for $N = 8 = 2^3$. For $N = 8$, (4.32), (4.35), and (4.36) become

$$X_d[m] = x[0] + x[1]W^m + x[2]W^{2m} + x[3]W^{3m} + x[4]W^{4m} + x[5]W^{5m}$$

$$+ x[6]W^{6m} + x[7]W^{7m}, \quad m = 0, 1, \ldots, 7$$

$$X_{d1}[m] = x[0] + x[2]W^{2m} + x[4]W^{4m} + x[6]W^{6m}, \quad m = 0, 1, 2, 3$$

$$X_{d2}[m] = x[1] + x[3]W^{3m} + x[5]W^{5m} + x[7]W^{7m}, \quad m = 0, 1, 2, 3$$

with $W = e^{-j2\pi/8}$. Note that $X_d[m]$ is periodic with period 8 and $X_{di}[m]$ are periodic with period 4. Furthermore, we have

$$X_d[m] = X_{d1}[m] + W^m X_{d2}[m], \quad m = 0, 1, \ldots, 7 \tag{4.40}$$

Next we decompose $x_1[n]$ into two subsequences $\{x[0], x[4]\}$ and $\{x[2], x[6]\}$. They are sequences of length 2. Define $W_2 = e^{-j2\pi/2} = W^4$. Their DFT are

$$X_{d11}[m] = x[0]W_2^0 + x[4]W_2^m = x[0] + x[4]W^{4m}, \quad m = 0, 1$$

and

$$X_{d12}[m] = x[2]W_2^0 + x[6]W_2^m = x[2] + x[6]W^{4m}, \quad m = 0, 1$$

They are periodic with period 2. Substituting these into $X_{d1}[m]$ yields

$$X_{d1}[m] = X_{d11}[m] + W^{2m}X_{d12}[m], \quad m = 0, 1, 2, 3 \tag{4.41}$$

Likewise, $X_{d2}[m]$ can be computed from DFT of sequences of length 2. This process of computing DFT is shown in Fig. 4.7. We see that every entry is computed from an equation of the form shown in (4.39). For easy reference, we call (4.39) a *butterfly* equation.

Let us extend the procedure in Fig. 4.7 to a general N that is a power of 2, say, $N = 2^k$. The first step is to shuffle the input data $x[n]$ into a proper order (that will be discussed later). Then the DFT is carried out in $k = \log_2 N$ stages. In each stage, we compute N numbers. Each number is computed from a butterfly equation, which requires only one complex multiplication. Thus each stage requires a total of N complex multiplications. There are $\log_2 N$ stages. Therefore, the entire FFT requires a total of $N \log_2 N$ complex multiplications. However, because $W^0 = 1$, $W^{N/2} = -1$, and $W^{\pm N/4} = \mp j$, about half of the multiplications are not really needed. Therefore, the entire FFT requires only roughly $0.5N \log_2 N$ complex multiplications. Every butterfly equation requires one complex addition. Thus the entire FFT requires a total of $N \log_2 N$ complex additions. Table 4.2 lists the numbers of complex multiplications and complex additions needed for direct computation of DFT and for the FFT just introduced. For large N, the saving in FFT is very significant. For example, for $N = 2^{10} = 1024$, the saving is by a factor of 200 in multiplications and by a factor of 100 in additions. This means that a program that may take 2 or 3 minutes in direct computation takes only 1 second by using FFT.

Consider again Fig. 4.7. If the input data $x[n]$ are arranged as shown in the left-most column, then the computation does not require any additional memory. Once $\{X_{d11}[0], X_{d11}[1]\}$ are computed, $\{x[0], x[4]\}$ are no longer needed in subsequent computation. Thus the computed $\{X_{d11}[0], X_{d11}[1]\}$ can be stored in the memory location of $\{x[0], x[4]\}$. Similar remarks apply to all pairs of computation. This is called *in-place* computation.

Let us discuss the arrangement of the input data $x[n]$ in Fig. 4.7. It is achieved by *bit reversal*. In this process, we first express the index n in $x[n]$ in binary form as $x[k_2 \, k_1 \, k_0]$, where k_i is either 0 or 1, as shown in Fig. 4.8. We then reverse the order of the binary bits as $x[k_0 \, k_1 \, k_2]$. Rearranging the bit-reversed indices into its natural order yields the input sequence in Fig. 4.7. This process of bit reversal can be justified from the way the FFT in Fig. 4.7 is developed. If we write $x[n]$ as $x[k_2 \, k_1 \, k_0]$, then the decomposition of $x[k_2 \, k_1 \, k_0]$ into two subsequences $x_1[n]$ and $x_2[n]$ in (4.34) is dictated by whether k_0 is 0 or 1. The decomposition of $x_1[n]$ and $x_2[n]$ into subsequences is then dictated by k_1. This process of decomposition can therefore be illustrated as shown in Fig. 4.9. We see that the binary index $[k_2 \, k_1 \, k_0]$ is reversed in the process. This justifies the arrangement of the input data in Fig. 4.7.

The FFT introduced is achieved by first rearranging the time data and is called a *decimation-in-time* algorithm. It is possible to develop its dual, called *decimation-in-frequency*, which does

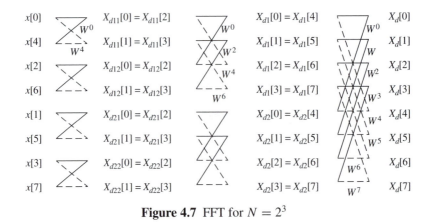

Figure 4.7 FFT for $N = 2^3$

Table 4.2 Comparison of Complex Operations in DFT and FFT

	DFT		FFT	
	multi ($\times$)	addi ($+$)	multi ($\times$)	addi ($+$)
$N = 2$	4	2	1	2
8	64	56	12	24
32	1024	922	80	160
64	4096	4022	192	384
1024	1048576	1048576	5120	10240
2^{20}	$\approx 10^{12}$	$\approx 10^{12}$	$\approx 10^7$	$\approx 2 \cdot 10^7$

not rearrange the time data but yields a rearranged frequency data. Other combinations are also possible.

4.4.1 Other FFT and DSP Processors

The FFT discussed in the preceding section is just one of many possible efficient ways of computing DFT. Its basic idea is to decompose a long sequence into the shortest possible subsequences. If N is a power of two, then the shortest subsequence has length two. The same idea can be extended to any positive integer N. For example, if $N = 15 = 5 \times 3$, then we can decompose a 15-point sequence into five 3-point subsequences and then develop an efficient algorithm. More generally, if N can be factored as

$$N = N_1 N_2 \cdots N_\nu$$

for some prime numbers N_i, for $i = 1, 2, \ldots, \nu$, then an efficient algorithm can be developed. If N is a power of 2, then $N_i = 2$ for all i, and the corresponding algorithm is called a radix-2 algorithm and is very efficient. See Refs. 8, 21, and 25. There are many efficient FFT programs,

Binary coding		Bit reversal		Rearrangement		Decoding	
$x[0]$	$\longrightarrow$ $x[000]$	$\longrightarrow$	$x[000]$	$\longrightarrow$	$x[000]$	$\longrightarrow$	$x[0]$
$x[1]$	$\longrightarrow$ $x[001]$	$\longrightarrow$	$x[100]$		$x[001]$	$\longrightarrow$	$x[4]$
$x[2]$	$\longrightarrow$ $x[010]$	$\longrightarrow$	$x[010]$		$x[010]$	$\longrightarrow$	$x[2]$
$x[3]$	$\longrightarrow$ $x[011]$	$\longrightarrow$	$x[110]$		$x[011]$	$\longrightarrow$	$x[6]$
$x[4]$	$\longrightarrow$ $x[100]$	$\longrightarrow$	$x[001]$		$x[100]$	$\longrightarrow$	$x[1]$
$x[5]$	$\longrightarrow$ $x[101]$	$\longrightarrow$	$x[101]$		$x[101]$	$\longrightarrow$	$x[5]$
$x[6]$	$\longrightarrow$ $x[110]$	$\longrightarrow$	$x[011]$		$x[110]$	$\longrightarrow$	$x[3]$
$x[7]$	$\longrightarrow$ $x[111]$	$\longrightarrow$	$x[111]$	$\longrightarrow$	$x[111]$	$\longrightarrow$	$x[7]$

Figure 4.8 Process of bit reversal.

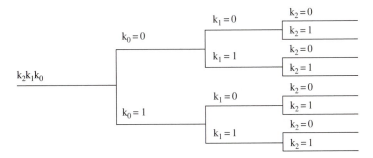

Figure 4.9 Justification of bit reversal.

most likely written in C language, that are well tested and incorporate all features to make them easy to use. For example, the FFT in MATLAB can take any integer N that may or may not be a power of 2. Therefore, there is no need to write your own FFT program unless for the sake of practice.

Inverse FFT The DFT of $x[n]$ is

$$X_d[m] = \sum_{m=0}^{N-1} x[n] W^{mn}$$

and the inverse DFT of $X_d[m]$ is

$$x[n] = \frac{1}{N} \sum_{m=0}^{N-1} X_d[m] W^{-mn} \tag{4.42}$$

They have similar forms and are dual to each other. Taking the complex conjugate of (4.42) yields

$$x^*[n] = \left[\sum_{0}^{N-1} \frac{X_d[m]}{N} W^{-mn}\right]^* = \sum_{0}^{N-1} \frac{X_d^*[m]}{N} W^{mn}$$

The summation after the last equality is the DFT of $X_d^*[n]/N$. Thus we have

$$\mathcal{D}^{-1}[X_d[n]] = \left(\mathcal{D}[X_d^*[n]/N]\right)^*$$

Thus the same FFT algorithm, with minor modification, can be used to compute DFT and inverse DFT. From now on, when we refer to FFT, we mean both FFT and inverse FFT.

In MATLAB, the function fft computes DFT and the function ifft computes inverse DFT. To be more specific, if x has N points, then fft(x) or fft(x,N) computes its N-point DFT. If $N > M$, the function fft(x,M) computes the DFT of the first M points of x. If $N < M$, fft(x,M) computes the DFT of x extended by $M - N$ number of zeros. If N equals a power of 2, then the computation will be very efficient. For example, let x=ones(1,1023). This sequence consists of 1023 number of ones. Typing

tic;fft(x);toc

where tic and toc start and stop the stopwatch, yields 0.11 s. However, tic;fft(x,1024);toc yields 0.05 s. Thus computing a 2^{10}-point FFT takes less than half the time of computing 1023-point FFT. Thus in application, we shall select, if possible, N to be a power of 2.

DSP Processors FFT carries out repeatedly sums of products or multiply-accumulates (MACs). A PC that uses a general-purpose processor, such as the Intel Pentium, requires a relatively large number of instruction cycles to carry out each MAC. Specialized processors, called DSP processors, have been developed to carry out more efficiently MACs and other computations often encountered in DSP processing. For example, recent DSP processors can carry out each MAC in one instruction cycle and implement bit reversal in hardware. This speeds up FFT considerably. Fixed-point DSP processors with 16 bits or more and floating-point DSP processors with 32 bits are now widely available commercially.

4.4.2 Real Sequences[2]

The FFT and DFT are developed for complex sequences. If given sequences are real valued, it is possible to save some computing time by simple manipulations, as we will discuss next. However, because of the tremendous speed of present-day digital computers, these manipulations probably are rarely employed.

Given two real sequences $x_1[n]$ and $x_2[n]$, both of length N, we form the complex-valued sequence

---

[2] This subsection may be skipped without loss of continuity.

$$x[n] = x_1[n] + jx_2[n] \tag{4.43}$$

for $n = 0, 1, \ldots, N - 1$. Then it is straightforward to verify

$$x_1[n] = \frac{x[n] + x^*[n]}{2} \quad \text{and} \quad x_2[n] = \frac{x[n] - x^*[n]}{2j} \tag{4.44}$$

where $x^*[n]$ is the complex conjugate of $x[n]$ or $x^*[n] = x_1[n] - jx_2[n]$. Because DFT is a linear operator, we have

$$X_{1d}[m] = \mathcal{D}[x_1[n]] = \frac{\mathcal{D}[x[n]] + \mathcal{D}[x^*[n]]}{2} \tag{4.45}$$

and

$$X_{2d}[m] = \mathcal{D}[x_2[n]] = \frac{\mathcal{D}[x[n]] - \mathcal{D}[x^*[n]]}{2j} \tag{4.46}$$

for $m = 0, 1, \ldots, N - 1$. Let us compute the DFT of $x^*[n]$. Define

$$X_d[m] := \mathcal{D}[x[n]] = \sum_{n=0}^{N-1} x[n]W^{nm}$$

with $W = e^{-j2\pi/N}$. Then we have

$$\mathcal{D}[x^*[n]] = \sum_{n=0}^{N-1} x^*[n]W^{nm} = \left(\sum_{n=0}^{N-1} x[n]W^{-nm} \right)^*$$

$$= X_d^*[-m] = X_d^*[N - m]$$

where we have used the fact that $X_d[m]$ is periodic with period N. Thus we have

$$X_{1d}[m] = \frac{1}{2}(X_d[m] + X_d^*[N - m]) \tag{4.47}$$

and

$$X_{2d}[m] = \frac{1}{2j}(X_d[m] - X_d^*[N - m]) \tag{4.48}$$

We see that once $X_d[m]$ is computed, both $X_{1d}[m]$ and $X_{2d}[m]$ can be readily obtained from (4.47) and (4.48). Thus, instead of computing two DFTs, the DFTs of two real sequences can be obtained by computing one DFT, and the computing time can be cut almost in half.

The preceding idea can be used to compute the DFT of a real sequence of length $2N$ by using an N-point DFT. Consider a real sequence $y[n]$ of length $2N$ or $n = 0, 1, \ldots, 2N - 1$. We define

$$x_1[n] := y[2n] \quad \text{and} \quad x_2[n] := y[2n + 1]$$

for $n = 0, 1, \ldots, N - 1$. They consist of even and odd parts of $y[n]$ and are of length N. We use them to form the complex sequence

$$x[n] = x_1[n] + jx_2[n]$$

Then the DFT of $x_1[n]$ and $x_2[n]$ can be computed from the DFT of $x[n]$ as in (4.47) and (4.48). Next we will express the DFT of $y[n]$ in terms of $X_{1d}[m]$ and $X_{2d}[m]$. By definition, the DFT of $y[n]$ is

$$Y_d[m] = \sum_{n=0}^{2N-1} y[n] W_{2N}^{nm} \tag{4.49}$$

for $m = 0, 1, \ldots, 2N - 1$, where

$$W_{2N} = e^{-2\pi/2N}$$

Clearly, we have

$$W := W_N = e^{-2\pi/N} = W_{2N}^2$$

Thus (4.49) can be written as

$$Y_d[m] = \sum_{n=0}^{N-1} y[2n] W_{2N}^{2nm} + \sum_{n=0}^{N-1} y[2n + 1] W_{2N}^{(2n+1)m}$$

$$= \sum_{n=0}^{N-1} x_1[n] W_N^{nm} + W_{2N}^m \sum_{n=0}^{N-1} x_2[n] W_N^{nm}$$

$$= X_{1d}[m] + W_{2N}^m X_{2d}[m]$$

for $m = 0, 1, \ldots, 2N - 1$. Using the periodicity of $X_{id}[m]$ and

$$W_{2N}^{N+m} = W_{2N}^N W_{2N}^m = -W_{2N}^m$$

we conclude

$$Y_d[m] = X_{1d}[m] + W_{2N}^m X_{2d}[m]$$

$$Y_d[N + m] = X_{1d}[m] - W_{2N}^m X_{2d}[m] \tag{4.50}$$

for $m = 0, 1, \ldots, N - 1$. Thus the DFT of a $2N$-point real sequence can be computed using a single N-point DFT.

4.5 Spectral Computation of Finite Sequences

This section discusses how to use FFT to compute the spectrum of DT signals of finite length. The procedure of FFT computation discussed in Chapter 2 for computing DT Fourier series is applicable here. We list first some pertinent points in the following.

- Consider a finite sequence $x[n] := x(nT)$ ranging from n_1 to n_2. Its spectrum $X_d(\omega)$ is defined as its DTFT, and is bounded and continuous in the Nyquist frequency range $(-\pi/T, \ \pi/T]$.
- Let $N = n_2 - n_1 + 1$. Then N equally spaced samples of $X_d(\omega)$ can be obtained by computing the DFT of $x[n]$ for $n = n_1, n_1 + 1, \ldots, n_2$. Let $\tilde{x}[n]$ be the periodic extension of $x[n]$ with period N. Then the N samples of $X_d(\omega)$ can also be obtained by computing the DFT of $\tilde{x}[n]$ for $n = 0, 1, \ldots, N - 1$. In computing DFT, the sampling period T does not play any role. It is used only to locate the samples of $X_d(\omega)$ at $\omega = m(2\pi/NT)$ with $m = 0, 1, \ldots, N - 1$ or

$$m = \overline{-(N-1)/2 : (N-1)/2} \tag{4.51}$$

Note that if N is even, then (4.51) can be replaced by

$$m = -N/2 + 1 : N/2$$

- DFT can be computed efficiently using FFT. In using FFT, however, we cannot use $x[n]$ from $n = n_1$ to n_2. We must use the data $\tilde{x}[n]$, for $n = 0, 1, \ldots, N-1$. The function $\mathsf{fft}(\tilde{x}, \mathsf{N})$ generates N data. They are to be located at $\omega = m(2\pi/NT)$ for $m = 0, 1, \ldots, N-1$. If they are to be arranged to lie inside the Nyquist frequency range, we apply shift as $\mathsf{shift}(\mathsf{fft}(\tilde{x}, \mathsf{N}))$. Then the N output data are located at the discrete frequencies $\omega = m(2\pi/NT)$ with m in (4.51). They are the N frequency samples, in $(-\pi/T, \pi/T]$, of the spectrum $X_d(\omega)$ of $x[n]$. Because $X_d(\omega)$ is defined for all ω, we must interpolate $X_d(\omega)$ from the N frequency samples.

We now give some examples.

◆ Example 4.4

Consider the finite sequence of length 4, with sampling period 0.5, in Fig. 4.3(a); that is, $x[0] = 2$, $x[1] = -1$, $x[2] = 1$, and $x[3] = 1$. Its spectrum was computed in Example 4.3 and plotted in Fig. 4.3(b). Now we will use FFT to compute its spectrum. Consider the program that follows:

Program 4.2
```
N=4;T=0.5;D=2*pi/(N*T);
```

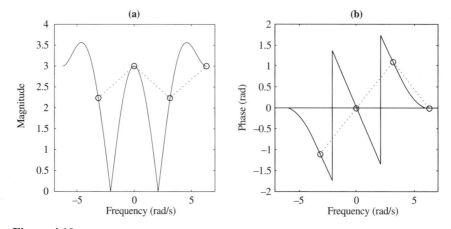

Figure 4.10 (a) Exact (solid line) and four-point FFT computed (hollow dots) magnitude spectra of the finite sequence in Fig. 4.3(a). (b) Corresponding phase spectra.

```
x=[2 -1 1 1];
X=shift(fft(x,N))
m=ceil(-(N-1)/2):ceil((N-1)/2);
w=-2*pi:0.01:2*pi; (% Frequencies to generate actual spectrum.)
X1=2-exp(-j*0.5*w)+exp(-j*w)+exp(-j*1.5*w); (% Actual spectrum)
subplot(1,2,1)
plot(w,abs(X1),m*D,abs(X),'o:'),title('(a)')
subplot(1,2,2)
plot(w,angle(X1),m*D,angle(X),'o:'),title('(b)')
```

The first four lines were explained earlier. The next two lines followed by % are used to generate the actual spectrum, computed in Example 4.3, for comparison with the one computed using FFT. The program yields the magnitude plot in Fig. 4.10(a) and the phase plot in Fig. 4.10(b). The solid lines are the actual spectra for all ω in the Nyquist frequency range (−6.28, 6.28]. The four hollow dots are the results of FFT. The hollow dots are spaced equally in (−6.28, 6.28]; therefore, the left-most dot is at −3.14 rad/s, and the right-most dot is at 6.28 rad/s. FFT indeed computes the samples of the spectrum. The MATLAB function **plot** automatically carries out interpolation by connecting neighboring dots by straight dotted lines as shown.

4.5.1 Padding with Zeros

FFT computes samples of the frequency spectrum of a finite sequence. The number of frequency samples must equal the number of data points. For the sequence of length 4 in Example 4.4, FFT computes four frequency samples as shown in Fig. 4.10. The frequency interval between two immediate samples is

$$D := \frac{2\pi}{NT} = \frac{2\pi}{4 \times 0.5} = 3.14 \, \text{rad/s}$$

This is called the frequency *resolution*. This resolution is very large or more often said to be very poor. Thus the straight-dotted-line interpolations of the four samples in Fig. 4.10 are very different from the actual magnitude and phase spectra (solid lines).

The way to improve the resolution is to pad zeros to the sequence. For example, we may pad twelve zeros to the sequence as

$$x_e[n] = \begin{cases} x[n] & \text{for } n = 0, 1, 2, 3 \\ 0 & \text{for } n = 4, 5, \cdots, 15 \end{cases}$$

This is a sequence of length 16. It is clear that the frequency spectrum of $x[n]$ equals the frequency spectrum of $x_e[n]$. If we use $x_e[n]$ as the input of FFT, then the frequency resolution becomes $2\pi/NT = 2\pi/(16 \times 0.5) = 0.39$. The results are plotted in Figs. 4.11(a) and (b) with hollow dots. They are obtained using Program 4.2 by changing only N from 4 to 16. No other changes are necessary. Recall that in using fft(x,N), if the length of x is less than N, trailing zeros are automatically added to x. We see that the frequency resolution in Figs. 4.11(a) and 4.11(b) is better than the one in Fig. 4.10, and its straight-line interpolation is quite close to the actual

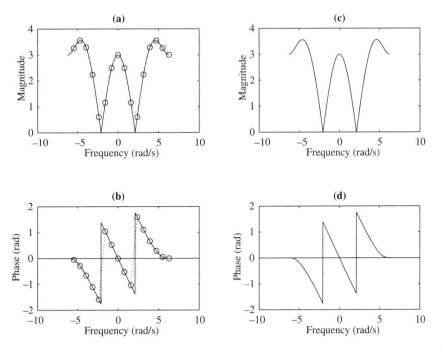

Figure 4.11 (a) Exact (solid line) and 16-point FFT computed (hollow dots) magnitude spectra of the finite sequence in Fig. 4.3(a). (b) Corresponding phase spectra in radians. (c, d) Exact and 1024-point FFT computed spectra.

spectrum. Next we pad 1020 zeros to the sequence; that is, we select $N = 1024$. The results are plotted in Figs. 4.11(c) and (d) with dotted lines without circles. We see that the magnitude and phase spectra computed using FFT are indistinguishable from the actual spectra (solid lines).

◆ Example 4.5

Consider the DT rectangular window $w_d[n]$ in Fig. 3.12(a). Its spectrum in closed form was developed in (3.47) and plotted in Fig. 3.12(b). Now we will use FFT to compute its spectrum. The sequence is real and even, and has length $N = 9$. If we use 9-point FFT to compute its spectrum, then the frequency resolution will be poor. Therefore, we will select a larger N. Suppose we select $N = 20$. Then we must pad 11 zeros to the sequence. In using DFT, we may pad 11 trailing zeros to $w_d[n]$ and compute the spectrum from $n = -4$ to 15. However, in using FFT, we must use the data from $n = 0$ to 19 of the periodic extension, with period $N = 20$, of $w_d[n]$. Thus the input of FFT should be

$$x = [1\ 1\ 1\ 1\ 1\ \text{zeros}(1, 11)\ 1\ 1\ 1\ 1]$$

They are the first 20 points, starting from $n = 0$, of the periodic extension of the rectangular window. Note that $w_d[n] = 1$ for $n = -4, -3, -2, -1$ are shifted to $n = 16, 17, 18, 19$. Then the program that follows will plot its spectrum.

Program 4.3
```
N=20;T=0.5;D=2*pi/(N*T);
x=[1 1 1 1 1 zeros(1,N-9) 1 1 1 1];
X=shift(fft(x,N));
```

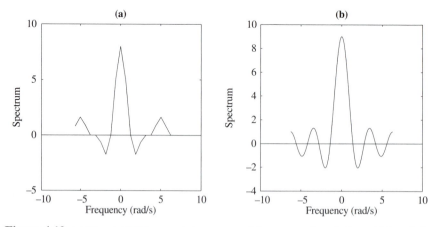

Figure 4.12 (a) 20-point FFT computed spectrum of the rectangular window with length 9. (b) 1024-point FFT computed spectrum.

m=-N/2+1:N/2;
plot(m*D,X)

The result is plotted in Fig. 4.12(a). Although the plot has a rough shape of the spectrum in Fig. 3.12(b), the frequency resolution is poor. We repeat Program 4.3 by selecting $N = 1024$, and the result is plotted in Fig. 4.12(b). It is indistinguishable from the actual spectrum in Fig. 3.12(b) in the Nyquist frequency range $(-2\pi, \ 2\pi]$.

In the preceding example, we have extended the time sequence periodically and then use the first N points as an input of FFT. If we use

$$x1 = [\text{ones}(1, 9) \ \text{zeros}(1, N - 9)]$$

as an input of FFT, then we are computing the spectrum of the *shifted* rectangular window or $w_d[n - 4]$. As discussed in Chapter 3, time shifting will not affect the magnitude spectrum but will introduce a linear phase into the phase spectrum. To verify this, we use Program 4.3 with $N = 1024$ to plot the magnitude spectra of $w_d[n]$ (solid line) and $w_d[n - 4]$ (dotted line) in Fig. 4.13(a). The two magnitude spectra are indeed identical. Figure 4.13(b) shows the phase spectra of $w_4[n]$ (solid line) and $w_4[n - 4]$ (dotted line). Because $w_4[n]$ is real and even, so is its spectrum. Thus the phase spectrum of $w_4[n]$ is either 0 or $\pm\pi$ radians. However, because of computer roundoff errors, the phase changes rapidly between $\pm\pi$ as shown. The phase spectrum of $w_4[n - 4]$ differs from the phase spectrum of $w_4[n]$ by $-\omega n_0 T = -2\omega$ (Problem 3.18). In MATLAB, phases are plotted only inside the range $\pm\pi$; therefore, instead of one straight line, we see many sections of straight lines lying inside the range. This verifies the assertion that time shifting introduces a linear phase into a phase spectrum.

We summarize what has been discussed in the preceding two sections. The spectrum of a finite sequence $x[n]$ of length N can be easily computed using an N-point FFT. The only problem is that the spectrum is defined for all ω in the Nyquist frequency range, whereas FFT computes only its frequency samples with frequency resolution $D = 2\pi/NT$. Thus we must interpolate the spectrum from the computed samples. It is possible to interpolate the spectrum by using interpolation formulas. It is however much simpler to pad trailing zeros to improve the frequency resolution. See Problems 4.9 and 4.10. If a finite sequence is not positive-time in the sense that $x[n] \neq 0$ for all $n < 0$, then the negative-time part must be moved after the trailing zeros before using FFT. Otherwise we will obtain an incorrect phase spectrum.

4.5.2 Spectral Computation of Infinite Sequences

This subsection discusses FFT computation of the spectrum of infinite sequences. For finite sequences, we may have to pad zeros to improve the frequency resolution or, equivalently, to eliminate the interpolation problem. This problem will not arise for infinite sequences, because we can select the number of data points as large as desired. However, we have a different problem. Every infinite sequence must be truncated to a finite sequence in computer computation.

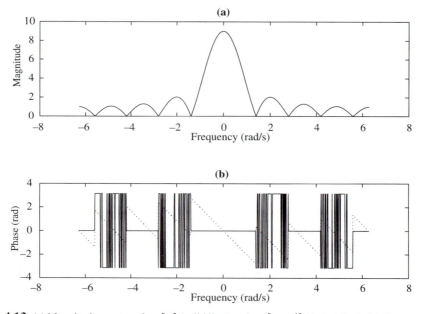

Figure 4.13 (a) Magnitude spectra of $w_d[n]$ (solid line) and $w_d[n-4]$ (dotted line). (b) Corresponding phase spectra.

Truncation, as discussed in Section 3.7, will introduce leakage and ripples. Thus the spectrum of a truncated sequence can only approximate the spectrum of the infinite sequence, and the issue of accuracy will arise.

We use an example to illustrate the issue. Consider the sequence

$$x[n] = x(0.5n) = 0.6^n$$

for $n \geq 0$, $x[n] = 0$ for $n < 0$, and $T = 0.5$. It is an infinite sequence, and its spectrum was computed analytically in Example 3.6 as $X_d(\omega) = 1/(1 - 0.6e^{-j0.5\omega})$. It is easy to verify that if we use $N = 16384$ data points[3] to compute its spectrum, then the computed spectrum is indistinguishable from the exact one.

It is natural to ask: Is it necessary to use such a large N? In fact, even for $N = 32$, the computed spectrum is indistinguishable from the exact one. The program that follows

```
T=0.5;N=32;D=2*pi/(N*T);
n=0:N-1;
x=0.6.^n;
X=fft(x);
```

---

[3] $N = 2^{14} = 16384$ is the largest number permitted in the Student Edition of MATLAB, Version 5.

```
mm=max(abs(X(n+1)))
e=max(abs(abs(X(n+1))-abs(1.0./(1-0.6.*exp(-j*0.5.*n*D)))))
pe=e/mm*100
```

computes $N = 32$ frequency samples of the spectrum of $(0.6)^n$ for $n = 0 : 31$, its peak magnitude mm=2.5, and the maximum error $e = 1.99 \times 10^{-7}$ between the computed and exact magnitude at frequencies mD for $m = 0 : 31$. Note that the index of X starts from 1 instead of 0. The percentage error pe is 7.96×10^{-6}. Thus for the infinite sequence 0.6^n, for $n \geq 0$, we can obtain a satisfactory result by using only 32 data points instead of using 16384. Therefore, for infinite sequences, it may not be necessary to use a very large N to compute their spectra.

Now let us pose the problem: Find the smallest possible N so that the computed spectrum differs, less than 1% of its peak magnitude, from the exact one. This problem is meaningful only if the exact spectrum is known. On the other hand, if the exact spectrum is known, there is no more need to compute it. Therefore, following Section 2.7.2, we will pose the problem differently as: Find the smallest possible integer a in N^a such that the spectrum computed using N data points differs, less than β percent of its peak magnitude, from the one computed using $N/2$ data points.

Before proceeding, we discuss how to compare the spectra computed using $N_1 = N$ and $N_2 = 2N$. For a DT infinite sequence with sampling period T, the spectrum X1 using N_1 points is located at

$$\omega_1[m] = m\frac{2\pi}{N_1 T}$$

and the spectrum X2 using N_2 points is located at

$$\omega_2[m] = m\frac{2\pi}{N_2 T} = m\frac{2\pi}{2N_1 T}$$

We see that $\omega_1[m] = \omega_2[2m]$. Thus we can compare the magnitudes of X1 and X2 at those frequencies. Furthermore, because magnitudes are even, we compare them only for $0 \leq m \leq N_1/2$. Note that if $N_2 \neq 2N_1$, generally X1 and X2 compute spectra at different frequencies; thus it is less useful to compare them. Following Program 2.7, we develop the program that follows

Program 4.4
```
T=0.5;a=1;b=100;beta=1;
while b>beta
        N1=2^a;
        n1=0:N1-1;x1=0.6.^n1;
        X1=fft(x1);
        N2=2*N1;
        n2=0:N2-1;x2=0.6.^n2;
```

```
    X2=fft(x2);
    m1p=0:N1/2;
    d=max(abs(abs(X1(m1p+1)))-abs(X2(2*m1p+1))));
    mm=max(abs(X1(m1p+1)));
    b=d/mm*100
    a=a+1;
end
N2,b
m=-N2/2+1:N2/2;D=2*pi/(N2*T);
subplot(1,2,1)
plot(m*D,abs(shift(X2))),title('(a) Magnitude spectra')
axis square
subplot(1,2,2)
plot(m*D,angle(shift(X2))),title('(b) Phase spectra')
axis square
```

The while loop searches the least $N = 2^a$ so that the computed result differs from its previous computation in less than β % of its peak magnitude. The remainder of the program uses that N to plot the magnitude and phase spectra of the sequence. If $\beta = 1$, the program yields $N2 = 32$, $b = 0.028$, and the magnitude and phase spectra in Fig. 4.14. It uses 32 data points of 0.6^n. The plot should be close to the spectrum of the infinite sequence.

Note that the magnitude spectrum in Fig. 4.14(a) is not smooth in the neighborhood of $\omega = 0$. It is due to the straight-line interpolation scheme used in plot. The magnitude spectrum also has a gap at -2π. This is due to the fact that the frequency range is $(-\pi/T, \ \pi/T]$ and

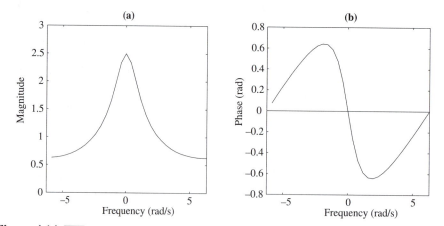

Figure 4.14 FFT computed frequency spectrum of $x[n] = 0.6^n$ for $n \geq 0$ and $T = 0.5$.

the left-most point computed is $D = 2\pi/NT$ from $-\pi/T$. To improve the interpolation and to reduce the gap, we may use $N = 64$ or 128. There is certainly no need to use $N = 16384$.

4.6 Spectral Computation of CT Signals

Consider the CT signals in Examples 3.1 and 3.2. Their frequency spectra were computed there by direct integration. Because of their simple forms, their integrations can be carried out analytically, and their exact spectra can then be plotted. This approach cannot be employed for most signals encountered in practice or for signals obtained by measurement. Thus we develop a numerical method.

Let x(t) be a CT signal and let $x(nT)$ be its sampled sequence. If T is selected to be sufficiently small, then we have, as discussed in Section 3.8.1,

$$X(\omega) = \mathcal{F}[x(t)] \approx \begin{cases} T\mathcal{F}_d[x(nT)] = TX_d(\omega) & \text{for } |\omega| \leq \pi/T \\ 0 & \text{for } |\omega| > \pi/T \end{cases} \tag{4.52}$$

Because $X_d(\omega)$ can be computed using FFT, so can the spectrum of CT signals. We discuss in this section only CT signals that are absolutely integrable.

We consider first positive-time signals. A signal $x(t)$ is defined to be positive time if its value is identically zero for $t < 0$. The time instant $t = 0$ is a relative one; it may be selected as the time instant we first study a signal. Therefore, in practical application, all signals can be considered to be positive time. We first discuss the relationships among time record length L, sampling period T, and the number of samples N. If a positive-time signal $x(t)$ does not contain impulses at $t = 0$ or L, then the spectra of $x(t)$ in $(0, L)$, $[0, L]$, $(0, L]$, and $[0, L)$ will all be the same. However, the spectra computed from its samples may be different. To see the reason, let $T = L/N$. Then the number of the samples of $x(t)$ in the four intervals will be, respectively, $N - 1$, $N + 1$, N, and N. Thus the four spectra computed from their samples will be different. Clearly, if N is very large, then the four spectra will be close. For convenience, time record length will be selected as $[0, L)$. Under this assumption, we have $L = TN$.

Computing the frequency spectrum of $x(t)$ from (4.52) may involve three issues. They are discussed in the following.

1. *Frequency aliasing:* If $x(t)$ is not bandlimited, we must select a sampling period T so that frequency aliasing due to time sampling is negligible. The discussions in Section 2.7.2, for computing CTFS coefficients, and in Section 3.8.1 are applicable here.
2. *Frequency resolution:* The spectrum $X(\omega)$ of $x(t)$ is defined for all ω. When we use FFT to compute $TX_d(\omega)$ in (4.52), we compute only its frequency sample $X_d(2m\pi/NT)$. Thus we must interpolate $X(\omega)$ from $TX_d(2m\pi/NT)$. Clearly, the smaller the frequency resolution $D = 2\pi/NT$, the better the interpolation as discussed in Section 4.5.1. Thus even if $x(t)$ is time limited, we may want to select a large $L = NT$ to improve its frequency resolution.

3. *Effects of truncation:* If $x(t)$ is of infinite duration, we must truncate it to a finite record length $L = NT$. This will introduce leakage and ripples and create the accuracy problem, as discussed in Section 4.5.2.

Because of the preceding three issues, care must be exercised in using FFT to compute the frequency spectrum of $x(t)$. Clearly, we should select the sampling period T small to reduce frequency aliasing and the time record length $L = NT$ large to reduce the effects of truncation and to improve the frequency resolution. If there is no limit on N, we can always find a sufficiently small T and a sufficiently large N so that the spectrum of $x(t)$ can be computed from (4.52). However, in practice, N may be limited. For example, N is limited to $10^{14} = 16384$ in the Student Edition of MATLAB, Version 5. For a fixed N, T small (to reduce frequency aliasing) and NT large (to reduce the effects of truncation and to improve the frequency resolution) are two conflicting requirements. Furthermore, a very small T and a very large N may not be necessary as discussed in Sections 2.7.2 and 4.5.2. Therefore, we suggest the following procedure for selecting T and N in computing the spectrum $X(\omega)$ of $x(t)$. In this computation, we assume no prior knowledge of $X(\omega)$.

1. Select first a time record length L such that $[0, \; L)$ includes most of the $x(t)$ that are significantly different from 0. We then use the procedure in Section 2.7.2 to select a sampling period T so that frequency aliasing due to time sampling is negligible. That is, we select a small N_1 and compute $T_1 = L/N_1$. We then use an N_1-point FFT to compute the spectrum $X_d(m2\pi/(N_1T_1))$ of the sampled sequence $x(nT_1)$, for $n = 0 : N_1 - 1$. Multiplying X_d by T_1 and interpolating it using plot, we obtain an approximate spectrum of $x(t)$. We repeat the computation by using $T_2 = T_1/2$ and $N_2 = 2N_1$. Because $T_1N_1 = T_2N_2$, the two spectra are computed using the same time record length. Thus they have the same effects of truncation. Consequently, any difference between the two spectra is due to frequency aliasing. As discussed in Section 2.7.2, if the two spectra are very close or, more specifically, their largest magnitude difference is less than, for example, 1% of the peak magnitude in the Nyquist frequency range $(-\pi/T_1, \; \pi/T_1]$, the effect of frequency aliasing is probably negligible, and we may use T_2 in all subsequent computations. Otherwise, we repeat the process until we find such a T_2.
2. We compute the spectrum of $x(t)$ using T_2 and N_2 and compare it with the one using T_2 and $2N_2$. The two spectra use the same T_2, thus they have the same frequency aliasing and the same Nyquist frequency range $(-\pi/T_2, \; \pi/T_2]$. Therefore, any difference between the two spectra is due to the effects of truncation. If the two spectra differ considerably in $(-\pi/T_2, \; \pi/T_2]$, we repeat the computation by doubling N until two consecutive spectra are very close. We then stop the computation and the last magnitude spectrum will be close to the actual magnitude spectrum of $x(t)$.

If $x(t)$ is absolutely integrable, the magnitude spectrum, the real part, and imaginary part of the spectrum computed using the preceding procedure will converge to the actual corresponding spectrum of $x(t)$. The computed phase spectrum however will differ from the actual phase spectrum. This is illustrated by the examples that follow.

◆ **Example 4.6**

Consider the CT signal $x(t) = e^{-0.1t}$, for $t \geq 0$. Use FFT to compute its frequency spectrum.

The signal has peak magnitude 1 and magnitude less than 0.0067 for $t \geq 50$; thus if we select $L = 50$, then the time record length will cover most significant part of the signal. However, we will select $L = 10$ in order to illustrate better the procedure. Note that $e^{-0.1 \times 10} = 0.37$; thus $L = 10$ does not include all significant part of the signal. However, because the signal is given analytically, the time interval [0, 10) will reveal all significant frequency components of $e^{-0.1t}$ for $t \geq 0$ and can be used in our discussion.

Arbitrarily, we select $N = 5$. Then we have $T = L/N = 2$ or $L = NT = 10$. The program that follows

Program 4.5
```
T=2;N=5;D=2*pi/(N*T);
n=0:N-1;
x=exp(-0.1*n*T);
X=T*shift(fft(x));
m=ceil(-(N-1)/2):ceil((N-1)/2);
plot(m*D,abs(X))
```

yields the magnitude spectrum in Fig. 4.15(a) with a solid line. We repeat the computation using $T = 1$ and $N = 10$ and plot the result in Fig. 4.15(a) with a dotted line. Note that both plots have the same frequency resolution $D = 2\pi/10$. The two plots differ considerably in $(-\pi/2, \pi/2] = (-1.56, 1.56]$, and the dotted line has a fairly large nonzero magnitude at $\pm\pi/1 = 3.14$; thus frequency aliasing is not negligible for $T = 1$. We repeat a number of times the computations by cutting T in half and doubling N (thus we have the same L). The results for $T = 0.2$, $N = 50$ (solid line) and for $T = 0.1$, $N = 100$ (dotted line) are plotted in Fig. 4.15(b). In order to see better the two plots, we zoom in Fig. 4.15(c) the frequency range $[-3, 3]$ and in Fig. 4.15(d) the frequency range $[15, 30]$. They can be easily obtained by adding, respectively, **axis([-3 3 0 10])** and **axis([15 30 0 0.5])** at the end of Program 4.5. The two plots in Fig. 4.15(c) are indistinguishable, and the magnitude in Fig. 4.15(d) is less than 0.05 at $\pi/0.1$ or less than $0.05/6 = 0.8\%$ of the peak value 6. Thus if $T = 0.1$, frequency aliasing due to time sampling is negligible, and there is no need to reduce T further. Thus we use $T = 0.1$ in all subsequent computations.

Figure 4.15(e) plots the magnitude spectra obtained using $T = 0.1$, $N = 100$ (solid line) and using $T = 0.1$, $N = 200$ (dotted line). We plot only the frequency range $[-3, 3]$. They differ considerably; thus the effects of truncation are appreciable, and we must use a longer time record length L. Figure 4.15(f) shows the magnitude spectra obtained using $T = 0.1$, $N = 400$ (solid line) and $T = 0.1$, $N = 800$ (dotted line). They are indistinguishable in $[-3, 3]$. Thus we conclude that $T = 0.1$ is small enough (to avoid frequency aliasing) and $L = 0.1 \times 800 = 80$ is large enough (to avoid the effects of truncation) in computing the frequency spectrum of $x(t) = e^{-0.1t}$. To verify this assertion, we plot in Fig. 4.15(f) the exact magnitude spectrum using a dashed line. It is indeed identical to the one computed using FFT.

To conclude this example, we compare the magnitude spectrum denoted by the dotted line in Fig. 4.15(b), which is replotted in Fig. 4.16(a), with the magnitude spectrum in Fig. 3.10(a). Both spectra were computed using the same time record length $L = 10$. The spectrum in Fig. 3.10(a) was plotted from an exact equation using frequency resolution 0.01 and shows ripples. The one in Fig. 4.16(a) was computed using FFT with frequency resolution $2\pi/10 = 0.628$. Because of the poor resolution, no ripples appear in Fig. 4.16(a). Now if we pad $(800 - 100)$ trailing zeros to the sequence, the FFT generated magnitude response will be as shown in Fig. 4.16(b). It is indistinguishable from the one in Fig. 3.10(a). Thus we conclude that it is possible to obtain the exact plot in Fig. 3.10(a) by using FFT. The question is: When more data of $e^{-0.1t}$ are available, is there any reason to pad zeros instead of using available data? The answer is probably not. If we use 800 data points, then we will obtain the plot in Fig. 4.15(f), which is closer

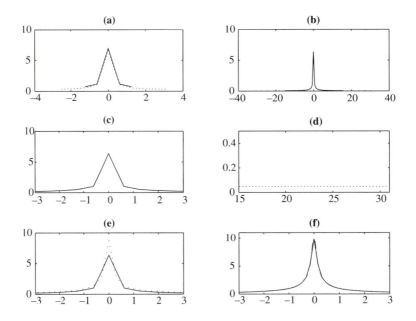

Figure 4.15 (a) Magnitude spectra of $e^{-0.1t}$, for $0 \le t < 10$, with $T = 2$ (solid line) and 1 (dotted line). (b) With $T = 0.2$ (solid line) and 0.1 (dotted line). (c) Zoom of (b) in the frequency range of $[-3, 3]$. (d) Zoom of (b) in the frequency range $[15, 30]$. (e) Magnitude spectra of $e^{-0.1t}$ using $T = 0.1$ and for $0 \le t < 10$ (solid line) and for $0 \le t < 20$ (dotted line). (f) For $0 \le t < 40$ (solid line) and for $0 \le t < 80$ (dotted line). All horizontal coordinates are frequency in rad/s.

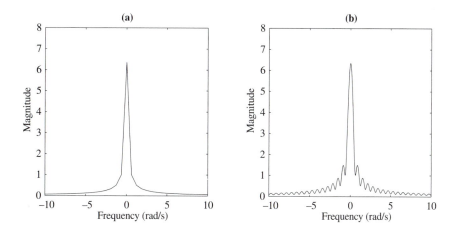

Figure 4.16 (a) FFT computed magnitude spectrum of $e^{-0.1t}$ with $0 \le t \le L = 10$. (b) FFT computed magnitude spectrum using $L = 10$ and trailing zeros.

to the actual magnitude spectrum than the one in Fig. 4.16(b). Therefore, for this example, it is unnecessary to pay any attention to the poor frequency resolution in Fig. 4.15(b) or 4.16(a). As a final remark, phase spectra computed using FFT are generally different from the actual phase spectrum. Thus it is not plotted here.

◆ **Example 4.7**

Consider the CT signal

$$x(t) = e^{-0.01t} \cos t + 2e^{-0.02t} \cos 1.1t \tag{4.53}$$

for $t \ge 0$ and $x(t) = 0$ for $t < 0$. We will use FFT to compute its spectrum. Because the signal is given in analytical form, any time record length will reveal its significant frequency components. Arbitrarily, we select $T_1 = 0.6$ and $N_1 = 256$. The program that follows

Program 4.6
```
T=0.6;N=256;D=2*pi/(N*T);
n=0:N-1;
x=exp(-0.01*n*T).*cos(n*T)+2*exp(-0.02*n*T).*cos(1.1*n*T);
X=T*shift(fft(x,N));
```

m=-N/2+1:N/2;
plot(m∗D,abs(X))

yields the plot in Fig. 4.17(a) with dashed line. We repeat the computation using $T = 0.3$ and $N = 512$, and plot the result in Fig. 4.17(a) with a solid line. They are indistinguishable in the range $(-\pi/0.6, \pi/0.6] = (-5.24, 5.24]$ and the second spectrum is practically zero outside the range, thus we conclude that the frequency aliasing is negligible for $T = 0.3$. Therefore, we may use $T = 0.3$ from now on with no need to reduce it further. We will increase only the time record length L or, equivalently, N.

Figure 4.17(a) also plots with a dotted line the magnitude spectrum for $T = 0.3$ and $N = 1024$. In order to see its difference from the solid line that is the magnitude spectrum for $T = 0.3$ and $N = 512$, we plot in Fig. 4.17(b) only the frequency interval $[0.2, 1.8]$. This is obtained by using the command

axis([0.2 1.8 0 60])

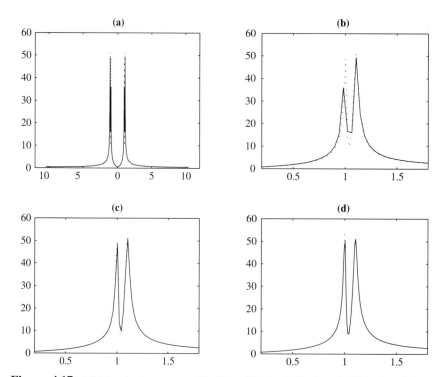

Figure 4.17 (a) Magnitude spectra with $N = 512$ (solid line) and 1024 (dotted line). (b) Zoom of (a). (c) Magnitude spectra using $N = 1024$ (solid line) and 2048 (dotted line). (d) Magnitude spectra using $N = 2048$ (solid line) and 4096 (dotted line). All horizontal coordinates are frequency in rad/s.

We see that the plot using $N = 512$ differs considerably from the plot using $N = 1024$. Thus $N = 512$ is not large enough. Whether or not $N = 1024$ is large enough must be determined by comparing it with a plot using a larger N. Figure 4.17(c) compares the magnitude spectra computed using $N = 1024$ and 2048. They are very close. We repeat in Fig. 4.17(d) using $N = 2048$ and 4096. They are almost indistinguishable. Thus we stop the computation. In conclusion, we stop the computation when the results of two consecutive computations are close.

The comparison in the preceding examples was done visually. It is possible to compare two computations mathematically by using the procedures in Sections 2.7.2 and 4.5.1. This will not be repeated here. See Problem 4.15. As a final example, we discuss how the plot in Fig. 3.9(a) was obtained using FFT.

◆ **Example 4.8**

Compute the frequency spectrum of

$$h(t) = \begin{cases} \dfrac{\sin 2t}{\pi t} & \text{for } |t| \leq a \\ 0 & \text{for } |t| > a \end{cases} \tag{4.54}$$

with $a = 5$. This signal is not positive time but is time limited. It is real and even; therefore, its spectrum is also real and even. When we use its sampled sequence to compute its spectrum, it is desirable to have the sampled sequence to be even so that the FFT computed spectrum will be real and even. In order to achieve this, we select the number of time samples to be odd.

First we select arbitrarily $T = 1$. Then there are a total of 11 samples of $h(nT)$ for $|t| \leq 5$. Consider the program

Program 4.7
```
T=1;M=5/T;N=2*M+1;D=2*pi/(N*T);
n=1:M;
hp=sin(2*n*T)./(pi*n*T);
h=[2/pi hp fliplr(hp)];
H=T*shift(fft(h));
m=ceil(-(N-1)/2):ceil((N-1)/2);
plot(m*D,H);
```

The five time samples for $n = 1$ to 5 are generated in hp. The sample of $h(t)$ at $t = 0$ is $2/\pi$, which cannot be generated from $(\sin 2t)/\pi t$ on a computer. Because of the evenness of $h(t)$, the five time samples for $n < 0$ are the left-right flip of hp. The input of FFT must be from $n = 0$ to $N - 1$ of the periodic extension of the

11 time samples; thus the input of FFT is h, as listed in the fourth line of Program 4.7. The program generates the solid line in Fig. 4.18(a). We plot in Fig. 4.18(a) with a dotted line the spectrum computed using $T = 0.5$ and $N = 21$. The second spectrum is practically zero in the neighborhood of $\pm\pi/0.5 = 6.28$; thus frequency aliasing due to time sampling with $T = 0.5$ is negligible, and we will use $T = 0.5$ in subsequent computation.

The frequency resolutions of the two spectra in Fig. 4.18(a) are, respectively, $2\pi/11 = 0.57$ and $2\pi/(0.5 \times 21) = 0.598$. They are very poor. Because we want to compute the spectrum of the truncated $h(t)$ with $|t| \leq 5$, the only way to improve its frequency resolution is by padding zeros. If we use a 1024-point FFT, then Program 4.7 must be modified as

Program 4.8

```
T=0.5;a=5;M=a/T;N=2*M+1;Nb=1024;D=2*pi/(Nb*T);
n=1:M;
hp=sin(2*n*T)./(pi*n*T);
h=[2/pi hp zeros(1,Nb-N) fliplr(hp)];
H=T*shift(fft(h));
m=ceil(-(Nb-1)/2):ceil((Nb-1)/2);
plot(m*D,H)
```

Note that there are only 21 data points for $T = 0.5$. Thus we must introduce $1024 - 21 = 1003$ zeros into the sequence h. These zeros are to be inserted as

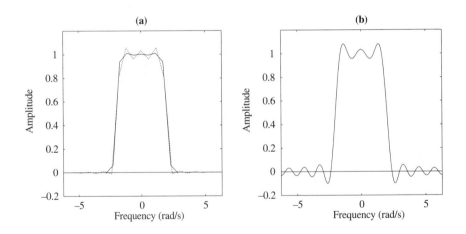

Figure 4.18 FFT computed spectra of (4.54) with $a = 5$: (a) Using $T = 1$ (solid line) and $T = 0.5$ (dotted line). (b) Using $T = 0.5$ and inserting 1003 zeros.

shown (not trailing zeros) for the same reason discussed in Example 4.5. Then the program will yield the frequency response in Fig. 4.18(b). This is the plot in Fig. 3.9(a). The frequency spectrum in Fig. 3.9(b) is for $a = 50$ and can be similarly obtained.

From the preceding examples, we conclude that FFT can indeed be used to compute the magnitude spectrum of CT signals.

4.6.1 Spectral Computation of CT Periodic Signals

If a CT signal is absolutely integrable, the spectra computed using the procedure in the preceding section will converge to the actual spectrum. However, if a CT signal is periodic, then the computed spectra will not converge. This is illustrated in the next example.

Consider $x(t) = \cos 3t$ defined for all t. Its spectrum is

$$X(\omega) = \pi[\delta(\omega - 3) + \delta(\omega + 3)]$$

It consists of two impulses with weight π at $\omega = \pm 3$ as shown in Fig. 3.8(d). In computer computation, the cosine function must be truncated to a finite length L. We use the shifted rectangular window $w_L(t)$ defined in (3.10) and consider $(\cos 3t)w_L(t)$. Its spectrum is, as derived in (3.36),

$$\bar{X}(\omega) = 0.5[W_L(\omega - 3) + W_L(\omega + 3)]$$

where $W_L(\omega)$ is the CTFT of $w_L(t)$ and is given in (3.11). The magnitude spectra of $\bar{X}(\omega)$ were plotted in Fig. 3.8(e) for $L = 8$ and in Fig. 3.8(f) for $L = 102.4$. The plots show the effects of truncation: leakage and ripples. Let us check whether we can obtain the same plots by using FFT.

The signal $x(t) = \cos 3t$ is bandlimited to 3 rad/s; thus, if the sampling period is smaller than $\pi/3 = 1.05$, then no frequency aliasing will occur. However, if we truncate $\cos 3t$ to a length of L, then its spectrum is no longer bandlimited. Thus we will select a smaller sampling period. Arbitrarily, we select $T = 0.2$. We then select $N = 40$, which yields $L = TN = 8$. The program that follows computes and plots the magnitude spectrum of $(\cos 3t)w_8(t)$.

Program 4.9
```
T=0.2;N=40;D=2*pi/(N*T);
n=0:N-1;
x=cos(3*n*T);
X=T*shift(fft(x));
m=-N/2+1:N/2;
plot(m*D,abs(X))
```

The result is shown in Fig. 4.19(a). Its magnitude spectrum shows two spikes at ± 3 with height slightly less than 4. As discussed in Section 3.5, truncation will introduce leakage and ripples. The plot in Fig. 4.19(a) shows leakage but no ripples. Figure 4.19(b) shows the magnitude spectrum for $T = 0.2$ and $N = 512$. Even though its frequency resolution is $D = 2\pi/NT = 0.06$, we still cannot see ripples. In fact, when we double N, the two spikes will double in height and become narrower as shown in Fig. 4.19(c). But there are still no ripples. This is in fact a blessing because the spectra plotted in Figs. 4.19(b) and (c) look more like impulses. It turns out that the only way for ripples to appear is, similar to the discussion in Fig. 4.15, by padding zeros. However, when more signal data are available, there seems no reason to pad zeros. In conclusion, if FFT computed magnitude spectra consist of narrow spikes whose heights double whenever N is doubled, we may conclude that the CT signal contains a periodic part.

Although the existence of periodic signals can be easily detected from the computed magnitude spectra, it is difficult to determine the θ in $\cos(\omega_0 t + \theta)$ from the corresponding phase spectra. We plot in Fig. 4.19(d) the phase spectra of $\cos(3t + \theta)$, using $T = 0.2$ and $N = 1024$, for $\theta = 0$ (solid line) and $-\pi/2$ (dotted line). It is indeed difficult to determine θ from the plot. We demonstrate this further using the example that follows.

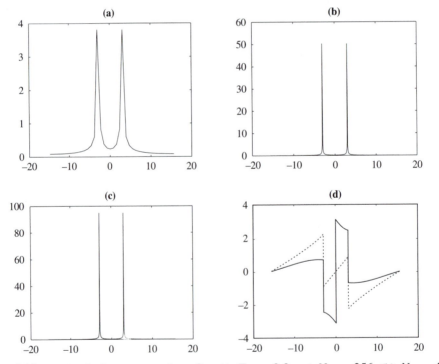

Figure 4.19 (a) Magnitude spectrum of $\cos 3t$ with $T = 0.2$ and $N = 256$. (b) $N = 512$. (c) $N = 1024$. (d) Phase spectra of $\cos(3t + \theta)$ for $\theta = 0$ (solid line) and $-\pi/2$ (dotted line). All horizontal coordinates are frequency in rad/s.

◆ **Example 4.9**

Consider the periodic signal in Example 2.4 or

$$x(t) = -1.2 + 0.8 \sin 0.6t - 1.6 \cos 1.5t \qquad (4.55)$$

Its highest frequency is 1.5 rad/s; thus the sampling period must be selected to be less than $\pi/1.5 = 2.09$. Arbitrarily, we select $T = 0.5$. We plot in Fig. 4.20(a) its magnitude spectrum (solid line) and phase spectrum (dotted line) using $N = 1024$, and in Fig. 4.20(b) using $N = 2048$. They show five spikes at $\omega = 0, \pm 0.6$, and ± 1.5. When we double N, the heights roughly double. Thus we conclude that the signal contains a dc part and two sinusoids at frequencies 0.6 and 1.5 in rad/s.

From Example 2.4 and the discussion in Section 3.2.1, we know that the ratio of the three different heights at $\omega = 0, \pm 0.6$, and ± 1.5 should be 3:1:2. It is almost the case in Fig. 4.20(a) but definitely not in Fig. 4.20(b). In conclusion, FFT spectral computation of periodic signals can be used to determine the frequencies of sinusoid components but not their phases and amplitudes.

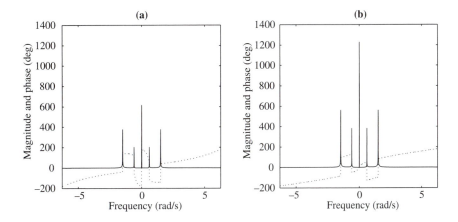

Figure 4.20 Magnitude (solid line) and phase (dotted line) spectra of (4.55): (a) $T = 0.5$ and $N = 1024$. (b) $T = 0.5$ and $N = 2048$.

We now compare FFT computations of frequency components (CT Fourier series) and of frequency spectra (CT Fourier transform) of CT periodic signals. In the former, we use only one period of a periodic signal; in the latter, we must truncate the signal and the computation will encounter the effects of truncation. Thus for periodic signals that last forever, it is simpler to use one period to compute its frequency components as in Section 2.7.2. The magnitudes of all frequency components can be obtained as accurate as desired, but the computed phases generally differ from the actual phases.

4.7 Computing DT Signals from Spectra

We discussed in the preceding sections FFT computation of frequency spectra of DT and CT signals. We now discuss the converse problem, namely, to compute DT and CT signals from their spectra. We discuss first the DT case.

Consider a frequency spectrum $X_d(\omega)$, for $-\pi/T < \omega \le \pi/T$. Its time sequence $x[n] = x(nT)$ can be computed uniquely from the inverse DT Fourier transform in (3.44). Unfortunately, the integration formula in (3.44) cannot be directly computed on a digital computer. In this section, we use inverse FFT or DFT to compute the time sequence. Ths procedure will be developed from the relationship between the inverse DTFT and inverse DFT discussed in Section 4.2.2.

First we discuss how to use inverse FFT to generate N data points from the N frequency samples of $X_d(\omega)$ or $X_d(2m\pi/NT)$ with

m=ceil(-(N-1)/2):ceil((N-1)/2)

As in FFT, the input of inverse FFT must be from $m = 0$ to $N - 1$. Thus the frequency samples of $x_d(2m\pi/NT)$ for $m < 0$ must be shited to the end of those for $m \ge 0$. Let mn=ceil(-(N-1)/2):-1 and mp=0:ceil((N-1)/2). Then the following program

Program 4.10
T= ;N= ;D=2*pi/(N*T);
mn=ceil(-(N-1)/2):-1;mp=0:ceil((N-1)/2);
X=[Xd(mp*D) Xd(mn*D)];
x=ifft(X):
stem(T*(0:N-1),x)

will yield N data $\hat{x}[n] = x(nT)$ for $n = 0, 1, \ldots, N - 1$. We now discuss the relationship between $\hat{x}[n]$ and the inverse DTFT $x[n]$ of $X_d(\omega)$. It involves two issues:

1. *Time aliasing*: As discussed in Section 4.2.2, frequency sampling may introduce time aliasing. If time aliasing is not negligible, then the computed $\hat{x}[n]$ is useless. The way to check time aliasing is as follows. If the computed $\hat{x}[n]$ are significantly different from zero for all n, then time aliasing is appreciable and we must select a larger N. We stop to increase N when the computed $\hat{x}[n]$ are practically zero for a range of n. This is dual to frequency aliasing. When the computed spectrum is practically zero in the neighborhood of $\pm\pi/T$, then frequency aliasing is negligible.
2. *Location*: If time aliasing is negligible, then (4.29) or (4.30) holds for some n_1. Now we discuss how to determine n_1 or the location of the time sequence. As discussed in

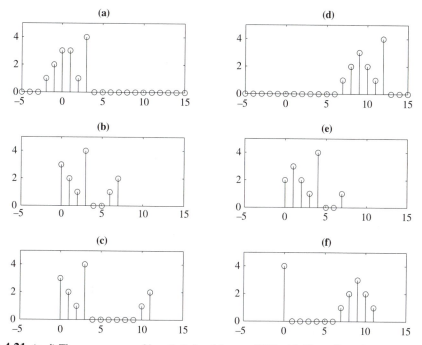

Figure 4.21 (a, d) Time sequences of length 6. (b, e) Inverse FFT with $N_1 = 8$. (c, f) Inverse FFT with $N_2 = 12$. All horizontal coordinates are time index.

Section 4.2, the DFT of $\hat{x}[n]$ can be considered to be defined for the periodic extension of $\hat{x}[n]$ with period N and any one period will yield the same DFT. Therefore, it is not possible to determine the location of the time sequence from one computation. To illustrate this point, consider the time sequence of length 6 shown in Fig. 4.21(a). If we select $N_1 = 8$ samples of its frequency spectrum and compute its inverse FFT, then the result will be as shown in Fig. 4.21(b). Clearly it is not possible to determine from the plot the location of the original sequence. We repeat the computation by selecting $N_2 = 12$ and plot the result in Fig. 4.21(c). Now we extend the plot in Fig. 4.21(b) periodically with period $N_1 = 8$ and the plot in Fig. 4.21(c) with period $N_2 = 12$. The two extensions are identical in the range $[-2, 5]$, $[-3, 4]$, and $[-4, 3]$ (they all have length $N_1 = 8$). Any one can be considered as the original time sequence. The same discussion applies to the sequences in Figs. 4.21(d)–(f). In conclusion, in order to determine the location of the original time sequence, we must carry out at least two computations with two different N.

Let N be the least common multiple of N_1 and N_2. For example, if $N_1 = 8$ and $N_2 = 12$, then we have $N = 24$. It is clear that the periodic extension of Fig. 4.21(b) with period $N_1 = 8$ and the periodic extension of Fig. 4.21(c) with period $N_2 = 12$ coincide not only in $[-2, 5]$ but also $[22, 29]$ and so forth. Therefore, in order to determine the original sequence uniquely, we must compute infinitely many different N_i. In practice, we will compute only a small number of different N_i. Furthermore, these N_i will be selected to have a large least common multiple.

Now we use examples to illustrate the procedure.

◆ Example 4.10

Given the frequency spectrum

$$X_d(\omega) = \frac{\sin 2.25\omega}{\sin 0.25\omega} \tag{4.56}$$

in the frequency range $(-6.28, \ 6.28]$, find its time sequence. The frequency range implies $T = 0.5$. Because $X_d(\omega)$ in (4, 56) is periodic with period 4π, we can use $X_d(\omega)$ directly for ω in $[0, 12.56]$ without carrying out the shifting of $X_d(\omega)$ as in Program 4.10. Let us first select $N_1 = 8$. Then we have $D = 2\pi/NT = \pi/2$ and the eight samples of $X_d(\omega)$ are $\sin(2.25mD)/\sin(0.25mD)$ for $m = 0:7$. Its value at $m = 0$ cannot be computed on a computer but can be computed, using l'Hôpital's rule, as 9. The program that follows

Program 4.11
```
T=0.5;N=8;D=2*pi/(N*T);
mb=1:N-1;
Xb=sin(2.75*mb*D)./sin(0.25*mb*D);
X=[9 Xb];
x=ifft(X);
stem(T*(0:N-1),x)
```

yields the plot in Fig. 4.22(a). Because $T = 0.5$, the $N = 8$ data are located at nT with $n = 0:7$. There is no range of n in which the computed x are zero; thus there is probably time aliasing due to frequency sampling. We repeat the computation by selecting $N_1 = 15$ and $N_2 = 22$, and the results are plotted in Figs. 4.22(b) and (c). There is a range of n in Fig. 4.22(b) in which the computed values are identically zero. Thus if N is 15 or larger, there is no time aliasing. This is confirmed in Fig. 4.22(c).

For $N_1 = 15$ and $N_2 = 22$, the computed sequences have exactly 9 nonzero entries; thus we may conclude that the inverse DTFT of the given spectrum has length 9. Their periodic extensions coincide in $n = -4:4$; thus we have

$$x(nT) = 1, \quad \text{for } n = -4, -3, \ldots, 3, 4$$

and zero for all other n. This is the DT rectangular window shown in Fig. 3.12(a). Thus, we obtain by using inverse FFT the correct time sequence or the inverse DT Fourier transform of (4.56).

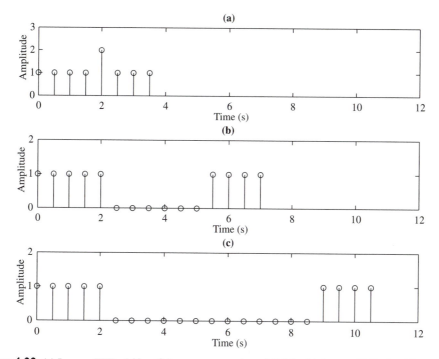

Figure 4.22 (a) Inverse FFT of $N = 8$ frequency samples of (4.56). (b) Inverse FFT with $N = 15$. (c) Inverse FFT with $N = 22$.

If the inverse DTFT of a spectrum is of infinite length, time aliasing due to frequency sampling always occurs. However, if the spectrum is bounded and continuous, then its inverse DTFT will be absolutely summable and, consequently, approaches zero as $|n| \rightarrow \infty$. In this case, if the number of frequency samples is selected to be sufficiently large, time aliasing will be negligible and the time sequence can still be computed using inverse FFT as the next example illustrates.

◆ **Example 4.11**

Consider the frequency spectrum

$$X(\omega) = \frac{1}{1 - 1.2e^{-j\omega}} \tag{4.57}$$

for ω in $(-\pi, \pi]$. Clearly, we have $T = 1$. Arbitrarily we select $N = 32$. Then Program 4.10 yields the plot in Fig. 4.23(a) with solid dots. The computed values for n small (roughly from $n = 0$ to 7) are close to zero; therefore, there is probably no time aliasing. We repeat the computation by selecting $N = 56$ and plot the

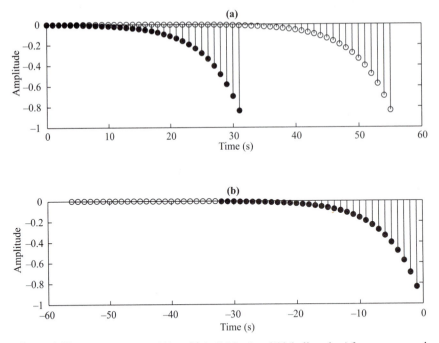

Figure 4.23 (a) Inverse FFT of $N = 32$ (solid dots) and 56 (hollow dots) frequency samples of (4.57). (b) Their periodic extensions.

result in Fig. 4.23(a) with hollow dots.[4] Its last 32 points are the same as the 32 solid dots, and the first 24 points are practically zero. Thus we may conclude that $N = 56$ is sufficiently large, and there is no appreciable time aliasing due to frequency sampling.

To find the location of the sequence, we extend periodically the solid dots with period $N = 32$ and the hollow dots with period $N = 56$. We see that they overlap for n in $[-32, -1]$. Thus the sequence is a negative-time sequence as shown in Fig. 4.22(b) with hollow dots. As we will discuss in Example 5.4 of the next chapter, the inverse DTFT of (4.57) is

$$x[n] = \begin{cases} 0 & \text{for } n \geq 0 \\ -(0.833)^{-n} & \text{for } n < 0 \end{cases} \tag{4.58}$$

Thus the inverse FFT yields a correct result.

---

[4] If we select $N = 64$, then the least common multiple of 32 and 64 is 64. Thus the periodic extensions of the two sequences will overlap every 64 samples. The least common multiple of 32 and 56 is 224. Thus their periodic extension will overlap every 224 samples, and it is more likely that we will obtain the correct location of the sequence.

We discuss an important situation to conclude this section. A signal is positive-time if its value is identically zero for all $n < 0$. Clearly, the DT signal in Fig. 4.21(d) is positive time and has length 13 counting from $n = 0$. If the inverse DTFT of a spectrum is known to be positive time and if the number of frequency samples N is larger than the length of the time sequence counting from $n = 0$, then the sequence generated by inverse FFT equals the original sequence. In this case, there is no need to extend the computed sequence periodically to find the location of the original sequence.

The way to determine whether or not the inverse DTFT of $X_d(\omega)$ is positive-time is simple. We carry out Program 4.10 using N_1 and $N_2 > N_1$. Let $\hat{x}_1(nT)$ denote the first output and $\hat{x}_2(nT)$ the second output. If

$$
\hat{x}_2(nT) \approx \begin{cases} \hat{x}_1(nT) & \text{for } 0 \leq n \leq N_1 - 1 \\ 0 & \text{for } N_1 \leq n \leq N_2 - 1 \end{cases}
$$

then the inverse DTFT of $X_d(\omega)$ is positive-time and roughly equals $\hat{x}_2(nT)$.

4.7.1 Computing CT Signals from Spectra

This section uses inverse FFT to compute the inverse CT Fourier transform, or to compute $x(t)$ from its frequency spectrum $X(\omega)$. Consider $X(\omega)$. We first select a T such that

$$
X(\omega) \approx 0 \qquad \text{for } |\omega| \geq \pi/T \tag{4.59}
$$

We then consider $X_d(\omega) := X(\omega)$, for ω in $(-\pi/T, \pi/T]$, to be the frequency spectrum of a discrete-time sequence with sampling period T. Let $\bar{x}(nT)$ be the inverse DT Fourier transform of $X_d(\omega)$. Because the spectra of $x(t)$ and its sampled sequence $x(nT)$ are related by $X(\omega) = TX_d(\omega)$, we must divide $\bar{x}(nT)$ by T to yield

$$
x(t) \approx \frac{\bar{x}(nT)}{T} \tag{4.60}
$$

where we have assumed implicitly that $\bar{x}(nT)$ has been interpolated to yield a continuous-time signal. The approximation becomes an equality if $X(\omega)$ is bandlimited to W, T is smaller than π/W, and $x(t)$ is obtained from $\bar{x}(nT)$ by using the ideal interpolation formula in (3.69).

As discussed in the preceding section, the DT sequence $\bar{x}(nT)$ can be computed from the frequency samples of $X_d(\omega)$ by using inverse FFT. Let $\hat{x}(nT)$ be the inverse DFT of the N frequency samples of $X_d(\omega)$. Then we have

$$
\hat{x}(nT) \approx \begin{cases} \bar{x}(nT) & \text{for } n_1 \leq n \leq n_1 + N - 1 \\ 0 & \text{for } n < n_1 \text{ and } n > n_1 + N - 1 \end{cases} \tag{4.61}
$$

for some n_1. The approximation becomes an equality if $x(t)$ or $\bar{x}(nT)$ is time limited and if N is sufficiently large. Combining (4.60) and (4.61) yields

$$x(t) \approx \begin{cases} \hat{x}(nT)/T & \text{for } n_1 T \leq t \leq (n_1 + N - 1)T \\ 0 & \text{for } t < n_1 T \text{ and } t > (n_1 + N - 1)T \end{cases} \tag{4.62}$$

where we have again assumed implicitly that $\hat{x}$ is interpolated to yield a continuous-time function. The approximation, however, can never become an equality because $x(t)$ cannot be both bandlimited and time limited, as discussed in Section 3.9. Even so, it is possible to obtain an FFT computed signal as close as desired to the actual CT signal. The program that follows

Program 4.12
```
T= ;N= ;D=2*pi/(N*T);
mn=ceil(-(N-1)/2):-1;mp=0:ceil((N-1)/2);
X=[Xd(mp*D) Xd(mn*D)];
x=ifft(X)/T;
plot(T*(0:N-1),x)
```

will yield an approximate $x(t)$ except its location. As T decreases and N increases, the computed result will become closer to the actual $x(t)$ except its location. This program differs from Program 4.10 only in two parts. First, we divide ifft(X) by T as is required by (4.62). Second, we use plot instead of stem, because $x(t)$ is a CT signal and plot will automatically interpolate two neighboring points by a straight line. We give two examples.

◆ **Example 4.12**

Consider the spectrum of the analog ideal low-pass filter shown in Fig. 3.3(a) with $\omega_c = 2$. The spectrum is 1 for $|\omega| \leq \omega_c = 2$ and 0 for $|\omega| > 2$. We will compute its CT signal or, equivalently, its inverse CT Fourier transform. The spectrum is bandlimited to ± 2 rad/s. If we select $T < \pi/2 = 1.57$, then frequency aliasing will not occur. However, because $x(t)$ will be interpolated from $\hat{x}(nT)$ by straight lines, instead of using the ideal interpolation formula in (3.69), errors will occur. Therefore, we use a smaller T. Arbitrarily, we select $T = 0.1$. Then the Nyquist frequency interval becomes $(-\pi/T, \pi/T] = (-31.4, 31.4]$. Let us select $N = 900$ equally spaced frequency samples in the interval. Then the frequency resolution is $D = 62.8/N = 0.0698$. Let $M = \text{floor}(2/D) = \text{floor}(28.65) = 28$, where the function floor rounds a number downward toward an integer. Thus the frequency sample of the spectrum in Fig. 3.3(a) is

$$X(mD) = 1, \quad \text{for } m = 0, \pm 1, \cdots, \pm M$$

and 0 otherwise. The next program

Program 4.13
```
T=0.1;N=900;D=2*pi/(N*T);M=floor(2/D);
```

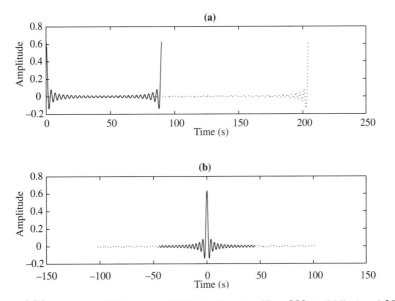

Figure 4.24 (a) Inverse FFT computed CT signals using $N = 900$ (solid line) and 2048 (dotted line) frequency samples of the spectrum shown in Fig. 3.3(a). (b) Their periodic extensions.

```
X=[ones(1,M+1) zeros(1,N-2*M-1) ones(1,M)];
x=ifft(X)/T;
n=0:N-1;
plot(n*T,x)
```

yields the CT signal shown in Fig. 4.24(a) with a solid line. We then repeat the program for $N = 2048$, and the result is plotted in Fig. 4.24(a) with a dotted line. We extend them periodically as shown in Fig. 4.24(b). We see that the dotted line equals the solid line for $|t| < 45$ and practically zero for $45 < |t| < 102.4$. Thus we may stop the computation and conclude that the inverse CT Fourier transform of the spectrum is as shown in Fig. 4.24(b) with a dotted line. The plot is very close to the one in Fig. 3.3(b), which was obtained analytically.

◆ **Example 4.13**

Consider a CT signal with spectrum $X(\omega) = 1/(j\omega + 0.1)$, as plotted in Fig. 3.1(b). The peak value of $X(\omega)$ is 10 as shown. If $T = 0.2$, then

$$|X(\omega)| = \left| \frac{1}{j\omega + 0.1} \right| \le 0.06 \quad \text{for } |\omega| \ge \frac{\pi}{T} = \frac{\pi}{0.2} = 15.7$$

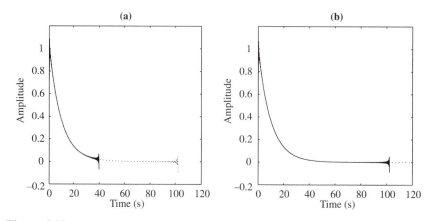

Figure 4.25 (a) Inverse FFT computed CT signals using $T = 0.2, N = 200$ (solid line) and 512 (dotted line) frequency samples of $1/(j\omega + 0.1)$. (b) Exact (dotted line) and computed (solid line) signals.

which is less than 1% of the peak value of $X(\omega)$. Thus selecting $T = 0.2$ will introduce negligible frequency aliasing. Program 4.12 with $N = 200$ will generate the plot in Fig. 4.25(a) with a solid line. We then repeat the computation with $N = 512$, and the result is plotted in Fig. 4.25(a) with a dotted line. We see that the dotted line equals the solid line for t in $[0, 200T]=[0, 40]$ and is practically zero for $t > 40$. Thus we conclude that the CT signal is positive-time and can be represented by the dotted line and zero thereafter. Note that there are ripples at the ends of both time intervals. They are probably due to truncation of the spectrum outside $\pm\pi/T$ and time aliasing due to frequency sampling of the spectrum. Figure 4.25(b) shows the computed time signal (solid line) and the exact signal (dotted line). They are indistinguishable if we disregard the ripples.

These examples show that CT signals can indeed be computed from their frequency spectra using inverse FFT.

4.8 Computing Energy Using FFT[5]

This section discusses FFT computation of energy for CT signals. Consider a CT signal $x(t)$. Its total energy is given by

$$E = \int_{t=-\infty}^{\infty} |x(t)|^2 dt = \frac{1}{2\pi} \int_{\omega=-\infty}^{\infty} |X(\omega)|^2 d\omega \qquad (4.63)$$

---

[5] This section may be skipped without loss of continuity

Thus the total energy can be computed directly in the time domain or indirectly from the frequency domain by computing first the frequency spectrum. If we compute the spectrum, then we can compute not only the total energy but also the energy in any nonzero frequency interval.

To compute the total energy directly in the time domain, we can approximate (4.63) as

$$E \approx \sum_{n=n_1}^{n_2} |x(nT)|^2 T \tag{4.64}$$

for some n_1 and n_2. In this equation, $x(t)$ is approximated by a stepwise signal as in Fig. 2.21 and is assumed to equal roughly zero for $t < n_1$ and $t > n_2$.

As discussed in Section 4.6, the spectrum $X(\omega)$ can be computed from the samples of $x(t)$ using FFT by selecting appropriate T and N. Let $X_d[m]$ be the FFT computed spectrum of $x(nT)$. Then we have

$$X(mD) \approx T X_d[m] \tag{4.65}$$

for m in (4.3) and with $D = 2\pi/NT$. Using (4.63) and (4.65), we can approximate the total energy as

$$E \approx \frac{1}{2\pi} \sum_{m=-(N-1)/2}^{\overline{(N-1)/2}} |T X_d[m]|^2 D$$

$$= \frac{T}{N} \sum_{m=0}^{N-1} |X_d[m]|^2 \tag{4.66}$$

This is the total energy of the CT signal $x(t)$. Once $X_d[m]$ is available, we can also compute the energy over any frequency interval by selecting an appropriate range of m, as the next example illustrates. Before proceeding, we mention that because energy depends only on magnitude spectra, which will not be affected by any time shifting, in using (4.66), there is no need to rearrange $x(t)$ before applying FFT. Furthermore, it is more convenient to use the spectrum computed in $[0, 2\pi/T)$ than in $(-\pi/T, \pi/T]$. Thus we will not rearrange the output of FFT.

◆ Example 4.14

Consider the reactangular window $w_2(t)$ shown in Fig. 3.2(a) with $a = 2$. Find the percentage of energy in the main lobe.

The total energy of the window can be computed directly in the time domain as

$$E = \int_{t=-\infty}^{\infty} |w_a(t)|^2 dt = \int_{t=-2}^{2} 1^2 dt = 1(2 - (-2)) = 4$$

To find distribution of energy in frequencies, we must compute its spectrum. Although its spectrum was computed analytically in (3.9) as $W_a(\omega) = 2\sin 2\omega/\omega$, we will use the procedure in Section 4.6 to compute it numerically. First we select the sampling period as $T = 0.01$. Then we have $M = 2/T = 200$ and

$$w_2(nT) = 1 \quad \text{for } -M/T \leq n \leq M/T$$

It is a finite sequence of length 201. We use FFT to compute its DTFT. In order to have a better frequency resolution, we will use 16384-point FFT. Because energy depends only on the magnitude spectrum, we will not rearrange the input of FFT. The program that follows

Program 4.14
```
T=0.01;N=16384;M=floor(2/T);D=2*pi/(N*T);
x=[ones(1,2*M+1)];
Xd=fft(x,N)
m=0:N-1;
E=T/N*sum(abs(Xd(m+1)).^2)
```

computes the total energy. The result is 4.01, which differs from the exact one by $(4.01 - 4)/4 = 0.0025$ or 0.25%.

The main lobe of the spectrum of $w_2(t)$ ranges, as discussed in Example 3.2, from $\omega = -\pi/2$ to $\pi/2$. We will compute the energy in $[0, \pi/2]$ and then double it. First we compute $m_1 = \text{floor}(\pi/2D)$; it is the largest integer m inside the range $[0, \pi/2]$. If we append the following

```
m1=0:floor(pi/(2*D));
E1=2*T/N*sum(abs(Xd(m1+1)).^2)
E1/E*100
```

at the end of Program 4.14, then the program will yield $E = 4.01$, $E_1 = 3.7185$, and 92.73%. The energy in the main lobe is 3.7185; it is $3.7185/4.01 = 0.9273$ or 92.73% of the total energy.

4.9 Concluding Remarks

In signal analysis, we often consider signals to be defined for positive- and negative-time. For example, the spectrum of $\sin \omega_0 t$ for all t in $(-\infty, \infty)$ is zero everywhere except the two impulses at $\pm\omega_0$. The spectrum of $\sin \omega_0 t$, for $t \geq 0$ however is nonzero for all ω. Thus including negative time will simplify the discussion of frequency spectra of periodic signals.

However, in using FFT to compute the spectrum of a two-sided signal, we must shift its negative-time part to the end of its positive-time part. In using inverse FFT to compute a time signal from a spectrum, we must compute two or more time signals and then extend them periodically to determine the location of the signal. Thus negative-time is most inconvenient in using FFT.

When we study a CT signal, the time instant the signal starts to appear can be considered to be $t = 0$. Thus it is fair to say that all signals encountered in practice are positive-time signals. For any positive-time signal $x(t)$, the program that follows

Program 4.15

```
T= ;N= (even) ;D=2*pi/(N*T);
n=0:N-1;
x=x(n*T);
X=T*fft(x);
m=-N/2+1:N/2;
w=m*D;
plot(w,abs(shift(X)));
```

computes and plots its magnitude spectrum in $(-\pi/T, \pi/T]$. The way to select T and N was discussed in Section 4.6. As discussed in Chapter 2, if we replace shift by fftshift and replace m=-N/2+1:N/2 by m=-N/2:N/2-1; then the preceding program will plot the magnitude spectrum in $[-\pi/T, \pi/T)$. The resulting program is Program 1.1 in Section 1.6.1.

Conversely, consider a frequency spectrum $X(\omega)$ of a CT signal. First select a T to meet (4.59) and then an N. The program that follows

Program 4.16

```
T= ;N= (even);D=2*pi/(NT);
mn=-N/2+1:-1;mp=0:N/2;
X=[X(mp*D) X(mn*D)];
x=ifft(X)/T;
x=0:N-1;
plot(n*T,x)
```

will generate a CT signal. If the generated $\bar{x}(t)$ is practically zero in the neighborhood of $(N - 1)T$, then $x(t)$ is positive time and roughly equals $\bar{x}(t)$, for $0 \leq t \leq (N - 1)T$ and zero elsewhere. Thus for positive-time signals, the use of FFT is fairly simple.

As a final remark, we discuss some limitations of Fourier analysis of signals. A signal can be expressed in the time domain as $f(t)$ or in the ferquency domain as $F(\omega)$. Because either one can be obtained from the other, the two descriptions contain the same amount of information. If we can combine the two descriptions, we may gain more insight about the signal. This however is difficult because the Fourier transform of a signal is developed from the entire time duration of the signal by using the same weight due to $|e^{-j\omega t}| = 1$ for all t. Thus the local property of a time signal becomes a global frequency property of its Fourier transform and vice versa. For example, if a signal is discontinuous at a time instant, no matter where it occurs, its frequency spectrum will be nonzero for all frequencies. Because time shifting does not affect the magnitude spectrum, the time location must be contained in the phase spectrum. However, phases may be sensitive to computational errors as shown in Example 2.12 or severely aliased due to time sampling in computer computation as discussed in (3.75). Thus it is difficult to obtain a time location or, more generally, the correlation between the time and frequency descriptions in Fourier analyses.

One way to remedy this problem is to divide a time signal into sections and then compute short-time or windowed Fourier transforms. Clearly, the narrower each time section, the more accurate its time location. However the computed transforms will be less accurate. Recall from Section 3.5 that a narrow time window has a wider main lobe in its transform and the resulting spectrum will convolve over a wider range and becomes less accurate. A better method is to use wavelets and wavelet transforms. Using wavelet transforms, the time and frequency correlation of a signal can be better established. See Refs. 2, 13.

PROBLEMS

4.1   Compute the DFT of the sequence $x[0] = 2, x[1] = 1$, and $x[2] = 1$ with $T = 0.5$. Compare your result with the DT Fourier series computed in Problem 2.12.

4.2   Compute the DFT of the sequence $x[-1] = -1, x[0] = 2, x[1] = 1$, and $x[2] = 1$ with $T = 1$. Compare your result with the DT Fourier series computed in Problem 2.13.

4.3   Find the DFT of the following sequences: (a) $x[-1] = 1, \ x[0] = 0, \ x[1] = -1$; (b) $x[0] = 1, \ x[1] = -2, \ x[2] = 1$; (c) $x[0] = -1, \ x[1] = 2, \ x[2] = 2, \ x[3] = -1$. If the sample period is $T = 0.5$, where are their DFT located? Verify that they are the frequency samples of the frequency spectrum computed in Problem 3.13.

4.4   Compute four frequency samples of the spectra of the sequences of length 3 in Problem 4.3(a) and (b) by padding one zero.

4.5   Compute six frequency samples of the frequency spectrum of the sequence in Problem 4.3(c) by padding two zeros.

4.6   Consider the two sequences in Problems 4.3(b) and (c). Can you use only one four-point DFT to compute four frequency samples of the frequency spectra of the two sequences.

4.7   Use two-point DFT to compute the DFT of the sequence of length 4 in Problem 4.3(c).

4.8   Use 1024-point FFT to compute and plot the frequency spectra of the three finite sequences in Problem 4.3. Are the results close to the ones obtained in Problem 3.13?

4.9   Show that the spectrum and three-point DFT of the sequence in Example 4.1 can be related by

$$X_d(\omega) = \frac{1}{3} \sum_{m=0}^{2} X_d[m] \frac{1 - e^{-j3(0.5\omega - 2\pi m/3)}}{1 - e^{-j(0.5\omega - 2\pi m/3)}}$$

This equation can be used to interpolate the spectrum from its sample $X[m]$. Can it be used to interpolate the spectrum of the sequence in Example 4.2? If not, develop an interpolation formula for the sequence.

4.10 Use the interpolation formula in Problem 4.9 and the method of padding zeros to compute 16 frequency samples of the spectrum of the sequence in Example 4.1. Which method is simpler?

4.11 Consider the infinite sequence $x[n] = 0.9^n$ for $n \geq 0$ and $T = 1$. Find the smallest a in $N = 2^a$ such that the spectrum of the first N data points differs from the spectrum of the infinite sequence in less than 1% of its peak magnitude. Compare magnitudes only at $2m\pi/N$ with $m = 0{:}N/2$.

4.12 Consider the infinite sequence $x[n] = 0.9^n$ for $n \geq 0$ and $T = 1$. Find the smallest a in $N = 2^a$ such that the spectrum of the first N data points differs from the spectrum of the first $N/2$ data points in less than 1% of its peak magnitude. Compare magnitudes only at $2m\pi/N$ with $m = 0{:}N/2$. Compare this a with the one in Problem 4.11.

4.13 Repeat Problem 4.12 for $x[n] = (-0.9)^n$ for $n \geq 0$.

4.14 Use FFT to generate the two plots in Fig. 3.15 for the truncated sequences of $h_d[n] = \sin n/\pi n$ for $|n| \leq 5$ and $|n| \leq 50$.

4.15 Consider the CT signal $x(t) = 2e^{-t}$ for $t \geq 0$. What is its spectrum in closed form? Use FFT to compute its spectrum following these steps: (a) Find L so that the time record length $[0, \; L)$ excludes only the part of the signal whose magnitude is less than 1% of its peak value. Use this L to find a sampling period T and the smallest integer a in $N = 2^a$ so that the effect of frequency aliasing is negligible. By this, we mean that the magnitude spectrum computed using T and $N = 2^a$ differs from the one using $2T$ and $N/2$ by less than 1% of its peak magnitude. (b) Use the T and N obtained in (a) to find the smallest integer c in $\bar{N} = 2^c N$ so that the effect of truncation is negligible. By this, we mean that the magnitude spectrum computed using T and $\bar{N}$ differs from the one using T and $\bar{N}/2$ by less than 1% of its peak magnitude. (c) Compare the magnitude spectrum obtained in (b) with the exact one. Compare also the phase spectrum with the exact one.

4.16 Repeat Problem 4.15 for the signal in Example 4.7.

4.17 Use FFT to compute the spectrum of

$$x(t) = e^{-0.01t} \sin 2t - 3e^{-0.02t} \cos 5t$$

4.18   Use FFT to compute the spectrum of

$$x(t) = e^{-0.01t} \sin 3t - 3e^{-0.02t} \cos 3.05t$$

Is the final time record length the same as the one in Problem 4.17?

4.19   Use FFT to compute the frequency spectrum of the signal in Problem 3.19. Which method is simpler: FFT or the method in Problem 3.19?

4.20   Consider the frequency spectrum

$$X_d(\omega) = e^{-j0.5\omega} \, (2\cos 0.5\omega - 2) \quad \text{for } |\omega| \le 2\pi$$

Compute the inverse DFT of its three frequency samples at $m(4\pi/3)$, $m = -1, 0, 1$. Compute the inverse DFT of its four frequency samples at $m\pi$, $m = 0, 1, 2, 3$. What is the inverse DTFT of $X_d(\omega)$?

4.21   Repeat Problem 4.20 for

$$X_d(\omega) = 2j \sin 0.5\omega \quad \text{for } |\omega| \le 2\pi$$

4.22   Consider a discrete frequency spectrum

$$X_d(\omega) = \frac{e^{j2\omega}}{e^{j2\omega} + 0.8e^{j\omega} - 0.09}$$

for $|\omega| \le \pi$. Use inverse FFT to compute and plot its time sequence. Is it a positive-time, negative-time, or two-sided sequence?

4.23   Consider

$$X(\omega) = \frac{2}{(j\omega + 1)^2 + 4}$$

for $-\infty < \omega < \infty$. Use inverse FFT to compute its time function.

4.24   Compute the energy in the first pair of the side lobes in Example 4.14.

PART 2

DIGITAL FILTER DESIGN

CHAPTER **5**

Linear
Time-Invariant
Lumped Systems

5.1 Introduction

In this chapter we study systems that will be used to process signals. There are many types of systems. For example, every block in Fig. 1.11 is a system. The first block from left is a transducer that changes a signal from one form to another such as from pressure to voltage. The second block may consist of amplifiers and limiters. The third block is an analog filter that attenuates high-frequency components and so forth. The last four blocks in Fig. 1.11 can also be considered as a single system as represented by the A/D converter in Fig. 1.17. Thus a system can be a single device or a collection of devices; it can be hardware or a computer program. In this text, every system will be modeled as a black box with at least one input terminal and one output terminal, as shown in Fig. 5.1. Note that a terminal does not necessarily mean a physical terminal such as a wire sticking out of the box. It merely indicates that a signal may be applied or measured from the terminal. We assume that if an excitation or input is applied to the input terminal, a *unique* response or output will be measurable or observable at the output terminal. This unique relationship between the excitation and response, input and output, or cause and effect is essential in defining a system.

A system is defined as a *continuous-time* (CT) or *analog* system if it accepts continuous-time signals as its input and generates CT signals as its output. We use $x(t)$ and $y(t)$ to denote, respectively, its input and output. Likewise, a system is defined as a *discrete-time* (DT) or *digital* system if it accepts DT signals as its input and generates DT signals as its output. We use $x[n]$ and $y[n]$ to denote its input and output sequences. This definition is independent of the sampling period and is applicable for any $T > 0$ with some minor restriction, as we will discuss later.

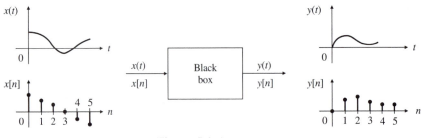

Figure 5.1 System.

Thus all DT signals in this and later chapters will be plotted with respect to a time index. We study first DT systems and then CT systems because the mathematics involved is simpler for the former.

DT systems are classified as follows: linear or nonlinear, time invariant or time varying, causal or noncausal, and lumped or distributed. This chapter studies only linear time-invariant (LTI) causal and lumped systems. We develop first the convolution description for LTI systems and difference equations for LTI systems that have the additional lumpedness property. We then introduce the z-transform and transfer functions. A basic concept in systems is stability. If a system is not stable, the output of the system may grow unbounded when an input is applied, and the system will overflow, saturate, disintegrate, or burn out. Thus systems designed to process signals must be stable. We then introduce the concept of frequency response for stable systems.

A DT system is said to be *initially relaxed* at $n = n_0$ if its output for all $n \geq n_0$ is identically zero when no input $x[n]$ with $n \geq n_0$ is applied. If a system is relaxed at $n = n_0$, then the output $y[n]$, for $n \geq n_0$, will be excited exclusively by the input applied on and after n_0. Under this assumption, the input and output relationship can be denoted by

$$\{x[n]\} \rightarrow \{y[n]\}$$

and we call $\{x[n], y[n]\}$ an input-output pair. In this text, unless stated otherwise, every system will be assumed to be initially relaxed.[1] If a system is initially relaxed at $n = n_0$, we can assume, without loss of generality, that $x[n] = 0$ and $y[n] = 0$ for all $n < n_0$.

5.2 Linearity and Time Invariance

A system is said to be linear if for any two input-output pairs $\{x_i[n], y_i[n]\}$, $i = 1, 2$, we have

$$\{x_1[n] + x_2[n]\} \rightarrow \{y_1[n] + y_2[n]\} \quad \text{(additivity)} \tag{5.1}$$

and, for any constant α,

---

[1] If the reader is familiar with the concept of state, this assumption is the same as assuming every initial state to be zero, or all initial conditions to be zero. Thus we study only zero-state responses in this text. See Ref. 6.

$$\{\alpha x_i[n]\} \rightarrow \{\alpha y_i[n]\} \quad \text{(homogeneity)} \tag{5.2}$$

It means that the output excited by the sum of any two inputs equals the sum of the two outputs excited by individual input. If we amplify an input sequence by α, the output sequence will be amplified by the same factor. The first property is called the *additivity property*; the second, the *homogeneity property*. Jointly, they are called the *principle of superposition*. If a system does not have the superposition property, it is called a *nonlinear* system.

A system is said to be *time invariant* or *shift invariant* if the characteristic of the system does not change with time. If a system is time invariant, no matter at what time an input sequence is applied, the output waveform will always be the same. Mathematically, this can be expressed as: For any $\{x[n] \rightarrow y[n]\}$ and any n_1, we have

$$\{x[n - n_1]\} \rightarrow \{y[n - n_1]\} \quad \text{(shifting)} \tag{5.3}$$

In other words, if the input sequence is shifted by n_1 samples, the generated output sequence is the original output sequence shifted by n_1 samples. If a system is not time invariant, it is said to be *time varying*. For time-invariant systems, we may assume, without loss of generality, that the initial time n_0 is 0 and the input sequence is applied from time 0 on. Note that $n_0 = 0$ is a relative one; it is the instant when we start to apply an input sequence.

This text studies only linear time-invariant systems. The topic of time-varying and/or nonlinear systems is very complex and is outside the scope of this text.

Before proceeding, we discuss the question of time interval. In signal analysis, signals are often defined for n in $(-\infty, \infty)$. The main reason is probably to simplify the discussion of frequency spectra of periodic signals. See Section 3.2.1. In system analysis, time interval is often limited to $[n_0, \infty)$ for some finite n_0, in particular, to $[0, \infty)$. There is no need to consider $(-\infty, \infty)$ because no input can be applied from $-\infty$.

5.2.1 LTI Systems—Convolutions

Using only the conditions of additivity, homogeneity, and shifting, we can develop a general description for linear time-invariant systems. First we define the *impulse sequence* as, for any given integer k,

$$\delta[n - k] = \begin{cases} 1 & \text{if } n = k \\ 0 & \text{if } n \neq k \end{cases} \tag{5.4}$$

It is zero everywhere except at $n = k$, where it equals 1. Thus $\delta[n] = \delta[n-0]$ is zero everywhere except at $n = 0$. We plot in Figs. 5.2(a)–(c) the three impulse sequences $2\delta[n]$, $3\delta[n - 1]$, and $-\delta[n - 2]$. Consider the sequence $x[n]$ in Fig. 5.2(d) with $x[0] = 2$, $x[1] = 3$, $x[2] = -1$, $x[3] = 0$, $x[4] = 0.5, \ldots$. From Fig. 5.2, we see that $x[n]$ can be expressed in terms of impulse sequences as

$$x[n] = 2\delta[n] + 3\delta[n - 1] - \delta[n - 2] + 0 \times \delta[n - 3] + 0.5\delta[n - 4] + \cdots$$

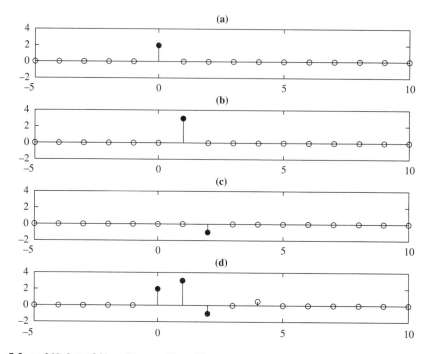

Figure 5.2 (a) $2\delta[n]$. (b) $3\delta[n-1]$. (c) $-\delta[n-2]$. (d) Decomposition of $x[n]$. All horizontal coordinates are time index.

In general, every time sequence $x[n]$ can be expressed as

$$x[n] = x[0]\delta[n] + x[1]\delta[n-1] + x[2]\delta[n-2] + \cdots$$

$$= \sum_{k=0}^{\infty} x[k]\delta[n-k] \tag{5.5}$$

This is an important equation.

Consider an LTI discrete-time system. Its output excited by the input $x[n] = \delta[n]$, an impulse sequence applied at $n = 0$, is called the (DT) *impulse response*. The impulse response is a very important concept because it describes the system completely in the sense that whatever information about the system we want to know, we can obtain it from the impulse response.

Let us use the impulse response to develop an equation to describe LTI discrete-time systems. Let $h[n]$ be the impulse response of a system; that is, the input $x[n] = \delta[n]$ excites the output $y[n] = h[n]$. If the system is linear and time invariant, then we have

$$\delta[n] \rightarrow h[n] \quad \text{(impulse response)}$$

$$\delta[n-k] \rightarrow h[n-k] \quad \text{(shifting)}$$

$$x[k]\delta[n-k] \rightarrow x[k]h[n-k] \quad \text{(homogeneity)}$$

$$\sum_{k=0}^{\infty} x[k]\delta[n-k] \rightarrow \sum_{k=0}^{\infty} x[k]h[n-k] \quad \text{(additivity)}$$

The left-hand-side term of the last row is the input sequence in (5.5). Thus we conclude that the output $y[n]$ excited by the input $x[n]$ is given by

$$y[n] = \sum_{k=0}^{\infty} x[k]h[n-k] \tag{5.6}$$

This is called the *discrete-time convolution*. Using this equation, we can compute the output excited by any input; there is no need to measure it from the actual system. More generally, using the impulse response, we can predict all properties of the system. Thus designing a system is the same as designing an impulse response.

Before discussing an alternative form of (5.6), we give two examples.

◆ **Example 5.1**

Consider a savings account in a bank. The account is time varying if the interest rate changes with time; nonlinear, if the interest rate depends on the amount of money in the account. If the interest rate is fixed and is the same no matter how much money is in the account, then it is a linear time-invariant system. We consider here only the LTI case with interest rate $r = 0.015\%$ per day and compounded daily. Let $x[n]$ denote the amount of money deposited into the account in the nth day and $y[n]$ be the total amount of money in the account at the end of the nth day. If we withdraw money, then $x[n]$ is negative.

If we deposit one dollar in the first day (that is $x[0] = 1$) and nothing thereafter ($x[n] = 0$ for $n = 1, 2, \ldots$), then $y[0] = x[0] = 1$ and $y[1] = 1 + 0.00015 = 1.00015$. Because the money is compounded daily, we have

$$y[2] = y[1] + y[1] \cdot 0.00015 = y[1] \times 1.00015 = (1.00015)^2$$

and, in general,

$$y[n] = (1.00015)^n$$

for $n \geq 0$. Because the input $\{1, 0, 0, \ldots\}$ is an impulse sequence, its output, by definition, is the impulse response or

$$h[n] = (1.00015)^n \tag{5.7}$$

for $n = 0, 1, 2, \ldots$. Thus we have

$$y[n] = \sum_{k=0}^{\infty} x[k]h[n-k] = \sum_{k=0}^{\infty} x[k](1.00015)^{n-k} \tag{5.8}$$

For example, if $x[0] = 100$, $x[1] = 50$, $x[3] = -30$, and $x[n] = 0$ for $n \geq 4$, then at the end of the eleventh day, we have

$$y[10] = x[0] \times (1.00015)^{10} + x[1] \times (1.00015)^{10-1} + x[3] \times (1.00015)^{10-3}$$

$$= 100 \times 1.0015 + 50 \times 1.00135 - 30 \times 1.00105 = 120.19$$

Using (5.8), we can compute the amount of money in the account at any day.

◆ **Example 5.2**

Consider a DT system with impulse response

$$h[n] = \begin{cases} 1/3 & \text{for } n = 0, 1, 2 \\ 0 & \text{for } n < 0 \text{ and } n > 2 \end{cases} \tag{5.9}$$

Its output is given by

$$y[n] = \sum_{k=0}^{\infty} x[k]h[n-k]$$

We compute this equation explicitly for $n = 0, 1, 2, 3$:

$$y[0] = \sum_{k=0}^{\infty} x[k]h[0-k] = x[0]h[0] + x[1]h[-1] + x[2]h[-2] + \cdots$$

$$= x[0]h[0] = x[0]/3$$

$$y[1] = \sum_{k=0}^{\infty} x[k]h[1-k] = x[0]h[1] + x[1]h[0] + x[2]h[-1] + \cdots$$

$$= x[0]h[1] + x[1]h[0] = (x[0] + x[1])/3$$

$$y[2] = x[0]h[2] + x[1]h[1] + x[2]h[0] = (x[0] + x[1] + x[2])/3$$

$$y[3] = x[0]h[3] + x[1]h[2] + x[2]h[1] + x[3]h[0] = (x[1] + x[2] + x[3])/3$$

In general, we have

$$y[n] = (x[n] + x[n-1] + x[n-2])/3$$

for $n \geq 2$. We see that the current output is the average of the current and two previous inputs. Thus the system is called a *moving-average filter*.

The moving-average filter can be used to compute the trend of a fast-changing signal and is widely used in stocks markets. If we want to see the trend over 1 month, we may use a 5-day moving average. The trend over 10 years, however, can be better seen using a 30-day moving average.

Causality A system is said to be *causal* if its current output depends only on its current and past inputs. If the output depends on future inputs, the system is said to be *noncausal* or *anticipatory*. A noncausal system can predict what input will be applied in the future. No physical system has such a capability. Thus every physical system is causal, and *causality is a necessary condition for a system to be realizable in the real world*.

The impulse responses in Examples 5.1 and 5.2 have the property $h[n] = 0$ for $n < 0$ and their outputs depend only on current and past inputs. If $h[n] \neq 0$ for some $n < 0$, such as $h[-1] = h[0] = h[1] = 1/3$, then (5.6) becomes, at $n = 0$,

$$y[0] = \sum_{k=0}^{\infty} x[k]h[0-k] = x[0]h[0] + x[1]h[-1] + x[2]h[-2] + \cdots$$

$$= \frac{1}{3}(x[0] + x[1])$$

We see that $y[0]$ depends on the future input $x[1]$. Thus we conclude that a system is causal if and only if $h[n] = 0$ for $n < 0$, denoted as

$$\text{causal} \iff h[n] = 0 \quad \text{for } n < 0 \tag{5.10}$$

This fact can also be deduced from the definition of impulse responses. The impulse response is the output excited by the input $\delta[n]$ applied at $n = 0$. If the system is initially relaxed and causal, no output will appear before an input is applied. Thus we have $h[n] = 0$ for $n < 0$.

If a system is causal, then $h[n-k] = 0$ for $n - k < 0$ or $n < k$. In this case, (5.6) can be reduced to

$$y[n] = \sum_{k=0}^{n} x[k]h[n-k]$$

which can also be written as, by defining $\bar{k} := n - k$,

$$y[n] = \sum_{\bar{k}=n}^{0} x[n-\bar{k}]h[\bar{k}] = \sum_{\bar{k}=0}^{n} x[n-\bar{k}]h[\bar{k}]$$

After dropping the overbar, we obtain

$$y[n] = \sum_{k=0}^{n} x[k]h[n-k] = \sum_{k=0}^{n} h[k]x[n-k] \tag{5.11}$$

Because of (5.11), the convolution is said to have the commutative property. If an LTI system is not causal, we must use (5.6). If an LTI system is causal, we can use either (5.6) or (5.11).

FIR and IIR Filters A digital system is called a finite-impulse-response (FIR) system or filter if its impulse response has a finite number of nonzero entries. It is an infinite-impulse-response (IIR) filter if its impulse response has infinitely many nonzero entries. The moving-average filter in Example 5.2 is clearly an FIR system, and the savings account in Example 5.1 is an IIR system.

A causal FIR filter is said to have length N if $h[N-1] \neq 0$ and $h[n] = 0$ for all $n > N - 1$. Such a filter has impulse response $h[n]$ for $n = 0, 1, \dots, N-1$, and has at most N nonzero entries in its impulse response. If $N = 1$, or $h[0] = a \neq 0$ and $h[n] = 0$ for $n \neq 0$, then (5.11) reduces to

$$y[n] = ax[n] \tag{5.12}$$

where a is a constant. Such a system is said to be *memoryless*. The memoryless system is more often called a *multiplier* with gain a. Other than multipliers (memoryless systems), the current output of every causal system must be computed from a convolution using current and past inputs.

5.3 LTIL Systems—Difference Equations

In computing $y[n]$ from (5.11), the number of operations (multiplications and additions) increases as n increases. This is not desirable. In this section, we develop a different description. Consider the savings account studied in Example 5.1. Let $x[n]$ be the amount of money deposited in the nth day and $y[n]$ be the total amount of money in the account at the end of the nth day. If the interest rate is $r = 0.00015$ per day and is compounded daily, then we have

$$y[n] = y[n-1] + y[n-1]\,0.00015 + x[n]$$
$$= 1.00015y[n-1] + x[n] \tag{5.13}$$

or

$$y[n] - 1.00015y[n-1] = x[n] \tag{5.14}$$

This is called a first-order linear difference equation with constant coefficients. Thus the savings account can be described by the difference equation in (5.14) or the convolution

$$y[n] = \sum_{k=0}^{n} (1.00015)^{n-k} x[k] \qquad (5.15)$$

First we show that the impulse response can be obtained from the difference equation. If the system is initially relaxed or, equivalently, $x[n] = 0$ and $y[n] = 0$ for $n < 0$, and if $x[n] = \delta[n]$, then (5.13) implies

$$n = 0: \quad y[0] = 1.00015 y[-1] + x[0] = 1.00015 \cdot 0 + 1 = 1$$

$$n = 1: \quad y[1] = 1.00015 y[0] + x[1] = 1.00015 \cdot 1 + 0 = 1.00015$$

$$n = 2: \quad y[2] = 1.00015 y[1] + x[2] = 1.00015 \cdot 1.00015 = (1.00015)^2$$

and, in general,

$$y[n] = (1.00015)^n$$

for $n \geq 0$. This output is, by definition, the impulse response of the system and equals (5.7).

Next we develop the difference equation in (5.14) from the convolution in (5.15). Replacing n by $n - 1$ in (5.15) yields

$$y[n - 1] = \sum_{k=0}^{n-1} (1.00015)^{n-1-k} x[k] \qquad (5.16)$$

Subtracting from (5.15) the product of (5.16) and 1.00015 yields

$$y[n] - 1.00015 y[n - 1]$$

$$= \sum_{k=0}^{n} (1.00015)^{n-k} x[k] - (1.00015) \sum_{k=0}^{n-1} (1.00015)^{n-1-k} x[k]$$

$$= \sum_{k=0}^{n-1} (1.00015)^{n-k} x[k] + 1 \cdot x[n] - \sum_{k=0}^{n-1} (1.00015)^{n-k} x[k] = x[n]$$

which is (5.14). Thus the two descriptions are equivalent.

We compare the numbers of multiplications required in the two descriptions. There are $(n+1)$ terms in (5.15), and each terms requires one multiplication. Thus the number of multiplications in the direct computation of (5.15) equals $n + 1$.[2] Thus computing $y[n]$, for $n = 0, 1, \ldots, 9$, requires

$$1 + 2 + \cdots + 10 = 55$$

---

[2] Equation (5.15) can also be computed using FFT, as we will discuss in Chapter 9.

multiplications. On the other hand, the difference equation in (5.13) requires only one multiplication for each n; thus computing $y[n]$, for $n = 0, 1, \ldots, 9$, requires only 10 multiplications. Thus the difference equation requires less computation than the convolution.

In computing $y[n]$, the convolution in (5.11) requires the storage of $h[k]$ and $x[k]$ for all $k < n$, whereas the difference equation requires the storage of only the coefficient 1.00015 and $y[n - 1]$. Thus the difference equation requires less memory location than the convolution.

Block diagram Furthermore, the difference equation can be implemented using the types of elements shown in Fig. 5.3(a). The element denoted by a box is a unit-delay element; its output $y[n]$ equals the input $x[n]$ delayed by one sample; that is,

$$y[n] = x[n - 1] \qquad \text{(unit-delay element)}$$

The element denoted by a line with an arrow and a real number a is called a multiplier with gain a; its input and output are related by

$$y[n] = ax[n] \qquad \text{(multiplier)}$$

If $a = 1$, it is a direct connection and the arrow and a are often not shown. In addition, a signal may branch out to two outputs with gains a and b, respectively, as shown. The point A will be called a *branch-out point*. The element denoted by a small circle with a plus sign is called an adder or summer. Every adder has two inputs $x_i[n]$, $i = 1, 2$, and one output $y[n]$. They are related by

$$y[n] = x_1[n] + x_2[n] \qquad \text{(adder)}$$

These three elements can easily be implemented in digital systems and are called DT basic elements. Any diagram that consists of only these three types of elements is called a block diagram. Using these elements, we can implement (5.13) as shown in Fig. 5.3(b). In fact, every difference equation, as we will show in Chapter 9, can be implemented as a block diagram. In conclusion, difference equations require less computation and less memory than convolutions; they can also directly be implemented using DT basic elements. Thus *difference equations are preferable to convolutions in describing discrete-time systems.*

Every LTI discrete-time system can be described by a convolution, but not necessarily by a difference equation. In order to be describable by a difference equation as well, the system must have one additional property, called *lumpedness*. An LTI system is defined to be lumped if it can be implemented using a finite number of unit-delay elements in Fig. 5.3(a); it is *distributed* if its implementation requires infinitely many unit-delay elements. Clearly, a distributed system cannot be easily implemented in practice. Thus we study in this text only LTI lumped (LTIL) systems or systems that can be described by difference equations. *Every LTIL systems is describable by a convolution and a difference equation.* But we prefer the latter.

We discuss some general forms of difference equations. Consider

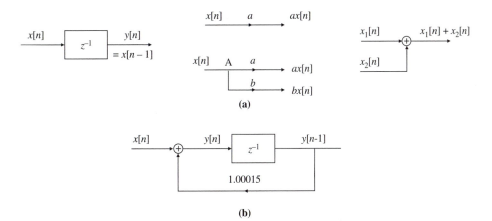

Figure 5.3 (a) DT basic elements. (b) Block diagram of (5.14).

$$a_1 y[n] + a_2 y[n-1] + \cdots + a_{N+1} y[n-N]$$
$$= b_1 x[n] + b_2 x[n-1] + \cdots + b_{M+1} x[n-M] \tag{5.17}$$

where a_i and b_i are real constants, and M and N are positive integers.[3] If $a_1 = 0$, $a_2 \neq 0$, and $b_1 \neq 0$, then the output $y[n-1]$ depends on $x[n]$, a future input; and the equation is not causal. If $a_1 \neq 0$, $y[n]$ depends only on current and past data for any zero or nonzero a_i and b_i and for any positive integers N and M. Thus (5.17) with $a_1 \neq 0$ is a most general form of difference equations to describe LTIL and causal systems. If $a_{N+1} \neq 0$ and $b_{M+1} \neq 0$, the difference equation is said to have order $\max(M, N)$. Equation (5.17) is said to be in the *delayed form*

In addition to the delayed form, the following *advanced form*

$$\bar{a}_1 y[n + \bar{N}] + \bar{a}_2 y[n + \bar{N} - 1] + \cdots + \bar{a}_{\bar{N}+1} y[n]$$
$$= \bar{b}_1 x[n + \bar{M}] + \bar{b}_2 x[n + \bar{M} - 1] + \cdots + \bar{b}_{\bar{M}+1} x[n] \tag{5.18}$$

can also be used to describe LTIL systems. To be causal, we require $\bar{a}_1 \neq 0$, $\bar{b}_1 \neq 0$, and $\bar{N} \geq \bar{M}$. The difference equation in (5.18) with $\bar{N} \geq \bar{M}$ is said to have order $\bar{N}$. Subtracting $\bar{N}$ from the indices inside all brackets in (5.18) yields the delayed form

$$\bar{a}_1 y[n] + \bar{a}_2 y[n-1] + \cdots + \bar{a}_{\bar{N}+1} y[n - \bar{N}]$$
$$= \bar{b}_1 x[n-d] + \bar{b}_2 x[n-d-1] + \cdots + \bar{b}_{\bar{M}+1} x[n - \bar{N}] \tag{5.19}$$

where $d = \bar{N} - \bar{M} \geq 0$. Thus delayed-form difference equations can be easily obtained from

---

[3] There is no standard way to assign coefficients a_i and b_i. We assign them to agree with those used in MATLAB.

advanced form and vice versa. This text uses both forms, although the delayed form is used almost exclusively in most DSP texts.

5.3.1 Recursive and Nonrecursive Difference Equations

Consider the difference equation in (5.17) with $a_1 \neq 0$. For convenience, we normalize a_1 to 1. If $a_i = 0$ for all $i > 1$, then the equation reduces to

$$y[n] = b_1 x[n] + b_2 x[n-1] + \cdots + b_{M+1} x[n-M] \tag{5.20}$$

This is called a *nonrecursive* difference equation. The current output of this equation depends only on current and past inputs. If $a_1 = 1$ and $a_i \neq 0$ for some $i > 1$, then (5.17) is called a *recursive* difference equation and can be written as

$$y[n] = -a_2 y[n-1] - \cdots - a_{N+1} y[n-N]$$
$$+ b_1 x[n] + b_2 x[n-1] + \cdots + b_{M+1} x[n-M] \tag{5.21}$$

Its current output depends not only on current and past inputs but also on past outputs. In this equation, because past outputs are needed in computing current output, the equation must be computed in the order of $y[0]$, $y[1]$, $y[2]$, and so forth or must be computed recursively from $n = 0$ on. There is no such restriction in computing the nonrecursive equation in (5.20).

We discuss the impulse response of the nonrecursive difference equation in (5.20). We assume $x[n] = 0$ and $y[n] = 0$ for all $n < 0$. Then the output of (5.20) excited by $x[n] = \delta[n]$ is, by definition, the impulse response of the equation. Direct substitution yields

$$h[0] = y[0] = b_1, \quad h[1] = y[1] = b_2, \quad \ldots, \quad h[M] = y[M] = b_{M+1}$$

and $h[n] = y[n] = 0$ for all $n > M + 1$. This is an FIR filter of length $(M + 1)$. Because the difference equation in (5.20) has order M, it is called an Mth-order FIR filter. Thus *an Mth-order FIR filter has length $M + 1$.*

Conversely, the convolution description of an FIR filter is in fact a nonrecursive difference equation. For example, consider an FIR filter of length 4; that is, $h[n] = 0$ for $n < 0$ and $n > 3$. Then we can write the second equation of (5.11) or

$$y[n] = \sum_{k=0}^{n} h[k] x[n-k]$$

as

$$y[n] = h[0] x[n] + h[1] x[n-1] + h[2] x[n-2] + h[3] x[n-3]$$

This is indeed a third-order nonrecursive difference equation. Thus there is no difference in using convolutions or nonrecursive difference equations to describe FIR filters.

Impulse responses of most recursive difference equations have infinitely many nonzero entries. The reason is that the impulse input will yield some nonzero output. This nonzero output will then propagate to all output because current output depends on past output. Thus *difference equations describing IIR filters must be recursive*. Although every FIR filter can be described by a nonrecursive difference equation, in some special situation, we may develop a recursive difference equation to describe an FIR filter to save the number of operations, as the next example illustrates.

◆ **Example 5.3**

Consider the 10-day moving-average filter or the filter with impulse response

$$h[n] = \begin{cases} 1/10 & \text{for } n = 0, 1, \ldots, 9 \\ 0 & \text{for } n < 0 \text{ and } n > 9 \end{cases}$$

Its convolution and nonrecursive difference equation both equal

$$y[n] = 0.1 \, (x[n] + x[n-1] + x[n-2] + \cdots + x[n-8] + x[n-9]) \quad (5.22)$$

Its computation requires nine additions and one multiplication.
Now we reduce the indices of (5.22) by one to yield

$$y[n-1] = 0.1 \, (x[n-1] + x[n-2] + x[n-3] + \cdots + x[n-9] + x[n-10])$$

Subtracting this equation from (5.22) yields

$$y[n] = y[n-1] + 0.1 \, (x[n] - x[n-10]) \quad (5.23)$$

This is a recursive difference equation of order 10. It requires one multiplication and two additions; thus it is more efficient than the nonrecursive difference equation in (5.22). Both (5.22) and (5.23) describe the same system.

5.3.2 Sampling Period and Real-Time Processing

In our discussion up to this point, we have not yet said anything regarding the sampling period. We use an example to discuss the issues involved. Consider

$$y[n] = -a_2 y[n-1] - a_3 y[n-2] + b_2 x[n-1] + b_3 x[n-2] \quad (5.24)$$

Note that $b_1 x[n]$ is not included in (5.24). Computing $y[n]$ requires four multiplications and three additions. To simplify the discussion, we disregard the time required to fetch data from the input port and registers and the time to send the data to the output port. Suppose one multiplication takes 1 μs (10^{-6} second), and one addition takes 0.2 μs. If we compute the four multiplications sequentially using one multiplier, then it will take 4 μs to obtain the four products. Two additions can be performed as soon as the first three products are available. The last addition takes 0.2 μs. Thus it takes a total of 4.2 μs to compute $y[n]$ sequentially. If four multipliers are available, we may compute the four products simultaneously. The summation of the four products, however, takes two steps; thus it needs a total of 1.4 μs to compute $y[n]$. Therefore, the time required to compute (5.24) depends on the hardware used and how it is programmed.

In digital processing of telephone conversation, appreciable time delay is not desirable. Such processing requires real-time processing in the sense that as soon as an input signal at time t_0 is received, the corresponding output must appear at the same t_0 or immediately thereafter. If the output is obtained from (5.24), then the sampling period must be larger than 1.4 μs. If the sampling period is less than 1.4 μs, then (5.24) cannot be computed in real time. Thus the complexity of a digital system and the hardware used will impose a fundamental restriction on the sampling period in real-time processing. We mention that if $y[n]$ also depends on $x[n]$ such as

$$y[n] = -a_2 y[n-1] - a_3 y[n-2] + b_1 x[n] + b_2 x[n-1] + b_3 x[n-2]$$

with $b_1 \neq 0$, then we can compute $y[n]$ only after $x[n]$ is received. Therefore, $y[n]$ can never appear at the same instant as $x[n]$. In this case, $y[n]$ is automatically delayed by one sampling period.

In non-real-time processing, the situation is different. In real-time processing, we can use only past and current inputs to compute current output. Thus the system must be causal. In non-real-time processing or in processing stored signals, causality is not an issue. For example, to reconstruct the missing signal in Fig. 1.13(a) at $t = 4.37$, we may use all available signal before and after $t = 4.37$. Furthermore, there is no restriction on the sampling period. We may take as much time as desired to compute the signal at $t = 4.37$. In conclusion, a DT system receives a stream of input data, carries out computations, generates a stream of output data, displays them immediately, or stores them in memory. In this process, the sampling period does not play any role. If the system is used in a real-time processing, a control logic will send out the output data at sampling instants, provided the sampling period is large enough to compute each $y[n]$ in time. Thus the sampling period T has nothing to do with the actual computation of the output. For example, $y[n]$ in (5.24) can be computed in 1.4 μs after $x[n-1]$ arrives, and then stored in a register. It is sent out to the output terminal at time nT. Thus, in DT system analysis and design, we may assume the sampling period T to be 1 or any other value. For convenience, we select $T = 1$, and the Nyquist frequency range becomes $(-\pi, \pi]$.

5.4 z-Transform

In this section we introduce an important tool, called the z-transform, to study DT systems. Its role in DT system analysis is the same as the role of the DT Fourier transform in DT signal analysis. Before proceeding, we introduce some terminology. A DT sequence is a *positive-time* sequence if it is identically zero for $n < 0$; a *negative-time* sequence if it is identically zero for

$n > 0$. It is a two-sided sequence if it is neither a positive-time nor a negative-time sequence.[4] Clearly, the sequences in Figs. 5.2 and 3.11 are positive time and the sequence in Fig. 4.23(b) is negative-time. The sequence in Fig. 3.12(a) is two sided.

Consider a DT signal $x[n]$ that can be positive-time, negative-time, or two sided. The two-sided z-transform of $x[n]$ is defined as

$$X(z) := Z_2[x[n]] := \sum_{n=-\infty}^{\infty} x[n]z^{-n} \tag{5.25}$$

where z is a complex variable. The one-sided z-transform or, simply, z-transform of $x[n]$ is defined as

$$X(z) := Z[x[n]] := \sum_{n=0}^{\infty} x[n]z^{-n} \tag{5.26}$$

It is defined only for the positive-time part of $x[n]$; the negative-time part is simply disregarded. For example, if $x[-2] = 1.5, x[-1] = 0.5, x[0] = 2, x[1] = -2, x[2] = 3, x[3] = -1$ or

$$x[n] = 1.5\delta[n+2] + 0.5\delta[n+1] + 2\delta[n]$$
$$- 2\delta[n-1] + 3\delta[n-2] - \delta[n-3]$$

then its two-sided z-transform is

$$X[z] = 1.5z^2 + 0.5z + 2z^0 - 2z^{-1} + 3z^{-2} - z^{-3}$$

and its (one-sided) z-transform is

$$X[z] = 2z^0 - 2z^{-1} + 3z^{-2} - z^{-3}$$

We see that the one- and two-sided z-transforms are obtained by multiplying $x[k]$ simply by z^{-k}, for each k. Thus z^k can be considered to indicate the sampling instant $n = -k$. In general, if $X(z)$ consists of only zero and negative powers of z, then its time sequence is positive-time. If $X(z)$ consists of zero and positive powers of z, then its time sequence is negative-time. If $X(z)$ consists of both positive and negative powers of z, then its time sequence is two sided. Clearly, if $x[n]$ is positive-time, then its z-transform equals its two-sided z-transform. If $x[n]$ is negative-time, then its z-transform is zero.

The inverse two- and one-sided z-transforms of $X(z)$ are given by

$$x[n] := Z^{-1}[X(z)] := \frac{1}{2\pi j} \oint X(z)z^{n-1}dz \tag{5.27}$$

---

[4] Some texts call positive-time sequences causal sequences and negative-time sequences anticausal sequences.

where the integration is around a circular contour in the complex z-plane lying inside the region of convergence, which we will discuss shortly. The integration is to be carried out in the counterclockwise direction. In particular, we may select $z = ce^{j\omega}$ in (5.27), where c is a constant so that the circle $ce^{j\omega}$ lies inside the region of convergence. In this case, we have $dz = cje^{j\omega}d\omega$ and (5.27) becomes

$$x[n] := Z^{-1}[X(z)] := \frac{c^n}{2\pi} \int_{\omega=0}^{2\pi} X(ce^{j\omega})e^{jn\omega}d\omega \tag{5.28}$$

If $c = 1$, this is the inverse DTFT discussed in (3.44) with $T = 1$. Before proceeding, we use an example to discuss the region of convergence and the constant c.

◆ **Example 5.4**

Consider the positive-time sequence

$$x[n] = \begin{cases} 1.2^n & \text{for } n = 0, 1, 2, \ldots \\ 0 & \text{for } n < 0 \end{cases} \tag{5.29}$$

Because $1.2^n = e^{0.1823n}$, the sequence grows *exponentially* to infinity, as shown in Fig. 5.4(a). Note that the DTFT of (5.29) diverges and is not defined. The two-sided and one-sided z-transforms of (5.29) are the same and equal

$$X(z) = \sum_{n=0}^{\infty} 1.2^n z^{-n} = \sum_{n=0}^{\infty} (1.2z^{-1})^n \tag{5.30}$$

This is an infinite-power series. As such, it is not very useful in analysis and design. Most z-transforms encountered in this text can be expressed in closed form by using

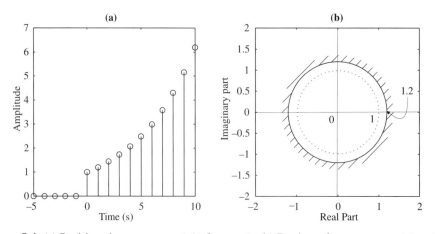

Figure 5.4 (a) Positive-time sequence 1.2^n for $n \geq 0$. (b) Region of convergence: $|z| > 1.2$.

$$1 + r + r^2 + r^3 + \cdots = \sum_{n=0}^{\infty} r^n = \frac{1}{1-r} \tag{5.31}$$

where r is a real or complex constant with magnitude less than 1 or $|r| < 1$. Using the formula, we can write (5.30) as

$$X(z) = Z[1.2^n] = \frac{1}{1 - 1.2z^{-1}} = \frac{z}{z - 1.2} \tag{5.32}$$

This equation holds only if $|1.2z^{-1}| < 1$ or $1.2 < |z|$. The region $1.2 < |z|$ or the region outside the circle denoted by solid line in Fig. 5.4(b) is called the *region of convergence.* If z is not in the region, the infinite summation in (5.30) diverges and does not equal (5.32). For example, if $z = 1$, then $X(1) = \infty$ in (5.30) and $X(1) = 1/(1 - 1.2) = -5$ in (5.32). However, in application, we may disregard the region of convergence and consider (5.32) to be defined for all z except at $z = 1.2$.

One may then wonder why we bother to introduce the region of convergence. To see the reason, we expand (5.32) into a negative power of z and a positive-power series of z as

$$\begin{aligned}
X(z) &= \frac{1}{1 - 1.2z^{-1}} \\
&= 1 + 1.2z^{-1} + (1.2)^2 z^{-2} + (1.2)^3 z^{-3} + \cdots \tag{5.33} \\
&= -0.833z - (0.833)^2 z^2 - (0.833)^3 z^3 - \cdots \tag{5.34}
\end{aligned}$$

where $0.833 = 1/1.2$. They are obtained by direct division as

$$
1 - 1.2z^{-1} \overline{\smash{\big)}\ \begin{array}{l} 1 + 1.2z^{-1} + (1.2)^2 z^{-2} + \cdots \\ 1 \\ \underline{1 - 1.2z^{-1}} \\ 1.2z^{-1} \\ \underline{1.2z^{-1} - (1.2)^2 z^{-2}} \\ (1.2)^2 z^{-2} \\ \vdots \end{array}}
$$

and, using $1/1.2 = 0.8333$,

$$
-1.2z^{-1} + 1 \overline{\smash{\big)}\ \begin{array}{l} -0.833z - (0.833)^2 z^2 - (0.833)^3 z^3 - \cdots \\ 1 \\ \underline{1 - 0.833z} \\ 0.833z \\ \underline{0.833z - (0.833)^2 z^2} \\ (0.833)^2 z^2 \\ \vdots \end{array}}
$$

Thus the expansion of $X(z)$ into (5.33) and (5.34) is straightforward. Let us identify (5.33) with (5.25). Then we can obtain

$$x[n] = \begin{cases} 0 & \text{for } n < 0 \\ (1.2)^n & \text{for } n \geq 0 \end{cases}$$

Thus the negative-power series in (5.33) yields a positive-time sequence. Identifying (5.34) with (5.25) yields

$$x[n] = \begin{cases} -(0.833)^n & \text{for } n < 0 \\ 0 & \text{for } n \geq 0 \end{cases} \tag{5.35}$$

Thus the positive-power series in (5.34) yields a negative-time sequence.

The region of convergence of (5.33) is $1.2 < |z|$; the region of convergence of (5.34) is $|0.833z| < 1$ or $|z| < 1/0.833 = 1.2$. Thus the inverse of $X(z)$ is not uniquely defined unless its region of convergence is specified. For example, if we select $c = 1$ in (5.28), then the circle $ce^{j\omega}$ lies inside the region of convergence of (5.34). If we use (5.28) with $c = 1$ to compute the inverse z-transform of $z/(z-1.2)$, then it will yield the negative-time sequence in (5.34). Indeed, this is the case as shown in Fig. 4.22(b), which was computed using FFT. On the other hand, if we select $c = 2$, then the circle $2e^{j\omega}$ lies inside the region of convergence of (5.33) and (5.28) with $c = 2$ yields, as we will show in the next subsection, the positive-time sequence in (5.29). Thus the specification of the region of convergence is essential in computing the inverse z-transform.

In studying LTI systems, the instant at which we start to apply an input can be considered as $n_0 = 0$. Thus the input is positive-time. If a system is causal, then its impulse response sequence and output are also positive-time. Thus in system analysis, we deal exclusively with positive-time signals, and all z-transforms will be assumed to be z-transforms of positive-time signals. There is a simple way to ensure that their inverse z-transforms are positive-time, as we will discuss later. Thus from now on, we pay no attention to the region of convergence and study only the (one-sided) z-transform.

We first develop some transform pairs. The z-transform of the impulse sequence in (5.4) is, by definition,

$$Z[\delta[n - k]] = 0 + 0 \cdot z^{-1} + \cdots + 1 \times z^{-k} + 0 \times z^{-(k+1)} + \cdots = z^{-k}$$

In particular, we have $Z[\delta[n]] = 1$ and $Z[\delta[n - 1]] = z^{-1}$. Because $\delta[n - 1]$ is the delay of $\delta[n]$ by one sample, z^{-1} is called the *unit-delay element*. In contrast, z is called the *unit-advance element*. Using (5.31), we have

$$Z[b^n] = \sum_{n=0}^{\infty} b^n z^{-n} = \frac{1}{1 - bz^{-1}} = \frac{z}{z - b} \tag{5.36}$$

where b can be real or complex such as $b = 2$ or $e^{j\omega}$. The z-transform of $b^n \sin \omega_0 n$, for $n \geq 0$, is

$$
\begin{aligned}
Z[b^n \sin \omega_0 n] &= Z\left[\frac{(be^{j\omega_0})^n - (be^{-j\omega_0})^n}{2j}\right] \\
&= \frac{1}{2j}\left(\frac{1}{1 - be^{j\omega_0}z^{-1}} - \frac{1}{1 - be^{-j\omega_0}z^{-1}}\right) \\
&= \frac{b(\sin \omega_0)z^{-1}}{1 - 2b(\cos \omega_0)z^{-1} + b^2 z^{-2}}
\end{aligned}
\tag{5.37}
$$

Its positive-power form can be easily obtained by multiplying its numerator and denominator by z^2. Proceeding similarly, we can establish the pairs in Table 5.1

To conclude this section, we discuss the z-transforms of the shiftings of $x[n]$. Let $x[n]$ be a positive-time sequence; that is, $x[n] = 0$ for $n < 0$. Let $X(z)$ be its z-transform or

$$
X(z) = \sum_{n=0}^{\infty} x[n] z^{-n}
$$

Then for any positive integer k, we have

$$
\begin{aligned}
Z[x[n-k]] &= \sum_{n=0}^{\infty} x[n-k] z^{-n} = \sum_{n=0}^{\infty} x[n-k] z^{-(n-k)} z^{-k} \\
&= z^{-k} \sum_{\bar{n}=0}^{\infty} x[\bar{n}] z^{-\bar{n}} = z^{-k} X(z)
\end{aligned}
\tag{5.38}
$$

where we have defined $\bar{n} = n - k$ and used $x[\bar{n}] = 0$, for $\bar{n} < 0$. Note that $x[n-k]$, with k positive, is the delay (right shifting) of $x[n]$ k samples and its z-transform equals $X(z)$ multiplied by z^{-k}. Note that (5.38) does not hold if $x[n]$ is not positive time. See Problem 5.7.

For $k > 0$, the sequence $x[n+k]$ is the advance (left shifting) of $x[n]$ by k samples. We compute the z-transform of $x[n+1]$:

$$
\begin{aligned}
Z[x[n+1]] &= \sum_{n=0}^{\infty} x[n+1] z^{-(n+1)} z = z \sum_{\bar{n}=1}^{\infty} x[\bar{n}] z^{-\bar{n}} \\
&= z\left(\sum_{\bar{n}=0}^{\infty} x[\bar{n}] z^{-\bar{n}} - x[0]\right) \\
&= z(X(z) - x[0])
\end{aligned}
\tag{5.39}
$$

where we have defined $\bar{n} = n + 1$. The equation can also be obtained as follows: Advancing $x[n]$ one sample is equivalent to multiplying $X(z)$ by z. It also shifts $x[0]$ to $n = -1$; therefore, $x[0]$ must be subtracted from $X(z)$ as in (5.39). Using the same argument, we have

Table 5.1 Some z-Transform Pairs

$x[n]\ n \geq 0$	$X(z)$	$X(z)$
$\delta[n]$	1	1
$\delta[n-k]$	z^{-k}	z^{-k}
b^n	$\dfrac{1}{1-bz^{-1}}$	$\dfrac{z}{z-b}$
nb^n	$\dfrac{bz^{-1}}{(1-bz^{-1})^2}$	$\dfrac{bz}{(z-b)^2}$
$b^n \sin \omega_0 n$	$\dfrac{b(\sin \omega_0)z^{-1}}{1-2b(\cos \omega_0)z^{-1}+b^2 z^{-2}}$	$\dfrac{b(\sin \omega_0)z}{z^2-2b(\cos \omega_0)z+b^2}$
$b^n \cos \omega_0 n$	$\dfrac{1-b(\cos \omega_0)z^{-1}}{1-2b(\cos \omega_0)z^{-1}+b^2 z^{-2}}$	$\dfrac{1-b(\cos \omega_0)z}{z^2-2b(\cos \omega_0)+b^2}$

$$Z[x[n+2]] = z^2 \left(X(z) - x[0] - x[1]z^{-1} \right) \tag{5.40}$$

and, in general,

$$Z[x[n+k]] = z^k \left(X(z) - \sum_{i=0}^{k-1} x[i]z^{-i} \right) \tag{5.41}$$

for $k \geq 0$. Equations (5.38) and (5.41) will be used to study the difference equations discussed in Section 5.3.

5.4.1 Inverse z-Transform

Given a z-transform $X(z)$, there are several ways to compute its inverse z-transform: (1) direct division, (2) partial fraction expansion and table lookup, and (3) the inverse z-transform formula in (5.27), which can be computed analytically using the Cauchy residue theorem or numerically using FFT. We discuss all of them except the one of using the Cauchy theorem.

Direct Division We use an example to illustrate the procedure. Consider

$$X(z) = \frac{2z^{-1} - z^{-2}}{1 - 1.6z^{-1} - 0.8z^{-2}} \tag{5.42}$$

This can be expanded into a positive- or negative-power series of z as in Example 5.4. If we expand it into a positive-power series, then it will yield a negative-time sequence. Thus in direct division, we must expand $X(z)$ into a negative-power series. We carry out the division as

$$1 \quad - 1.6z^{-1} \quad - 0.8z^{-2} \quad \overline{\smash{\big)}\ 2z^{-1} \quad - z^{-2}} \quad \overset{\displaystyle 2z^{-1} \quad + 2.2z^{-2} \quad + 5.12z^{-3} \quad + \cdots}{}$$

$$\begin{array}{r}
2z^{-1} \quad - 3.2z^{-2} \quad - 1.6z^{-3} \\
\hline
2.2z^{-1} \quad + 1.6z^{-3} \\
2.2z^{-2} \quad - 3.52z^{-3} \quad - 1.76z^{-4} \\
\hline
5.12z^{-3} \quad + 1.76z^{-4} \\
\vdots
\end{array}$$

which implies

$$X(z) = 2z^{-1} + 2.2z^{-2} + 5.12z^{-3} + \cdots$$

Thus the inverse z-transform of (5.42) is

$$x[0] = 0, \ x[1] = 2, \ x[2] = 2.2, \ x[3] = 5.12, \ \ldots$$

This can also be expressed as

$$x[n] = 2\delta[n-1] + 2.2\delta[n-2] + 5.12\delta[n-3] + \cdots$$

This is a positive-time sequence and is the inverse z-transform of (5.42).

Partial Fraction Expansion and Table Lookup In this method, we express $X(z)$ as a sum of terms whose inverse z-transforms are available in a table. Consider the z-transform in (5.42). We express it as

$$X(z) = \frac{2z^{-1} - z^{-2}}{1 - 1.6z^{-1} - 0.8z^{-2}} = \frac{2z^{-1} - z^{-2}}{(1 - 2z^{-1})(1 + 0.4z^{-1})}$$

$$= k_1 + \frac{r_1}{1 - 2z^{-1}} + \frac{r_2}{1 + 0.4z^{-1}}$$

where r_i are called *residues*. The constants k_1 and r_i can be computed as

$$k_1 = X(z)|_{z^{-1}=\infty} = \frac{2z^{-1} - z^{-2}}{1 - 1.6z^{-1} - 0.8z^{-2}}\bigg|_{z^{-1}=\infty}$$

$$= \frac{2z - 1}{z^2 - 1.6z - 0.8}\bigg|_{z=0} = \frac{-1}{-0.8} = 1.25$$

$$r_1 = X(z)(1 - 2z^{-1})|_{z^{-1}=1/2} = \frac{2z^{-1} - z^{-2}}{(1 + 0.4z^{-1})}\bigg|_{z^{-1}=1/2}$$

$$= \frac{2(1/2) - (1/2)^2}{1 + 0.4(1/2)} = \frac{0.75}{1.2} = 0.625$$

$$r_2 = X(z)(1 + 0.4z^{-1})\big|_{z^{-1} = -1/0.4 = -2.5} = \frac{2z^{-1} - z^{-2}}{(1 - 2z^{-1})}\bigg|_{z^{-1} = -2.5}$$

$$= \frac{2(-2.5) - (-2.5)^2}{1 - 2(-2.5)} = \frac{-11.25}{6} = -1.875$$

Thus we have[5]

$$X(z) = 1.25 + \frac{0.625}{1 - 2z^{-1}} - \frac{1.875}{1 + 0.4z^{-1}} \tag{5.43}$$

The inverse z-transform of each term can be found in Table 5.1. Thus the inverse z-transform of $X(z)$ in (5.42) is

$$x[n] = 1.25\delta[n] + 0.625 \times 2^n - 1.875 \times (-0.4)^n \tag{5.44}$$

for $n \geq 0$.

To compare the result with the one obtained by direct division, we compute (5.44) explicitly for $n = 0, 1, 2, 3$:

$$x[0] = 1.25 + 0.625 - 1.875 = 0$$

$$x[1] = 0 + 0.625 \times 2 - 1.875 \times (-0.4) = 1.25 + 0.75 = 2$$

$$x[2] = 0 + 0.625 \times 4 - 1.875 \times 0.16 = 2.5 - 0.3 = 2.2$$

$$x[3] = 0 + 0.625 \times 8 - 1.875 \times (-0.064) = 5 + 0.12 = 5.12$$

They are indeed the same.

We discuss next the use of positive-power form of (5.42). Multiplying the numerator and denominator of (5.42) by z^2, we obtain

$$X(z) = \frac{2z - 1}{z^2 - 1.6z - 0.8} = \frac{2z - 1}{(z - 2)(z + 0.4)}$$

For the reason to be given shortly, we expand $X(z)/z$, instead of $X(z)$, by partial fraction expansion as

$$\frac{X(z)}{z} = \frac{2z - 1}{z(z - 2)(z + 0.4)} = \frac{r_1}{z} + \frac{r_2}{z - 2} + \frac{r_3}{z + 0.4} \tag{5.45}$$

---

[5] The region of convergence of $1/(1 - 2z^{-1})$ is $|z| > 2$. The point $z^{-1} = -2.5$ or $z = -0.4$ is not inside the region, and we still use it because we can disregard the region of convergence.

with

$$r_1 = \left.\frac{2z-1}{(z-2)(z+0.4)}\right|_{z=0} = \frac{-1}{(-2)(0.4)} = 1.25$$

$$r_2 = \left.\frac{2z-1}{z(z+0.4)}\right|_{z=2} = \frac{3}{2(2.4)} = 0.625$$

$$r_3 = \left.\frac{2z-1}{z(z-2)}\right|_{z=-0.4} = \frac{-1.8}{(-0.4)(-2.4)} = -1.875$$

Multiplying (5.45) by z, we obtain

$$X(z) = 1.25 + 0.625\frac{z}{z-2} - 1.875\frac{z}{z+0.4}$$

The inverse z-transform of each term can be found in Table 5.1. Thus the inverse z-transform of $X(z)$ is

$$x[n] = 1.25\delta[n] + 0.625 \times 2^n - 1.875 \times (-0.4)^n$$

which is the same as (5.44), as should be the case. Now if we expand $X(z)$ directly, its expanded term will not be in Table 5.1. If we expand $X(z)/z$ and then multiply z as we did, then each term is in Table 5.1. Thus in using positive-power form, we expand $X(z)/z$ instead of $X(z)$.

   For the example in (5.42), using positive-power or negative-power form requires roughly the same amount of computation. This may not always be the case. See Problem 5.6. For a more detailed discussion of the partial fraction method, see Ref. 6.

5.4.2 FFT Computation of the Inverse z-Transform[6]

We use (5.28) to compute the inverse z-transform. Consider

$$x[n] = \frac{c^n}{2\pi}\int_{\omega=0}^{2\pi} X(ce^{j\omega})e^{jn\omega}d\omega =: c^n\bar{x}[n]$$

where

$$\bar{x}[n] = \frac{1}{2\pi}\int_{\omega=0}^{2\pi} X(ce^{j\omega})e^{jn\omega}d\omega$$

The second equation is the inverse DTFT defined in (3.44) with $T = 1$ and can be computed using FFT. Thus the only question is how to select c so that the formula will yield a positive-time sequence. The rule is very simple. The constant c must be selected to be larger than the largest

---

[6] This subsection may be skipped without loss of continuity.

magnitude of all roots of the denominator of $X(z)$ or, equivalently, all roots must lie inside the circle with radius c. For example, if all roots of the denominator of $X(z)$ lie inside the unit circle, we may select $c = 1$. For the $X(z)$ in (5.42) or

$$X(z) = \frac{2z^{-1} - z^{-2}}{1 - 1.6z^{-1} - 0.8z^{-2}} = \frac{2z - 1}{z^2 - 1.6z - 0.8} = \frac{2z - 1}{(z + 0.4)(z - 2)}$$

we have $p_1 = -0.4$ and $p_2 = 2$. Thus c must be larger than 2. Arbitrarily we select $c = 2.2$. Then we have

$$X(ce^{j\omega}) = X(2.2e^{j\omega}) = \frac{2 \times 2.2e^{j\omega} - 1}{(2.2e^{j\omega})^2 - 1.6 \times 2.2e^{j\omega} - 0.8}$$

We will use the procedure in Section 4.7 to compute its inverse DTFT. In using FFT, we must select an N, the number of frequency samples of $X(2.2e^{j\omega})$ in $(-\pi, \pi]$ or $[0, 2\pi)$. If N is sufficiently large, time aliasing due to frequency sampling will be negligible, and the inverse FFT will yield the inverse DTFT $\bar{x}[n]$. In using inverse DFT or FFT, one problem is to determine the location of a time sequence. In computing the inverse z-transform, because we have agreed that the inverse is a positive-time sequence, the output of inverse FFT will simply be the inverse z-transform $\bar{x}[n]$ for $n = 0, 1, \ldots, N - 1$.

For the example, let us select $N = 128$ frequency samples in $[0, 2\pi)$. Then the program that follows

Program 5.1
```
N=128;D=2*pi/N;
m=0:N-1;w=m*D;
X=(2.0*2.2.*exp(j.*w)-1)./(2.2*2.2.*exp(j*2.*w)...
     -1.6*2.2.*exp(j.*w)-0.8);
xb=ifft(X);
x=(2.2.^ m).*xb;
n=0:4;
x(n+1) (%display only x[0], x[1], . . . , x[4])
```

yields

$$x[0] = 0, \ x[1] = 2, \ x[2] = 2.2, \ x[3] = 5.12, \ x[4] = 9.9521$$

They are the same as those computed using direct division and partial fraction expansion. Note that indices in MATLAB start from 1 not 0. Note also that if we use $N = 64$ in Program 5.1, then the result will be slightly different from the exact one because of appreciable time aliasing due to frequency sampling. The more frequency samples we use, the more accurate the result is.

5.5 Transfer Functions

Every linear time-invariant DT system that is causal and initially relaxed can be described by

$$y[n] = \sum_{k=0}^{n} h[n-k]u[k] = \sum_{k=0}^{\infty} h[n-k]u[k] \tag{5.46}$$

where $h[n]$ is the impulse response and is zero for $n < 0$. Substituting (5.46) into the z-transform of $y[n]$, we obtain

$$Y(z) = Z[y[n]] = \sum_{n=0}^{\infty} y[n]z^{-n}$$

$$= \sum_{n=0}^{\infty} \left(\sum_{k=0}^{\infty} h[n-k]u[k] \right) z^{-(n-k)} z^{-k}$$

Interchanging the order of summations, introducing a new index $\bar{n} := n - k$, and using the causality property $h[\bar{n}] = 0$, for $\bar{n} < 0$, we obtain

$$Y(z) = \sum_{k=0}^{\infty} \left(\sum_{n=0}^{\infty} h[n-k]z^{-(n-k)} \right) u[k]z^{-k}$$

$$= \sum_{k=0}^{\infty} \left(\sum_{\bar{n}=-k}^{\infty} h[\bar{n}]z^{-\bar{n}} \right) u[k]z^{-k}$$

$$= \left(\sum_{\bar{n}=0}^{\infty} h[\bar{n}]z^{-\bar{n}} \right) \sum_{k=0}^{\infty} u[k]z^{-k}$$

or

$$Y(z) = H(z)X(z) \tag{5.47}$$

where

$$H(z) = Z[h[n]] = \sum_{n=0}^{\infty} h[n]z^{-n} \tag{5.48}$$

is called the (discrete-time or digital) transfer function. It is the z-transform of the impulse response. Thus the z-transform transforms the convolution in (5.46) into the multiplication in (5.47). In the z-transform domain, the output is the inverse z-transform of the product of the transfer function and the z-transform of the input. Note that the transfer function $H(z)$ can also be defined as

$$H(z) = \frac{Z[\text{output}]}{Z[\text{input}]} = \frac{Y(z)}{X(z)} \tag{5.49}$$

In this definition, the system must be initially relaxed or, equivalently, we must assume $x[n] = y[n] = 0$, for $n < 0$.

The transfer function can be a rational or an irrational function of z. If an LTI system is also lumped, then its transfer function is a rational function. Indeed, every LTI lumped DT system can be described by a difference equation of the form in (5.17). Because $Z[x[n-k]] = z^{-k}X(z)$, applying the z-transform to (5.17) yields

$$a_1 Y(z) + a_2 z^{-1} Y(z) + \cdots + a_{N+1} z^{-N} Y(z)$$
$$= b_1 X(z) + b_2 z^{-1} X(z) + \cdots + b_{M+1} z^{-M} X(z) \tag{5.50}$$

and

$$\left(a_1 + a_2 z^{-1} + \cdots + a_{N+1} z^{-N} \right) Y(z)$$
$$= \left(b_1 + b_2 z^{-1} + \cdots + b_{M+1} z^{-M} \right) X(z)$$

Thus we have

$$H(z) = \frac{Y(z)}{X(z)} = \frac{b_1 + b_2 z^{-1} + \cdots + b_{M+1} z^{-M}}{a_1 + a_2 z^{-1} + a_3 z^{-2} + \cdots + a_{N+1} z^{-N}} =: \frac{B(z)}{A(z)} \tag{5.51}$$

This is a rational function of z^{-1} and is called a negative-power DT transfer function. If $a_1 \neq 0$, for any integers N and M, the transfer function describes a causal system. If $b_{M+1} \neq 0$, $a_{N+1} \neq 0$, and if $B(z)$ and $A(z)$ have no common factor, the transfer function is said to have degree $\max(N, M)$.

Next we consider the advanced-form difference equation in (5.18). Because of (5.41), applying the z-transform to (5.18) will yield an equation involving $x[n]$ and $y[n]$ for some $n \geq 0$. However, using the initial conditions $x[n] = 0$ and $y[n] = 0$ for $n < 0$, the equation can be reduced as

$$\bar{a}_1 z^{\bar{N}} Y(z) + \bar{a}_2 z^{\bar{N}-1} Y(z) + \cdots + \bar{a}_{\bar{N}+1} Y(z)$$
$$= \bar{b}_1 z^{\bar{M}} X(z) + \bar{b}_2 z^{\bar{M}-1} X(z) + \cdots + \bar{b}_{\bar{M}+1} X(z)$$

See Problem 5.9. Thus in computing the transfer function of the advanced-form difference equation in (5.18), we may use

$$Z[x[n + k]] = z^k X(z)$$

for $z = 0, 1, 2, \ldots$, instead of (5.41). Thus the transfer function of (5.18) is

$$H(z) = \frac{Y(z)}{X(z)} = \frac{\bar{b}_1 z^{\bar{M}} + \bar{b}_2 z^{\bar{M}-1} + \cdots + \bar{b}_{\bar{b}} z + \bar{b}_{\bar{M}+1}}{\bar{a}_1 z^{\bar{N}} + \bar{a}_2 z^{\bar{N}-1} + \bar{a}_3 z^{\bar{N}-2} + \cdots + \bar{a}_{\bar{N}} z + \bar{a}_{\bar{N}+1}} = \frac{\bar{B}(z)}{\bar{A}(z)} \qquad (5.52)$$

It is also a rational functions of z, and is said to be in *positive-power* form.

◆ **Example 5.5**

Consider the delayed-form difference equation

$$y[n] + 5y[n-1] = x[n-1] + 4x[n-3]$$

Its order is max(1, 3)=3. Applying the z-transform yields

$$Y(z) + 5z^{-1}Y(z) = z^{-1}X(z) + 4z^{-3}X(z)$$

Thus its transfer function is

$$H(z) = \frac{Y(z)}{X(z)} = \frac{z^{-1} + 4z^{-3}}{1 + 5z^{-1}}$$

This is a negative-power transfer function with degree 3.
Let us consider

$$y[n+3] + 5y[n+2] = x[n+2] + 4x[n]$$

It is obtained by adding 3 to all indices in the delayed-form difference equation. Applying the z-transform yields

$$z^3 Y(z) + 5z^2 Y(z) = z^2 X(z) + 4X(z)$$

Thus its transfer function is

$$H(z) = \frac{Y(z)}{X(z)} = \frac{z^2 + 4}{z^3 + 5z^2}$$

This is a positive-power transfer function with degree 3. Of course it can also be obtained directly from the negative-power transfer function by multiplying z^3 to its numerator and denominator.

This text studies only LTI and lumped systems. Thus we encounter only rational transfer functions that are ratios of two polynomials of z or z^{-1}. Consider the positive-power transfer function in (5.52), where we implicitly assume $\bar{a}_1 \neq 0$ and $\bar{b}_1 \neq 0$. The rest of the coefficients can be zero or nonzero. We call $H(z)$ proper if $\bar{N} \geq \bar{M}$, biproper if $\bar{N} = \bar{M}$, strictly proper if

$\bar{N} > \bar{M}$, and improper if $\bar{M} > \bar{N}$. Properness of a positive-power transfer function depends only on the relatives degrees of its numerator and denominator. If the numerator and denominator of (5.52) have no common factor, then the *degree* of (5.52) is defined as the larger of $\bar{M}$ and $\bar{N}$ or $\bar{N}$ if (5.52) is proper.

It is also possible to define properness for negative-power transfer functions. The condition for (5.51) to be proper is $a_1 \neq 0$, biproper if $a_1 \neq 0$ and $b_1 \neq 0$, strictly proper if $a_1 \neq 0$ and $b_1 = 0$, and improper if $a_1 = 0$ and $b_1 \neq 0$. Its properness is independent of whether $N > M$, $N = M$, or $N < M$. If the numerator and denominator of (5.51) have no common factor, then the degree of (5.51) is defined as the larger of M and N. For example, the rational function

$$H(z) = \frac{2}{z^4 - 2z^2 + z} = \frac{2z^{-4}}{1 - 2z^{-2} + z^{-3}}$$

is strictly proper and has degree 4. The rational function

$$H(z) = \frac{3z(z^2 - 1)}{(z + 1)(z^2 + z + 1)} = \frac{3z(z - 1)}{z^2 + z + 1} = \frac{3(1 - z^{-1})}{1 + z^{-1} + z^{-2}}$$

is biproper and has degree 2.

All systems to be designed in practice are required to be causal; otherwise they cannot be implemented in the real world. A system is causal if and only if its transfer function is proper. Thus we study in this text only proper rational transfer functions. The numerator and denominator of every proper transfer function will be assumed to have no common factor unless stated otherwise.

We mention that both positive- and negative-power transfer functions are used in MATLAB. To differentiate them, we use bn, an to denote, respectively, the numerator's and denominator's coefficients of (5.51) and bp, ap the corresponding coefficients of (5.52). For example, if

$$H(z) = \frac{5z}{z^3 - z^2 + 2z + 3} = \frac{5z^{-2}}{1 - z^{-1} + 2z^{-2} + 3z^{-3}}$$

then we have bn=[0 0 5], an=[1 -1 2 3] and bp=[5 0], ap=[1 -1 2 3]. Note that an always equals ap, and bn generally differs from bp. However, if we write $H(z)$ as

$$H(z) = \frac{0 \times z^3 + 0 \times z^2 + 5z + 0}{z^3 - z^2 + 2z + 3} = \frac{0 + 0 \times z^{-1} + 5z^{-2} + 0 \times z^{-3}}{1 - z^{-1} + 2z^{-2} + 3z^{-3}}$$

then we have b=bn=bp=[0 0 5 0] and a=an=ap=[1 -1 2 3]. By so doing, we do not have to differentiate positive- and negative-power coefficients. Whenever we use b and a for discrete-time transfer functions, they are assumed to have the same length. See also Problems 5.21–5.23.

5.5.1 Poles and Zeros

Consider a proper rational function $H(z) = \bar{B}(z)/\bar{A}(z)$. A real or complex number λ is called a *pole* if $|H(\lambda)| = \infty$, a *zero* if $H(\lambda) = 0$. For example, if

$$H(z) = \frac{\bar{B}(z)}{\bar{A}(z)} = \frac{3z + 6}{2z^3 - 0.8z^2 - 1.1z + 1.7}$$

$$= \frac{1.5(z + 2)}{(z + 1)(z - 0.7 + j0.6)(z - 0.7 - j0.6)} \tag{5.53}$$

then $H(z)$ has poles at -1 and $0.7 \pm j0.6$. The number -2 is clearly a zero. We also have $H(\infty) = 0$; thus ∞, by definition, is also a zero. If the degree difference of $\bar{A}(z)$ and $\bar{B}(z)$ is $d := \bar{N} - \bar{M}$, then $H(z)$ has d number of zeros at ∞. If we include zeros at ∞, then the number of poles equals the number of zeros. We study only finite zeros unless stated otherwise. We mention that if all coefficients of $H(z)$ are real, as is always the case in practice, then complex-conjugate poles and zeros must appear in pairs.

If $\bar{A}(z)$ and $\bar{B}(z)$ have no common factors, then all roots of $\bar{A}(z)$ are poles and all roots of $\bar{B}(z)$ are finite zeros. Poles and finite zeros are usually plotted on the z-plane using crosses and circles as shown in Fig. 5.5. Zeros at infinite are not shown. It is obtained in MATLAB by typing zplane(bp,ap) or zplane(b,a). Note that the last form in (5.53) is called the *zero-pole-gain* form and can be obtained in MATLAB as [z,p,k]=tf2zp(bp,ap), where tf2zp is an acronym for transfer function to zero pole. As in zplane, tf2zp can also use b,a. If we use bn and an in zplane and tf2zp, then some zeros at $z = 0$ may not appear.

Multiplying the numerator and denominator of (5.53) by z^{-3}, we obtain

$$H(z) = \frac{3z^{-2} + 6z^{-3}}{2 - 0.8z^{-1} - 1.1z^{-3} + 1.7z^{-3}}$$

$$= \frac{1.5z^{-2}(1 + 2z^{-1})}{[1 - (-1)z^{-1}][1 - (0.7 + j0.6)z^{-1}][1 - (0.7 - j0.6)z^{-1}]}$$

Using this form, we can also obtain the poles and zeros of $H(z)$. In this form, the two infinite zeros are given by $z^{-2} = (0 + z^{-1})^2$. We see that it is simpler to use positive-power transfer functions to define and to compute poles and zeros.

Now we use an example to discuss the significance of poles and zeros. Consider a DT system with transfer function

$$H(z) = \frac{2z - 1}{z^2 - 1.5z - 1} = \frac{2z^{-1} - z^{-2}}{(1 - 2z^{-1})(1 + 0.5z^{-1})} \tag{5.54}$$

This system has poles at 2 and -0.5 and one zero at 0.5. We will compute its output excited by the input $x[n] = 1$ for $n \geq 0$. Because this input is called a (unit) step sequence, the output is called the *unit step response* or, simply, *step response*. Clearly, we have

$$Y(z) = H(z)X(z) = \frac{2z^{-1} - z^{-2}}{(1 - 2z^{-1})(1 + 0.5z^{-1})} \frac{1}{1 - z^{-1}} \tag{5.55}$$

which can be expanded as, using partial fraction expansion,

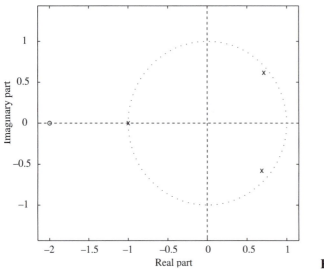

Figure 5.5 Poles and zeros.

$$Y(z) = k_1 + \frac{r_1}{1 - z^{-1}} + \frac{r_2}{1 - 2z^{-1}} + \frac{r_3}{1 + 0.5z^{-1}} \qquad (5.56)$$

Thus the output is, using Table 5.1,

$$y[n] = k_1 \delta[n] + r_1 \times 1^n + r_2(2)^n + r_3(-0.5)^n$$

for $n \geq 0$. We see that the terms 1^n, 2^n, and $(-0.5)^n$ are dictated by the poles of the input $X(z)$ and the transfer function $H(z)$. The zeros of $H(z)$ do not appear explicitly in (5.56); they affect only k_1 and residues r_i. Thus poles play a more important role than zeros in determining the response of systems.

We discuss some responses of poles. A pole λ_1 is called a *simple* pole if there is only one pole at λ_1; a *repeated* pole if there are two or more poles at λ_1. We plot in Fig. 5.6 time responses of some simple poles. The time responses of all simple poles lying inside the unit circle approach zero as $n \to \infty$. This is also true for repeated poles. For example, the time response or inverse z-transform of the repeated pole $1/(1 - \lambda_1 z^{-1})^2$ is $n(\lambda_1)^n$, for $n \geq 0$. Using l'Hôpital rule, we can show $|n(\lambda_1)^n| \to 0$, as $n \to \infty$, if $|\lambda_1| < 1$. In general, we have the following

- Simple or repeated poles lying inside the unit circle→time responses approach zero as $n \to \infty$.
- Simple or repeated poles lying outside the unit circle→time responses grow unbounded as $n \to \infty$.
- Simple poles lying on the unit circle→ constant or sustained oscillation as $n \to \infty$.
- Repeated poles lying on the unit circle→ time responses grow unbounded as $n \to \infty$.

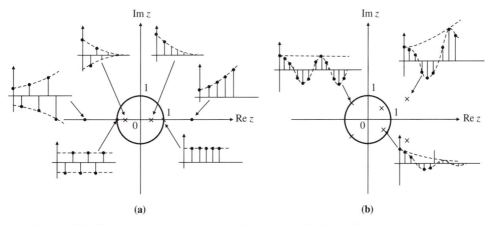

Figure 5.6 Time responses of poles: (a) Real poles. (b) Complex-conjugate poles.

It is important to know that *the time response of a pole, simple or repeated, approaches zero as* $n \to \infty$ *if and only if the pole lies inside the unit circle or has magnitude less than 1.*

To conclude this subsection, we mention a number of MATLAB functions. The impulse response and step response of the transfer function in (5.53) can be obtained, respectively, by using **dimpulse** and **dstep**, where the first character d denotes discrete time. In using these functions, we use the positive-power form of transfer functions. For example, the next program

```
bp=[3 6];ap=[2 -0.8 -1.1 1.7];
n=0:40;
yi=dimpulse(bp,ap,41);
ys=dstep(bp,ap,41);
subplot(1,2,1)
stem(n,yi),title('(a)')
hold on
plot(n,zeros(1,41)) (%draw the horizontal coordinate)
subplot(1,2,2)
stem(n,ys),title('(b)')
```

computes and plots in Figs. 5.7(a) and (b) the impulse response and step response of the transfer function in (5.53).

MATLAB contains the function **residuez**, which computes the partial fraction expansion as in (5.56). In using **residuez**, a transfer function must be expressed in negative-power form. For example, if we write (5.53) as

$$H(z) = \frac{3z + 6}{2z^3 - 0.8z^2 - 1.1z + 1.7} = \frac{3z^{-2} + 6z^{-3}}{2 - 0.8z^{-1} - 1.1z^{-2} + 1.7z^{-3}}$$

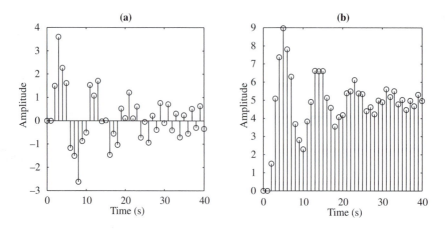

Figure 5.7 (a) DT impulse response. (b) DT step response.

then typing

bn=[0 0 3 6];an=[2 -0.8 -1.1 1.7];[r,p,k]=residuez(bn,an)

will yield

$$H(z) = 3.5294 + \frac{-0.4615}{1 + z^{-1}} + \frac{-1.53 - j1.41}{1 - (0.7 + j0.6)z^{-1}} + \frac{-1.53 + j1.41}{1 - (0.7 - j0.6)z^{-1}}$$

Its inverse z-transform can then be obtained using Table 5.1 and is the impulse response of (5.53). This method of computing impulses responses is prone to numerical error and should be avoided in computer computation. The functions **dimpulse** and **dstep** do not use **residuez**. Instead, they transform transfer functions into sets of first-order difference equations, called state-space equations, and then compute directly the responses. This approach is simpler and yields more accurate results. See Section 9.2.1 and Ref. 7.

5.5.2 Transfer Functions of FIR and IIR Filters

Consider an Nth-order FIR filter with impulse response $h[n]$ for $n = 0, 1, \ldots, N$. Note that it has length $N + 1$. Its transfer function is

$$\begin{aligned} H(z) &= h[0] + h[1]z^{-1} + \cdots + h[n]z^{-N} \\ &= \frac{h[0]z^N + h[1]z^{N-1} + \cdots + h[n]}{z^N} \end{aligned} \tag{5.57}$$

Thus every FIR filter has poles only at the origin of the z plane. Conversely, if a transfer function has poles only at $z = 0$, then it must describe an FIR filer.

◆ Example 5.6

Consider the 10-day moving average filter studied in Example 5.3. Its transfer function is

$$H(z) = \frac{z^9 + z^8 + \cdots + z + 1}{10z^9}$$

$$= \frac{(z^9 + z^8 + \cdots + z + 1)(z - 1)}{10z^9(z - 1)}$$

$$= \frac{z^{10} - 1}{10z^9(z - 1)} \tag{5.58}$$

where we have multiplied the numerator and denominator by $z-1$. Although $z = 1$ is a root of the denominator $z^9(z - 1)$, it is not a pole of $H(z)$. Using l'Hôpital's rule, we have

$$H(1) = \left.\frac{z^{10} - 1}{10(z^{10} - z^9)}\right|_{z=1} = \left.\frac{10z^9}{10(10z^9 - 9z^8)}\right|_{z=1} = 1 \neq \infty$$

Thus $z = 1$ is not a pole. We mention that applying the z-transform to (5.23) yields

$$Y(z) - z^{-1}Y(z) = 0.1[X(z) - z^{-10}X(z)]$$

which implies

$$H(z) = \frac{Y(z)}{X(z)} = \frac{1 - z^{-10}}{10(1 - z^{-1})}$$

This is the same as the one in (5.58).

While every FIR filter has no pole other than $z = 0$, every IIR filter has at least one pole other than $z = 0$. The reason is as follows. The inverse z-transform of $z/(z - a)$ has infinitely many nonzero entries if and only if $a \neq 0$. The inverse z-transform of $1/z^k$, for $k \geq 0$, is $\delta[n - k]$; it has only one nonzero entry. Thus if $H(z)$ has one or more poles other than $z = 0$, then its inverse z-transform has infinitely many nonzero entries. Thus, every IIR filter must have one or more poles other than $z = 0$.

5.5.3 DT Fourier Transform and z-Transform

This subsection compares the z-transform and the DT Fourier transform in system analysis. Consider (5.46), where $x[n]$, $h[n]$, and $y[n]$ are all positive-time sequences. Applying the DTFT to (5.46) yields

$$Y_d(\omega) = H_d(\omega)X_d(\omega) \tag{5.59}$$

where $X_d(\omega)$ is the DTFT of $x[n]$; that is,

$$X_d(\omega) = \mathcal{F}_d[x[n]] = \sum_{n=0}^{\infty} x[n]e^{-jn\omega}$$

and $H_d(\omega)$ and $Y_d(\omega)$ are similarly defined. The derivation of (5.59) is similar to that of $Y(z) = H(z)X(z)$ in (5.47) and will not be repeated. We mention that (5.59) is still applicable if $x[n]$, $h[n]$, and $y[n]$ are two-sided signals defined for all n in $(-\infty, \infty)$ so long as their DTFT are defined. Note that if we use the two-sided z-transform, then (5.47) can also be extended to two-sided signals.

In system analysis, the time instant we start to apply an input signal can be considered as $n = 0$. In this sense, no input can be applied from a negative time instant, not to mention $-\infty$. Thus in system analysis, we may consider the time interval to be $[0, \infty)$.

We now compare the DT Fourier transform and z-transform. The applicability of DTFT is limited. For example, if $x[n] = (-2)^n$ and $h[n] = 1.2^n$, then their DTFT are not defined and (5.59) cannot be used to study a system. If $h[n] = 1$ and $x[n] = 1$, for $n \geq 0$, then their DTFTs are defined; however, the output $y[n]$ of (5.46) will grow unbounded. Thus $Y_d(\omega)$ is not defined and (5.59) is again not applicable.

In using $Y(z) = H(z)X(z)$ in system analysis, there is no such restriction. For example, if $x[n] = h[n] = 1$, then we have

$$Y(z) = \frac{1}{1 - z^{-1}} \frac{1}{1 - z^{-1}} = \frac{1}{(1 - z^{-1})^2}$$

and $y[n] = n(1)^n = n$, for $n \geq 0$. Thus the z-transform is more general than the DTFT in system analysis.

As we will show in a later section, if $h[n]$ is absolutely summable and $x[n]$ has a DTFT, then (5.59) is applicable. However, even in this case, the z-transform is still simpler. For example, consider a DT system with impulse response $h[n] = 0.9^n$, for $n \geq 0$. Its output excited by $x[n] = 1$ for $n \geq 0$ can be computed from (5.47) as

$$Y(z) = \frac{1}{1 - 0.9z^{-1}} \frac{1}{1 - z^{-1}}$$

Its inverse z-transform is the output. If we use (5.59), then we can show

$$H_d(\omega) = \frac{1}{1 - 0.9e^{-j\omega}} \qquad X_d(\omega) = \pi\delta(\omega) + \frac{1}{1 - e^{-j\omega}}$$

(see Problem 5.25) and

$$Y_d(\omega) = \frac{1}{1 - 0.9e^{-j\omega}} \left(\pi\delta(\omega) + \frac{1}{1 - e^{-j\omega}} \right)$$

Computing its inverse DTFT is more complex than computing inverse z-transform of $Y(z)$. Thus we conclude that the z-transform is more general and simpler than DTFT in computing responses of DT systems. Furthermore, the z-transform can be used to develop transfer functions and general properties of DT systems. This is difficult, if not impossible, in using DTFT. Thus there seems no reason to use DTFT in system analysis.

5.6 Stability

A DT system is said to be *BIBO stable* or, simply, *stable* if every bounded input sequence excites a bounded output sequence. If a system is not stable, generally the output will grow unbounded when an input (no matter how small) is applied. Such a system will eventually saturate, overflow, or burn out and cannot be used in practice. Thus systems that are designed to process signals must be stable.

Consider an LTI causal system described by the convolution

$$y[n] = \sum_{k=0}^{n} h[n-k]x[k] = \sum_{k=0}^{n} h[k]x[n-k]$$

The system is stable if and only if $h[n]$ is absolutely summable; that is,

$$\sum_{n=0}^{\infty} |h[n]| < \infty \tag{5.60}$$

Indeed, if an input sequence is bounded or $|x[n]| < p$ for all $n \geq 0$ and for some constant p, then we have

$$|y[n]| = \left| \sum_{k=0}^{n} h[k]x[n-k] \right| \leq \sum_{k=0}^{n} |h[k]||x[n-k]|$$

$$\leq p \sum_{k=0}^{\infty} |h[k]| < \infty$$

for all $n \geq 0$. Thus if $h[n]$ is absolutely summable, every bounded input excites a bounded output.

Next we show the converse. Suppose $h[n]$ is not absolutely summable. Then for any arbitrarily large M, there exists n_1 such that

$$\sum_{n=0}^{n_1} |h[n]| \geq M$$

Let

$$x_1[n_1 - k] = \begin{cases} 1 & \text{if } h[k] \geq 0 \\ -1 & \text{if } h[k] < 0 \end{cases}$$

This input sequence is clearly bounded. The output excited by this input is

$$y[n_1] = \sum_{k=0}^{n_1} h[k]x[n_1 - k] = \sum_{k=0}^{n_1} |h[k]| \geq M$$

Because $y[n_1]$ can be arbitrarily large, we conclude that the system is not stable. This establishes the assertion that *a system is stable if and only if its impulse response is absolutely summable*.

Discrete-time LTIL systems can also be described by proper rational transfer functions. In terms of transfer functions, a system is stable if and only if every pole lies inside the unit circle on the z plane or, equivalently, every pole has a magnitude less than 1. This condition can be established using the fact that $h[n]$ is the inverse z-transform of $H(z)$. We expand $H(z)$, by partial fraction expansion, into the sum of its poles. If $H(z)$ has one or more poles on or outside the unit circle, then $h[n]$ contains at least one term that is not absolutely summable. If $H(z)$ has poles only inside the unit circle, then every term of $h[n]$ approaches 0 *exponentially* and is absolutely summable. Thus *a system with proper rational transfer function $H(z)$ is stable if and only if all its poles have magnitudes less than 1*.

If a system is stable, its transfer function $H(z)$ has finite magnitudes on the unit circle of the z-plane. Indeed, because $|e^{-jn\omega}| = 1$, for all n and ω, we have

$$|H(e^{j\omega})| = \left| \sum_{n=0}^{\infty} h[n]e^{-jn\omega} \right| \leq \sum_{n=0}^{\infty} |h[n]||e^{-jn\omega}|$$

$$= \sum_{n=0}^{\infty} |h[n]| < \infty$$

Furthermore, in using (5.28) to compute the inverse z-transform of $H(z)$, we may select $c = 1$. Thus, for stable systems, we have

$$\mathcal{Z}^{-1}[H(z)] = \mathcal{F}_d^{-1}[H(e^{j\omega})] \tag{5.61}$$

5.6.1 The Jury Test

Consider a DT system with transfer function

$$H(z) = \frac{(z + 10)(z - 5)}{2z^4 - 2.98z^3 + 0.17z^2 + 2.3418z - 1.5147} \tag{5.62}$$

One way to check its stability is to compute its poles. Typing d=[2 -2.98 0.17 2.3418 -1.5147];roots(d) in MATLAB will yield the four poles

$$0.7 + 0.6i, \quad 0.7 - 0.6i, \quad -0.9, \quad 0.99$$

The two real poles clearly have magnitudes less than 1. The magnitudes of the two complex poles equal $\sqrt{0.7^2 + (\pm 0.6)^2} = \sqrt{0.85} = 0.92 < 1$. Thus the DT transfer function is stable.

The preceding method of checking stability is not simple by hand if the denominator of $H(z)$ has a degree three or higher. Let us discuss a method of checking stability without computing poles.[7] A polynomial $D(z)$ is called a *Schur polynomial* if all its roots have magnitudes less than 1. We introduce a method, called the *Jury test*, to check whether a polynomial is Schur without computing its root. For convenience, we use a polynomial of degree five to discuss the method. Consider

$$D(z) = a_0 z^5 + a_1 z^4 + a_2 z^3 + a_3 z^2 + a_4 z + a_5 \quad \text{with } a_0 > 0 \qquad (5.63)$$

If $a_0 < 0$, we simply consider $-D(z)$, and if $-D(z)$ is Schur, so is $D(z)$. We form Table 5.2. The first row is the coefficients of $D(z)$ arranged in descending powers of z. The second row is the reverse of the first row. We then compute $k_1 = a_5/a_0$, the ratio of the last entries of the first two rows. We subtract from the first row the product of the second row and k_1 to yield the first b_i row as shown. For example, we have $b_0 = a_0 - k_1 a_5$, $b_1 = a_1 - k_1 a_4$, and so forth. Because the way k_1 is defined, the last entry of the b_i row is always zero and will be discarded. We then reverse the order of b_i, compute $k_2 = b_4/b_0$, and subtract from the first b_i row the product of the second b_i row and k_2 to obtain the first c_i row. The last entry of the c_i row is zero and is discarded. Proceeding repeatedly until we obtain the last entry f_0 as shown. This completes the Jury table.

The Jury Test The polynomial of degree five with a positive leading coefficient in (5.63) is Schur if and only if the five leading coefficients $\{b_0, c_0, d_0, e_0, f_0\}$ in Table 5.2 are all positive. If any one of them is zero or negative, then the polynomial is not Schur.

Table 5.2 The Jury Table

a_0	a_1	a_2	a_3	a_4	a_5	
a_5	a_4	a_3	a_2	a_1	a_0	$k_1 = a_5/a_0$
b_0	b_1	b_2	b_3	b_4	0	
b_4	b_3	b_2	b_1	b_0		$k_2 = b_4/b_0$
c_0	c_1	c_2	c_3	0		
c_3	c_2	c_1	c_0			$k_3 = c_3/c_0$
d_0	d_1	d_2	0			
d_2	d_1	d_0				$k_4 = d_2/d_0$
e_0	e_1	0				
e_1	e_0					$k_5 = e_1/e_0$
f_0						

---

[7] The method may be skipped without loss of continuity.

◆ **Example 5.7**

Consider the transfer function in (5.62). We apply the Jury test to its denominator:

2.0000	−2.9800	0.1700	2.3418	−1.5147	k_1	$= -1.5147/2$
−1.5147	2.3418	0.1700	−2.9800	2.0000		$= -0.7574$
0.8528	−1.2064	0.2987	0.0847	0	k_2	$= 0.0849/0.8528$
0.0847	0.2987	−1.2064	0.8528			$= 0.0996$
0.8443	−1.2361	0.4188	0		k_3	$= 0.4188/0.8443$
0.4188	−1.2361	0.8443				$= 0.4960$
0.6366	−0.6230	0			k_4	$= -0.6230/0.6366$
−0.6230	0.6366					$= -0.9786$
0.0269						

Because all four leading coefficients are positive, the denominator is Schur and the $H(z)$ in (5.62) is stable.

5.7 Frequency Response

Consider a DT system with transfer function $H(z)$. Its output excited by the input $x[n]$ with z-transform $X(z)$ is given by

$$Y(z) = H(z)X(z)$$

The inverse z-transform of $Y(z)$ yields the output $y[n]$ in the time domain. The output $y[n]$ as $n \to \infty$ is called the *steady-state output* or *steady-state response*, denoted as

$$y_{ss}[n] = \lim_{n \to \infty} y[n]$$

If a system is not stable, generally $y_{ss}[n]$ will grow unbounded no matter what input is applied. On the other hand, if the system is stable, the steady-state response is determined by the applied input and the frequency response of the system, as we will introduce in this section.

First we show that if a system is stable and if the input is a unit step sequence defined as $x[n] = 1$ for $n \geq 0$, then the output of the system approaches a step sequence with magnitude $H(1)$ or $y_{ss}[n] = H(1)$. Indeed, if $x[n] = 1$ for $n \geq 0$, then $X(z) = 1/(1 - z^{-1})$ and the output is given by

$$Y(z) = H(z)X(z) = H(z)\frac{1}{1 - z^{-1}}$$

We use partial fraction expansion to expand it as

$$Y(z) = \frac{r_1}{1 - z^{-1}} + [\text{terms due to poles of } H(z)] \qquad (5.64)$$

with

$$r_1 = H(z)\frac{1}{1 - z^{-1}}(1 - z^{-1})\Big|_{z=1} = H(1)$$

If $H(z)$ is stable, then all its poles lie inside the unit circle and their time responses all approach zero as $n \to \infty$. Thus we have

$$y_{ss}[n] = \lim_{n\to\infty} y[n] = H(1) \times 1^n = H(1)$$

This establishes the assertion. Note that, because the system is linear, if the input is a step sequence with magnitude a or $x[n] = a$ for $n \geq 0$, then the output approaches a step sequence with magnitude $aH(1)$.

Conversely, if $H(z)$ is not stable, its output excited by a step input sequence cannot approach a step sequence. Indeed, if $H(z)$ is not stable, it has at least one pole on or outside the unit circle. If the pole is outside the unit circle, its time response will grow unbounded. If it is at $z = 1$, then $Y(z)$ contains a repeated pole at $z = 1$ and its inverse contains the term $n(1)^n$, for $n \geq 0$, which grows unbounded. If $H(z)$ has a pair of complex-conjugate poles at $e^{\pm j\omega_0}$, then its time response contains $\sin \omega_0 n$, a sustained oscillation. Thus we have established the theorem that follows.

Theorem 5.1 A DT system is stable if and only if its output excited by a step input sequence approaches a constant (zero or nonzero) as $n \to \infty$.

This theorem can be used to check stability of a system by measurement. We apply a step input sequence. If the output approaches a constant, the system is stable. If the output grow unbounded or remains oscillatory, the system is not stable.

Next we study the steady-state response excited by sinusoidal sequences. Before proceeding, we discuss $H(z)$ with z replaced by $e^{j\omega_0}$. For example, if $H(z) = 0.5/(1 - 0.5z^{-1})$ and $z = e^{j1.2}$, then we have

$$H(e^{j1.2}) = \frac{0.5}{1 - 0.5e^{-j1.2}} = \frac{0.5}{1 - 0.5(\cos 1.2 - j\sin 1.2)}$$

$$= \frac{0.5}{1 - 0.5(0.36 - j0.93)} = \frac{0.5}{0.82 + j0.465}$$

$$= \frac{0.5e^{j0}}{0.943e^{j0.516}} = 0.53e^{-j0.516}$$

Thus $H(e^{j1.2})$ has magnitude 0.53 and phase -0.516 rad. In general, we can express $H(e^{j\omega_0})$ as $A(\omega_0)e^{j\theta(\omega_0)}$ with

$$A(\omega_0) = |H(e^{j\omega_0})| \quad \text{and} \quad \theta(\omega_0) = \sphericalangle H(e^{j\omega_0}) \tag{5.65}$$

where $A(\omega_0)$ and $\theta(\omega_0)$ denote, respectively, the magnitude and phase. If all coefficients of $H(z)$ are real, then $A(\omega_0)$ is even and $\theta(\omega_0)$ is odd; that is,

$$A(-\omega_0) = A(\omega_0) \quad \text{and} \quad \theta(-\omega_0) = -\theta(\omega_0)$$

With this preliminary, we are ready to discuss the steady-state response excited by a sinusoidal sequence. Let $x[n] = e^{j\omega_0 n}$, for $n \geq 0$. Then we have

$$X(z) = \frac{1}{1 - e^{j\omega_0}z^{-1}}$$

and

$$Y(z) = H(z)X(z) = \frac{H(z)}{1 - e^{j\omega_0}z^{-1}}$$

Its partial fraction expansion yields

$$Y(z) = \frac{r}{1 - e^{j\omega_0}z^{-1}} + [\text{terms due to poles of } H(z)]$$

where

$$r = H(z)|_{z=e^{j\omega_0}} = H(e^{j\omega_0}) =: A(\omega_0)e^{j\theta(\omega_0)}$$

If $H(z)$ is stable, the time responses of all terms due to the poles of $H(z)$ will approach zero as $n \to \infty$. Thus we have

$$y_{ss}[n] = \lim_{n \to \infty} y[n] = H(e^{j\omega_0})e^{j\omega_0 n} = A(\omega_0)e^{j[\omega_0 n + \theta(\omega_0)]} \tag{5.66}$$

$$= A(\omega_0)\{\cos[\omega_0 n + \theta(\omega_0)] + j \sin[\omega_0 n + \theta(\omega_0)]\} \tag{5.67}$$

This implies that if a system is stable and if we applied a complex exponential sequence as an input, the output will approach a complex exponential sequence with the same frequency but different amplitude and phase.

If $x[n] = \cos\omega_0 n = \text{Re}(e^{j\omega_0 n})$, where Re stands for the real part, then the steady-state response of $H(z)$ is given by

$$y_{ss}[n] = A(\omega_0)\cos[\omega_0 n + \theta(\omega_0)] \tag{5.68}$$

This follows directly from (5.67). Likewise, if $x[n] = \sin\omega_0 n = \text{Im}(e^{j\omega_0 n})$, where Im stands for the imaginary part, then its steady-state output is given by

$$y_{ss}[n] = A(\omega_0)\sin[\omega_0 n + \theta(\omega_0)] \tag{5.69}$$

We state these as a theorem.

Theorem 5.2 Consider a system with proper transfer function $H(z)$. If $H(z)$ is stable, then the output excited by $x[n] = \cos \omega_0 n$ [$\sin \omega_0 n$] will approach

$$y_{ss}[n] = A(\omega_0) \cos[\omega_0 n + \theta(\omega_0)]$$

$$[y_{ss}[n] = A(\omega_0) \sin[\omega_0 n + \theta(\omega_0)]]$$

as $n \to \infty$, and $H(e^{j\omega}) = A(\omega)e^{j\theta(\omega)}$ is called the *frequency response*. Its magnitude $A(\omega)$ is called the *magnitude response* and its phase $\theta(\omega)$, the *phase response*.

Because $e^{j(\omega + k2\pi)} = e^{j\omega} e^{jk2\pi} = e^{j\omega}$ for all integer k, the frequency response $H(e^{j\omega})$ is periodic with period 2π. Furthermore, because of the evenness of $|H(e^{j\omega})|$ and oddness of $\angle H(e^{j\omega})$ for $H(z)$ with real coefficients, we usually plot the frequency response only in the frequency range $[0, \pi]$. If a frequency response is plotted, the steady-state response of the system excited by a step or sinusoidal sequence can be read out directly from the plot. Note that $H(1)$ is the value of $H(e^{j\omega})$ at $\omega = 0$.

MATLAB function freqz computes the frequency response of DT transfer functions expressed in negative-power form. Let

$$H(z) = \frac{B(z)}{A(z)} = \frac{b_1 + b_2 z^{-1} + \cdots + b_{M+1} z^{-M}}{a_1 + a_2 z^{-1} + \cdots + a_{N+1} z^{-N}}$$

Then the program that follows

```
bn=[b(1) b(2) ··· b(M+1)];an=[a(1) a(2) ··· a(N+1)];
[h,w]=freqz(bn,an,n);
```

computes n equally spaced points of the frequency response in $[0, \pi)$. It is obtained by computing fft(bn,n)./fft(an,n). We note that freqz(bn,an), without the third argument, computes 512 points in $[0, \pi)$. The magnitude and phase responses can then be obtained as plot(w,abs(h)) and plot(w,angle(h)). We mention that stability is essential in Theorem 5.2 and in defining the frequency response, as the following examples illustrate.

◆ **Example 5.8**

Consider a DT systems with transfer functions

$$H_1(z) = \frac{0.3(1 + z^{-1})}{(1 - 0.9z^{-1})} = \frac{0.3(z + 1)}{(z - 0.9)}$$

We use MATLAB to plot its frequency response and the response excited by the input

$$x[n] = \cos 0.2n + \sin 3n$$

We use freqz(bn,an) to compute 512 samples of the frequency response in $[0, \pi)$, and dlsim(bp,ap,x) to compute the output excited by the input x, where dlsim is an acronym for discrete linear simulation. For $H_1(z)$, we have bn=bp=:b and an=ap=:a. The program that follows

Program 5.2
```
b=[0.3 0.3];a=[1 -0.9];
[h,w]=freqz(b,a);
n=0:1:100;
x=cos(0.2*n)+sin(3*n);
y=dlsim(b,a,x);
subplot(2,1,1)
plot(w,abs(h),w,angle(h),':'),title('(a)')
subplot(2,1,2)
plot(n,y),title('(b)')
```

generates the magnitude response (solid line) and the phase response (dotted line) in Fig. 5.8(a) and the output excited by $x[n]$ in Fig. 5.8(b). For easier viewing, we plot the *envelope* of the output by using plot instead of stem.
    To verify Theorem 5.2, we read from Fig. 5.8(a)

$$H_1(e^{0.2j}) = 2.8e^{-j1.1}; \quad H_1(e^{3j}) = 0.02e^{-j1.57}$$

Clearly, this reading cannot be very accurate. Then Theorem 5.2 implies

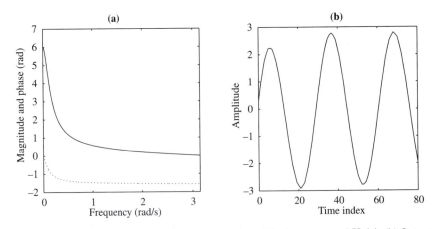

Figure 5.8 (a) Magnitude (solid line) and phase (dotted line) responses of $H_1(z)$. (b) Output excited by $x[n]$.

$$y_{ss}[n] = \lim_{n \to \infty} y[n] = 2.8\cos(0.2n - 1.1) + 0.02\sin(3n - 1.57)$$

We see that the sinusoidal sequence $\cos 0.2n$ is amplified and the sinusoidal sequence $\sin 3n$ is severely attenuated. Thus the output of the filter approaches $2.8\cos(0.2n - 1.1)$, as shown in Fig. 5.9(b). This verifies Theorem 5.2.

Before proceeding, we mention that the use of plot(n,y) in Program 5.2 is, strictly speaking, incorrect. The output y is a sequence of numbers, and we should use stem(n,y). However, the plot of stem(n,y) will become very crowded for n large. Thus we use the function plot.

◆ **Example 5.9**

Consider a DT system with transfer function

$$H_2(z) = \frac{0.33(1 + z^{-1})}{1 - 1.11z^{-1}} \tag{5.70}$$

Program 5.2 will yield the plot in Fig. 5.9 if b and a are replaced by b=[0.33 0.33] and a=[1 -1.11]. The plot of $|H_2(e^{j\omega})|$ is identical to the magnitude response in Fig. 5.8(a); the plot of $\measuredangle\, H_2(e^{j\omega})$, however, is different from the phase response in Fig. 5.8(a). Because $H_2(z)$ is not stable, its response excited by $x[n]$ grows to infinity as shown in Fig. 5.9(b). Thus the plot of $H_2(e^{j\omega})$ in Fig. 5.9(a) has no physical meaning.

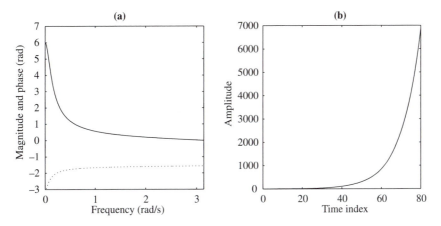

Figure 5.9 (a) Magnitude (solid line) and phase (dotted line) plots of $H_2(z)$. (b) Output excited by $x[n]$.

The MATLAB function filter can also be used to compute responses of transfer functions excited by any input. In other words, dlsim(bp,ap,x) can be replaced by filter(bn,an,x). Note that dlsim uses coefficients of positive-power transfer functions as its input and filter uses coefficients of negative-power transfer functions as its input.

5.7.1 Infinite Time

If a system with transfer function $H(z)$ is stable and if the input is a step or sinusoidal sequence, then the output, following (5.64) and (5.66), is of the form

$$y[n] = y_{ss}[n] + [\text{terms due to poles of } H(z)]$$

The terms enclosed by the parentheses are called collectively the *transient response*. Clearly, the output reaches steady state only after the transient response dies out completely. Mathematically speaking, the transient will die out only as $n \to \infty$. In practice, the transient, however, can be considered to have died out when its value decreases to less than 1% of its peak value. Using this convention, the response of a stable system can reach steady state in a finite time.

To simplify the discussion, we assume $H(z)$ to have a real pole at ρ_1 and a pair of complex-conjugate poles at $\rho_2 e^{j\omega_0}$ and $\rho_2 e^{-j\omega_0}$. Then the output of the system excited by a unit step sequence or the sinusoidal input sequence $\sin \omega_1 n$ with $\omega_1 \neq \omega_0$ is of the form

$$y[n] = k_1 \rho_1^n + k_2 \rho_2^n \sin(\omega_0 n + k_3) + \text{steady-state response} \tag{5.71}$$

If $H(z)$ is stable, then ρ_1 and ρ_2 have magnitudes less than 1 and the transient response

$$y_t[n] = k_1 \rho_1^n + k_2 \rho_2^n \sin(\omega_0 n + k_3) \tag{5.72}$$

will approach zero as $n \to \infty$. Clearly the smaller the magnitudes of ρ_1 and ρ_2, the faster the corresponding term approaches zero. Thus the larger magnitude will dictate the time for the response to decrease to 1% or less. Let ρ be the larger of $|\rho_1|$ and $|\rho_2|$. We define

$$\text{time constant} = -1/\ln \rho$$

where ln stands for natural logarithm. Because $0 \leq \rho < 1$, the time constant is a positive number. This definition is adopted from the CT case. See Ref. 6, p. 469. Define

$$\alpha = \overline{-4.5/\ln \rho}$$

where the overline rounds a number upward to an integer. We show by direct verification that it takes roughly α samples for (5.72) to reach 1% of its peak value or, equivalently, for (5.71) to reach steady state. For example, if $\rho = 0.95$, then we have $\alpha = 87.7 = 88$ and $\rho^n \leq 0.011 = 1.1\%$, for $n \geq 88$. Thus we may consider the response in (5.71) to have reached steady state in 88 samples. We list in Table 5.3 some ρ, α, and ρ^α. From the table, we see that it

Table 5.3 Time Constant and Response

ρ	α	ρ^α
0.99	448	0.011
0.95	88	0.011
0.9	43	0.011
0.8	21	0.009
0.5	7	0.007
0.3	4	0.008

takes about α samples for the transient in (5.71) to decrease to 1% of its peak value and for the response to reach steady state.

From Table 5.2, we see that if $\rho = 0.5$, it takes only 7 samples for the response to reach steady state. If $\rho = 0$, we have $\alpha = 0$. In this case, all three poles of (5.71) are located at $z = 0$, and the system introduces only a delay of three samples. There will be no transient response.

In conclusion, even though it takes an infinite time mathematically for a response to reach steady state, in practice, the response can be considered to have reached steady state in a finite time. For example, if the sampling period is $T = 0.01$ s, then for $\rho = 0.95$, it takes only 0.88 s for the response to reach steady state. Therefore, we often disregard transient responses in design. Nevertheless, it is important to know that the response cannot reach steady state instantaneously.

5.7.2 Frequency Response and Frequency Spectrum

We discuss first a relationship between the z-transform and DT Fourier transform. Let $x[n]$ be positive-time and absolutely summable. Then its z-transform and DTFT are, respectively,

$$X(z) = Z[x[n]] = \sum_{n=0}^{\infty} x[n]z^{-n}$$

and

$$X_d(\omega) = \mathcal{F}_d[x[n]] = \sum_{n=0}^{\infty} x[n]e^{-jn\omega}$$

Comparing these yields immediately

$$X_d(\omega) = \mathcal{F}_d[x[n]] = Z[x[n]]|_{z=e^{j\omega}} = X(e^{j\omega}) \tag{5.73}$$

It is important to mention that (5.73) holds only if $x[n]$ is positive-time and absolutely summable. If $x[n]$ is positive-time and periodic for $n \geq 0$, then its z-transform and DT Fourier transform are defined. However, (5.73) must be modified. See Problems 5.24 and 5.25. If $x[n]$ grows unbounded, (5.73) does not hold.

Next we show that if a DT causal system is stable and if its input is absolutely summable, so is its output. Indeed, we have

$$\left| y[n] \right| = \left| \sum_{k=0}^{\infty} x[k]h[n-k] \right| \leq \sum_{k=0}^{\infty} |x[k]||h[n-k]|$$

Thus we have

$$\sum_{n=0}^{\infty} |y[n]| \leq \sum_{n=0}^{\infty} \sum_{k=0}^{\infty} |x[k]||h[n-k]| = \sum_{k=0}^{\infty} \left(\sum_{n=0}^{\infty} |h[n-k]| \right) |x[k]|$$

$$= \sum_{k=0}^{\infty} \left(\sum_{\bar{n}=-k}^{\infty} |h[\bar{n}]| \right) |x[k]| = \left(\sum_{k=0}^{\infty} |x[k]| \right) \left(\sum_{\bar{n}=0}^{\infty} |h[\bar{n}]| \right)$$

where we have interchanged the order of summations, introduced a new index $\bar{n} = n - k$, and used the causality condition $h[\bar{n}] = 0$ for $\bar{n} < 0$. Thus if a DT system is stable (its impulse response is absolutely summable), and if the input is absolutely summable, so is the output.

Consider an LTI discrete-time system with transfer function $H(z)$. Its input $x[n]$ and output $y[n]$ are related by

$$Y(z) = H(z)X(z) \tag{5.74}$$

where $Y(z)$, $H(z)$, and $X(z)$ are, respectively, the z-transforms of $y[n]$, $h[n]$, and $x[n]$. This is a general equation, applicable whether or not the system is stable, and whether or not the frequency spectrum of the input signal is defined. Now if the system is stable and if the input is absolutely summable, then (5.74) and (5.73) imply

$$Y(e^{j\omega}) = H(e^{j\omega})X(e^{j\omega}) \tag{5.75}$$

and

$$Y_d(\omega) = H_d(\omega)X_d(\omega) = H(e^{j\omega})X_d(\omega) \tag{5.76}$$

This is an important equation and forms the basis of filtering. It states that if a system is stable, its output frequency spectrum excited by an input with frequency spectrum $X_d(\omega)$ is well defined and equals the product of the system frequency response and the input frequency spectrum. The product is carried out point-by-point at every frequency. For example, if the input frequency spectrum is shown in Fig. 5.10(a) with a solid line and the system frequency response is shown with a dotted line. Then their product is as shown in Fig. 5.10(b). The part of the input whose spectrum lying inside $[-1, 1]$ passes through the system without any attenuation; the part outside $|\omega| \leq 2$ is completely stopped by the system. The part in $1 \leq |\omega| \leq 2$ is partially attenuated. Equations (5.75) and (5.76) are basic in filter design. We mention that if $x[n]$ is positive time and periodic for $n \geq 0$, then (5.76) still holds. However, if $H(z)$ is not stable or if $x[n]$ grows

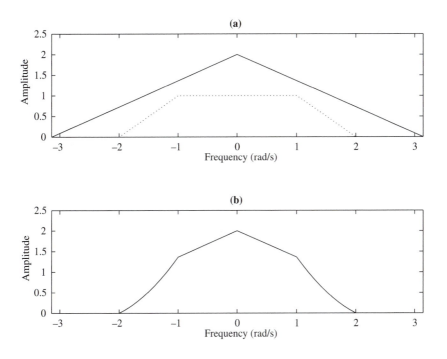

Figure 5.10 (a) Input frequency spectrum (solid line) and system frequency response (dotted line). (b) Output frequency spectrum.

unbounded, then (5.76) is meaningless.

To conclude this subsection, we mention that if $h[n]$ and $x[n]$ are positive time and absolutely summable, then their DT Fourier transforms are called, respectively, the frequency response and frequency spectrum. Thus the *frequency response and frequency spectrum are mathematically the same. The former is defined for systems, the latter for signals.*

5.7.3 Alternative Derivation of Frequency Responses

We now discuss a different way of developing the frequency response. Consider (5.6), which was developed by assuming that the system is linear, time-invariant, initially relaxed at $n = 0$, and that the input is applied from $n = 0$ on. Now if the system is initially relaxed at $n = -\infty$ and the input is applied from $-\infty$, then (5.6) must be modified as

$$y[n] = \sum_{k=-\infty}^{\infty} x[k]h[n-k] = \sum_{k=-\infty}^{\infty} h[k]x[n-k] \qquad (5.77)$$

where the last term can be obtained from the second term by defining $\bar{k} = n - k$ and then renaming $\bar{k}$ as k.

Let $x[n] = e^{j\omega_0 n}$ for all n in $(-\infty, \infty)$. In other words, the input $e^{j\omega_0 n}$ is applied from $-\infty$. Then (5.77) implies

$$y[n] = \sum_{k=-\infty}^{\infty} h[k]e^{j\omega_0(n-k)} = \left(\sum_{k=-\infty}^{\infty} h[k]e^{-j\omega_0 k} \right) e^{j\omega_0 n}$$

$$= H_d(\omega_0)e^{j\omega_0 n} \tag{5.78}$$

where $H_d(\omega)$ is the DTFT of $h[n]$ and is the frequency response. This is the approach adopted by most DSP texts and appeared to be simpler than the derivation in Section 5.7.

The derivation of (5.78), however, is not complete without discussing the condition of its validity. If $h[n] = 1.2^n$, for $n \geq 0$, then the infinite summation in (5.78) diverges and $H_d(\omega)$ is not defined. If $h[n] = 1$, for $n \geq 0$, then its DTFT is

$$H_d(\omega) = \frac{1}{1 - e^{j\omega}} + \pi\delta(\omega)$$

See Problem 5.25. It contains an impulse at $\omega = 0$. Thus $H_d(\omega)$ is not well defined at $\omega = 0$, and the meaning of (5.78) is not clear. If $h[n]$ is absolutely summable or, equivalently, the system is stable, then the DTFT $H_d(\omega)$ is finite at every ω and (5.78) is well defined. In this case, the DTFT $H_d(\omega)$ equals the frequency response $H(e^{j\omega})$, as shown in (5.73), and (5.78) becomes

$$y[n] = H(e^{j\omega_0})e^{j\omega_0 n} \tag{5.79}$$

for all n. This is similar to (5.66) and requires the same stability condition.

Equation (5.79) holds for all n in $(-\infty, \infty)$, whereas (5.66) holds only as $n \to \infty$. Can we reconcile this discrepancy? Note that the causality condition is not used in (5.77). Now suppose the system is causal and initially relaxed at n_0. If an input is applied from $n = n_0$, then (5.6) must be modified as

$$y[n] = \sum_{k=n_0}^{n} x[k]h[n-k] = \sum_{k=0}^{n-n_0} h[k]x[n-k] \tag{5.80}$$

The second equality can again be obtained as in (5.11) and (5.77). Note that (5.80) does not have the commutative property as in (5.11) and (5.77). Substituting $x[n] = e^{j\omega_0 n}$, for $n \geq n_0$, into (5.80), we obtain

$$y[n] = \sum_{k=0}^{n-n_0} h[k]e^{j\omega_0(n-k)}$$

$$\neq \sum_{k=0}^{\infty} h[k]e^{j\omega_0(n-k)} = H(e^{j\omega_0})e^{j\omega_0 n}$$

for all finite n. The inequality becomes an equality only as $n \to \infty$. Thus if a complex exponential is applied from $n = n_0$, where n_0 is any positive or negative finite integer, then the output cannot reach steady state immediately and will consist of a transient response. However, if the complex exponential input is applied from $-\infty$, the transient response somehow dies out at $-\infty$ and (5.80) reaches steady state instantaneously. This is possible only because of the ambiguity of $-\infty$. Moreover. no input can be applied from $-\infty$ in the real world.[8]

In conclusion, the derivation of frequency responses in (5.77) and (5.78) is deceivingly simple. Although the derivation in Section 5.7 is more complex, it reveals explicitly the necessity of the condition of stability. More important, it shows the transient and steady-state responses. This is what happens in the real world when an input is applied.

5.8 Continuous-Time LTIL Systems

In this and subsequent sections, we discuss briefly the CT counterparts of what has been discussed for DT systems. Although most concepts in the DT case can be applied here, the mathematics involved is much more complex. Fortunately, we need in this text only CT transfer functions, stability, and frequency responses. The concept of impulse responses, which is essential in defining FIR and IIR filters in the DT case, is not needed in this text. Thus the reader may glance through this section.

The concepts of linearity, time invariance, and initial relaxedness are directly applicable to CT systems. For every LTI system that is initially relaxed at $t = 0$, its output excited by the input $x(t)$, for $t \geq 0$, can be described by the convolution

$$y(t) = \int_{\tau=0}^{\infty} h(t - \tau)x(\tau)d\tau \tag{5.81}$$

where $h(t)$ is the output excited by the impulse $x(t) = \delta(t)$ and is called the *impulse response*. See the appendix. Unlike the DT case, where the impulse sequence $\delta[n]$ can easily be generated, the impulse $\delta(t)$, which has a zero width and infinite height, cannot be generated in practice.

If the current output of a CT system does not depend on future input, the system is said to be causal. The condition for a system to be causal is $h(t) = 0$ for $t < 0$, denoted as

$$\text{causal} \iff h(t) = 0, \quad \text{for } t < 0$$

For LTI causal systems, (5.81) can be modified as

$$y(t) = \int_{\tau=0}^{t} h(t - \tau)x(\tau)d\tau = \int_{\tau=0}^{t} h(\tau)x(t - \tau)d\tau \tag{5.82}$$

---

[8] We encountered both ∞ and $-\infty$. The concept $n \to \infty$ is simple and easily understandable, as discussed in Section 5.7.1. Applying an input from $-\infty$, however, is a strange concept. For example, if an input is applied from $n_0 = -100^{100}$, the output will still consist of a transient response because -100^{100} is still far away from $-\infty$.

This is the CT counterpart of (5.11).

If a CT LTI system has the additional lumpedness property, then it can also be described by a differential equation of the form

$$a_1 y^{(N)}(t) + a_2 y^{(N-1)}(t) + \cdots + a_N \dot{y}(t) + a_{N+1} y(t)$$
$$= b_1 x^{(M)}(t) + b_2 x^{(M-1)}(t) + \cdots + b_M \dot{x}(t) + b_{M+1} x(t) \tag{5.83}$$

where a_i and b_i are real constants, $\dot{y}(t) = dy(t)/dt$, and $y^{(k)}(t) = d^k y(t)/dt^k$. We study only the case where $a_1 \neq 0$ and $N \geq M$.

The CT counterpart of the z-transform is the Laplace transform. Consider a signal $x(t)$ defined for all t in $(-\infty, \infty)$. Its two-sided Laplace transform is defined as

$$X(s) = \mathcal{L}_2[x(t)] = \int_{t=-\infty}^{\infty} x(t) e^{-st} dt \tag{5.84}$$

and its one-sided Laplace transform, or, simply, Laplace transform is defined as

$$X(s) = \mathcal{L}[x(t)] = \int_{t=0}^{\infty} x(t) e^{-st} dt \tag{5.85}$$

where s is a complex variable. Their inverse Laplace transforms are both given by

$$x(t) = \frac{1}{2\pi j} \int_{c-j\infty}^{c+j\infty} X(s) e^{st} ds \tag{5.86}$$

where the integration is along a vertical line passing through the constant c in the complex s plane. The line must lie inside the region of convergence. As in the DT case, the inverse Laplace transform of $X(s)$ can, depending on the region of convergence, be a positive-time signal, a negative-time signal, or a two-sided signal. However, in CT system analysis and design, we study only positive-time signals. Therefore, we use only (one-sided) Laplace transform and require all inverse Laplace transforms to be positive time. Inverse Laplace transforms will be computed using partial fraction expansion and a Laplace transform table. Thus we may disregard the region of convergence. We list in Table 5.4 some Laplace transform pairs.

We discuss one property of the Laplace transform to conclude this section. Let $X(s)$ be the Laplace transform of $x(t)$. If $x(t) = 0$ for all $t < 0$, then from (5.85) we can show

$$\mathcal{L}\left[\frac{dx(t)}{dt}\right] = sX(s) \tag{5.87}$$

and, more generally,

$$\mathcal{L}[x^{(k)}(t)] = \mathcal{L}\left[\frac{d^k x(t)}{dt^k}\right] = s^k X(s) \tag{5.88}$$

Table 5.4 Some Laplace Transform Pairs

$x(t)$ $t \geq 0$	$X(s)$
$\delta(t - t_0)$	$e^{-t_0 s}$
e^{at}	$\dfrac{1}{s - a}$
$t e^{at}$	$\dfrac{1}{(s - a)^2}$
$e^{at} \sin \omega_0 t$	$\dfrac{\omega_0}{(s - a)^2 + \omega_0^2}$
$e^{at} \cos \omega_0 t$	$\dfrac{s - a}{(s - a)^2 + \omega_0^2}$

for any positive integer k. Thus differentiation in the time domain equals multiplication by s in the Laplace transform.

5.8.1 Laplace Transform and z-Transform

This subsection discusses the relationship between the z-transform and the Laplace transform. Let $x[n]$ be a positive-time sequence. If we apply the Laplace transform to $x[n]$, then the result will be identically zero. See the appendix. Now we define

$$x_s(t) := \sum_{n=0}^{\infty} x[n]\delta(t - n) \tag{5.89}$$

It is zero everywhere except at $t = n$, where it is an impulse with weight $x[n]$. The signal $x_s(t)$ can be considered as a continuous-time representation of the discrete-time sequence $x[n]$.

Applying the Laplace transform to $x_s(t)$ yields

$$X_s(s) := \mathcal{L}[x_s(t)] = \sum_{n=0}^{\infty} x[n]\mathcal{L}[\delta(t - n)]$$

$$= \sum_{n=0}^{\infty} x[n]e^{-ns} = \sum_{n=0}^{\infty} x[n]z^{-n}$$

where we have defined $z = e^s$. Thus we have

$$\mathcal{L}[x_s(t)] = Z[x[n]]|_{z=e^s} \tag{5.90}$$

This relates the Laplace transform and z-transform. Their variables are related by $z = e^s$. Clearly, we have

$$s = 0 \rightarrow z = 1$$

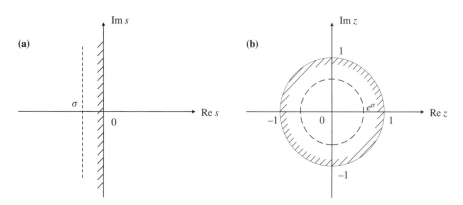

Figure 5.11 Relationship between (a) s plane and (b) z plane.

$$s = j\omega \quad \text{(imaginary axis)} \qquad\qquad \rightarrow |z| = 1 \quad \text{(unit circle)}$$
$$s = -\infty \qquad\qquad\qquad\qquad\qquad \rightarrow z = 0$$

More generally, the left-half s plane is mapped into the interior of the unit circle on the z-plane, as shown in Fig. 5.11. The imaginary axis of the s plane is mapped into the unit circle, and the right-half s plane is mapped into the exterior of the unit circle. The mapping is not one-to-one. For example, $s = 0, \pm 2\pi j, \pm 4\pi j, \ldots$ all map into $z = 1$.

5.9 CT Transfer Function, Stability, and Frequency Response

Applying the Laplace transform to (5.81) yields

$$Y(s) = H(s)X(s) \tag{5.91}$$

where

$$H(s) = \mathcal{L}[h(t)] = \int_{t=0}^{\infty} h(t)e^{-st}dt \tag{5.92}$$

which is called the CT *transfer function*. It is the Laplace transform of the impulse response. The transfer function $H(s)$ can be an irrational or rational function of s. The CT systems we will encounter in this text are all lumped, and their transfer functions are all rational functions of s. Indeed, if $x(t) = 0$ and $y(t) = 0$ for $t < 0$, applying the Laplace transform to (5.83) and using (5.88), we can obtain

$$H(s) = \frac{Y(s)}{X(s)} = \frac{b_1 s^M + b_2 s^{M-1} + \cdots + b_{M+1}}{a_1 s^N + a_2 s^{N-1} + \cdots + a_{N+1}} =: \frac{B(s)}{A(s)} \tag{5.93}$$

where we assume implicitly $a_1 \neq 0$ and $b_1 \neq 0$. Unlike the DT case, we use exclusively positive powers of s in expressing $H(s)$. We study only proper rational functions, that is, $N \geq M$. If $H(s)$ is improper or $N < M$, then $H(s)$ will amplify high-frequency noise that often exists in the real world. It is also difficult to implement improper $H(s)$. Thus we rarely encounter improper transfer functions in practice.

As in the discrete-time case, we can define poles and zeros for proper rational functions. If $A(s)$ and $B(s)$ in (5.93) have no common factors, then all roots of $A(s)$ are the poles of $H(s)$ and all roots of $B(s)$ are the finite zeros of $H(s)$. We plot time responses of some real poles in Fig. 5.12(a) and time responses of some pairs of complex-conjugate poles in Fig. 5.12(b). The time response of a pole approaches zero as $t \to \infty$ if and only if the pole lies inside the left-half s plane or, equivalently, has a negative real part.

A CT system is defined to be stable if every bounded input excites a bounded output. The necessary and sufficient condition for a system to be stable is that its impulse response is absolutely integrable; that is,

$$\text{stable} \iff \int_{t=0}^{\infty} |h(t)|dt < \infty$$

In terms of proper rational transfer functions, a system is stable if and only if every pole lies inside the left-half s plane or has a negative real part. If $H(s)$ is stable, and if $x(t) = a$, for $t \geq 0$, then we have

$$y_{ss}(t) := \lim_{t \to \infty} y(t) = aH(0) \tag{5.94}$$

If $x(t) = a \cos \omega_0 t$, for $t \geq 0$, then

$$y_{ss}(t) = a|H(j\omega_0)| \cos[\omega_0 t + \measuredangle H(j\omega_0)] \tag{5.95}$$

If $x(t) = e^{j\omega_0 t}$, for $t \geq 0$, then

$$y_{ss}(t) = H(j\omega_0)e^{j\omega_0 t} \tag{5.96}$$

As in the DT case, if $H(s)$ is stable, $H(j\omega)$ is called the *frequency response*, $|H(j\omega)|$ the *magnitude response*, and $\measuredangle H(j\omega)$ the *phase response*.

Mathematically it takes infinite time to reach steady state. In practice, if the transient response decreases to less than 1% of its peak value, then the response can be considered to have reached steady state. In continuous-time systems, it takes roughly five time constants to reach steady state, where the time constant is defined as

$$\text{time constant} = \frac{1}{\text{shortest distance from all poles to the imaginary axis}}$$

See Ref. 6. For example, if $H(s)$ has poles -2 and $-1.5 \pm j4$, then its time constant is $1/1.5$, and it takes roughly $5/1.5 = 3.3$ s for a response to reach steady state.

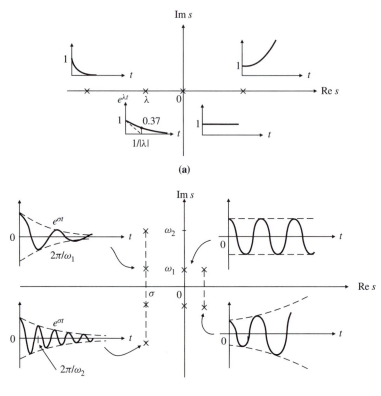

Figure 5.12 Time responses of some poles: (a) Real poles. (b) Complex-conjugate poles.

If a CT system is stable, then we have

$$H_a(\omega) = \mathcal{F}[h(t)] = \mathcal{L}[h(t)]|_{s=j\omega} = H(j\omega) \tag{5.97}$$

where $H_a(\omega)$ is the CT Fourier transform and $H(s)$ is the Laplace transform of $h(t)$. We also have

$$\mathcal{L}^{-1}[H(s)] = \mathcal{F}^{-1}[H(j\omega)] \tag{5.98}$$

If a CT system is not stable, its output may grow unbounded, and its frequency spectrum is not defined. On the other hand, if a system is stable, and if the input frequency spectrum is defined, so is the output frequency spectrum. In this case, we have

$$Y(j\omega) = H(j\omega)X(j\omega)$$

and

$$Y_a(\omega) = H_a(\omega)X_a(\omega)$$

They are the CT counterparts of (5.75) and (5.76).

5.9.1 Measuring CT Frequency Responses

We discuss how to obtain frequency responses of CT systems. If the transfer function of a system is given, its frequency response can be obtained by calling the MATLAB function freqs. For the CT transfer function in (5.93), we type

b=[b(1) b(2) $\cdots$ b(M+1)];a=[a(1) a(2) $\cdots$ a(N+1)];
[H,w]=freqs(b,a)

Then the function freqs automatically picks a set of 200 frequencies in $[0, \infty)$ and computes the frequency response at these frequencies.

   If the transfer function of a system is not available, the only way to obtain the frequency response is by measurement. In fact, once the frequency response is measured, we can then find a rational transfer function whose frequency response approximates the measured one. This is a widely used method of finding transfer functions in practice. We discuss in the following two methods of obtaining frequency responses from measured outputs.

Method 1 If we apply the input $x(t) = \sin \omega_0 t$ to a stable system, the output will approach $|H(j\omega_0)|\sin[\omega_0 t + \not\!\measuredangle\, H(j\omega_0)]$. Thus from the steady-state response, we can measure the frequency response at $\omega = \omega_0$. By sweeping ω_0 over a range such as $\omega_0 = m\Delta\omega$, for $m = 0, 1, 2, \ldots$, we can obtain the magnitude response $|H(jm\Delta\omega)|$ and the phase response $\not\!\measuredangle\, H(jm\Delta\omega)$. Instruments called spectrum analyzers, network or control system analyzers are commercially available to carry out this measurement and to generate transfer functions.

Method 2 Consider

$$Y(j\omega) = H(j\omega)X(j\omega) \tag{5.99}$$

If we can generate an impulse as an input or, $x(t) = \delta(t)$, then $X(j\omega) = 1$ and $Y(j\omega) = H(j\omega)$ for all ω as shown in Fig. 5.13(a). Thus the frequency response of the system equals the frequency spectrum of this output. The only problem is that no impulse having a zero width and infinite height can be generated in practice. Fortunately, frequency responses of most physical systems are bandlimited in the sense $H(j\omega) \approx 0$ for ω larger than some W, as shown in Fig. 5.13(a). Now if we can find an input with frequency spectrum roughly equal to 1 for $\omega < W$, as shown in Fig. 5.13(b), then $Y(j\omega) \approx H(j\omega)$ and the frequency response can be computed from the corresponding output. Let us consider the signal shown in Fig. 5.14(a). It is a pulse with height $1/a$ and width a. As in (3.10) and (3.11), its spectrum can be readily computed as[9]

---

[9] The spectrum can also be computed from its Laplace transform. Its Laplace transform is $(1 - e^{-as})/s$, which yields the spectrum after replacing s by $j\omega$.

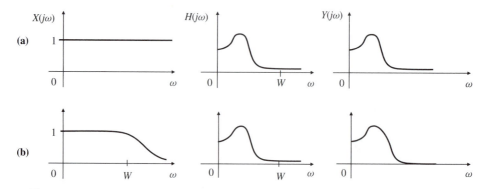

Figure 5.13 Measuring frequency response: (a) Impulse input. (b) Practical input.

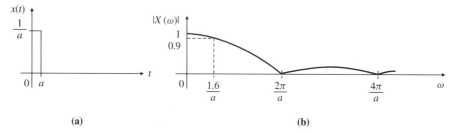

Figure 5.14 (a) Pulse. (b) Its frequency spectrum.

$$X(j\omega) = \frac{2e^{-j0.5a\omega} \sin 0.5a\omega}{a\omega}$$

Its magnitude spectrum is plotted in Fig. 5.14(b). Its magnitude is 1 at $\omega = 0$ and decreases to 0.9 at $\omega = 1.6/a$. Thus if a is selected such that $1.6/a > W$ or $a < 1.6/W$, then the frequency response of the system roughly equals the frequency spectrum of the output excited by the pulse. The spectrum of the output can be computed using FFT, as discussed in Chapter 4. This is the setup in Fig. 1.17. In inspecting the loudspeakers in Fig. 1.17, we are interested in only the waveform of frequency responses, thus the height of the pulse in Fig. 5.14(a) is not important. However, the height must be large enough to drive the system but not too large to drive the system outside the linear range or damage it.

Let us compare the preceding two methods. In Method 1, after applying $\sin \omega_0 t$, we must wait until the response reaches steady state before measuring. We then change ω_0 and repeat. If we measure 20 points of $H(j\omega)$, we need 20 runs. In Method 2, we apply a pulse, then $H(j\omega)$ can be computed from the samples of the output. This method requires only one run; thus it takes much less time to complete than the first method. Method 1, however, may yield a more accurate result than Method 2, which uses a pulse to approximate an impulse.

We mention that the frequency spectrum of an output can also be obtained by passing the output through a bank of bandpass analog filters and then compute the energy in each frequency band. This technique is less often used than the FFT method, and will not be discussed.

5.10 Concluding Remarks

This chapter introduced the class of DT and CT systems to be used in this text. We introduced the concept of impulse responses for DT systems and then classify DT systems as FIR (finite impulse response) or IIR (infinite impulse response). Although the same concept can be applied to CT systems, it is much more complex. Furthermore, most CT systems encountered in practice are IIR, and no such classification is used in CT systems.

The z-transform of DT impulse responses are called DT transfer functions. The Laplace transform of CT impulse responses are called CT transfer functions. In the transform domains, DT and CT systems can be described by

$$Y(z) = H(z)X(z) \quad \text{and} \quad Y(s) = H(s)X(s)$$

if they are initially relaxed. They are algebraic equations, and their algebraic manipulations are identical.

Systems are defined to be stable if every bounded input excites a bounded output. The condition for DT systems to be stable is that all poles of $H(z)$ lie inside the unit circle on the z-plane. The condition for CT systems to be stable is that all poles of $H(s)$ lie inside the left-half s plane.

In signal analysis, how to select a sampling period T is an important issue. In system analysis, however, we may assume, as discussed in Section 5.3.2, the sampling period to be 1 or any other value. For stable systems, $H(e^{j\omega})$, for $-\pi < \omega \le \pi$, by assuming $T = 1$, and $H(j\omega)$, for $-\infty < \omega < \infty$, are called frequency responses. From frequency responses, we can read out steady-state responses of systems excited by step or sinusoidal inputs. For unstable systems, frequency responses have no physical meaning and are not defined.

In CT systems, we use exclusively positive-power transfer functions. Consequently, all MATLAB functions for CT systems use positive-power coefficients. Many of them, such as tf2zp, zp2tf and residue, can be directly applied to DT systems. However, many DSP texts use exclusively negative-power transfer functions. Negative-power form is more convenient in using the MATLAB function freqz but is less convenient or less transparent in discussing properness of transfer functions and the roles of poles and zeros in shaping magnitude responses, as we will discuss in the next chapter. Thus this text uses both positive- and negative-power DT transfer functions.

Although residue can be directly applied to DT systems to yields positive-power partial fraction expansion, MATLAB contains residuez to carry out negative-power partial fraction expansion. However, if we apply dstep to negative-power transfer functions, we will obtain incorrect step responses. See Problems 5.21–5.23. Thus care must be exercised in using MATLAB functions for DT systems. One way to avoid this problem is to use the same length of numerator and denominator coefficients as discussed at the end of Section 5.5.

PROBLEMS

5.1   Find a convolution and a difference equation to describe a savings account with interest rate 0.0001 per day and compounded daily.

5.2   Develop a difference equation from the convolution in Problem 5.1. Also compute the impulse response sequence from the difference equation obtained in Problem 5.1.

5.3   Design a 5-day moving-average filter. Find a recursive and a nonrecursive difference equation to describe the filter. What are their orders? Which difference equation requires less computation?

5.4   Develop a block diagram for the nonrecursive difference equation in Problem 5.3. Does the number of unit-delay elements equal the order of the equation? Repeat the questions for the recursive equation in Problem 5.3.

5.5   (a) Compute the impulse response of the first-order difference equation

$$y[n] + y[n-1] = x[n]$$

   Is it an FIR or IIR filter?
   (b) Repeat the questions for the second-order difference equation

$$y[n] + y[n-1] = x[n] - x[n-2]$$

5.6   Use direct division to find the inverse z-transform of

$$X(z) = \frac{2z^2 + 3}{z(z+1)(z+2)}$$

   Also find its inverse using positive-power and negative-power partial fraction expansions. Which is simpler?

5.7   Show that if $x[n]$ is not positive time, then

$$Z[x[n-1]] = z^{-1}X(z) + x[-1]$$

   and

$$Z[x[n-2]] = z^{-2}X(z) + x[-1]z^{-1} + x[-2]$$

5.8   Consider

$$y[n] + 3y[n-1] + 2y[n-2] = x[n-1] + 2x[n-2]$$

What is its unit step response, that is, the output excited by $x[n] = y[n] = 0$, for $n < 0$, and $x[n] = 1$ for $n \geq 0$? Compute it directly and indirectly using the z-transform.

5.9   Consider

$$a_1 y[n+2] + a_2 y[n+1] + a_3 y[n] = b_1 x[n+2] + b_2 x[n+1] + b_3 x[n]$$

Use (5.39) and (5.40) and the initially relaxed condition ($x[n] = y[n] = 0$ for $n < 0$) to find its transfer function $Y(z)/X(z)$. Is the result the same if we use $Z[y[n+k]] = z^k Y(z)$ instead of (5.39) and (5.40)?

5.10   Discuss properness, degree, poles, and zeros of the following transfer functions.

$$H_1(z) = \frac{z^4}{(z+1)^2 (z^2 + 0.5z + 0.06)}$$

$$H_2(z) = \frac{1 - z^{-3}}{z^{-1} - z^{-2}}$$

$$H_3(z) = \frac{z^2 + z + 1}{z^3 - 1}$$

$$H_4(z) = \frac{1 - z^{-2}}{1 + z^{-1}}$$

What type of filter does each transfer function describe?

5.11   Consider a DT system with impulse response sequence $h[n] = e^{-0.1n} + 2(0.5)^n$, for $n \geq 0$. What is its transfer function? Where are its poles and zeros? What is the general form of its unit step response? What is its steady-state response?

5.12   Find the output of a DT system with impulse resonse sequence $h[n] = 1$ for $n \geq 0$ excited by the input $x[n] = 1$ for $n \geq 0$. Is the output bounded?

5.13   Consider a DT system with transfer function

$$H(z) = \frac{z^2 + 1}{(z-1)(z+0.5)}$$

Find its convolution, delayed-form, and advanced-form difference-equation descriptions.

5.14   Use MATLAB to write a program to compute and to plot the output $y[n]$ for $n = 0:15$ of the system in Problem 5.13 excited by the input $x[n] = n + 1$ for $n = 0:3$, $x[n] = n - 8$ for $n = 4:7$, $x[n] = 0$ for $n \geq 7$, and the initial conditions

$x[-2] = x[-1] = 0$, $y[-2] = y[-1] = 0$. Use the most convenient description among transfer function, convolution, and difference equation, to carry our the task.

5.15 Determine the stability of the following DT systems:

(a) $h[n] = \begin{cases} 10^n & \text{for } n = 0, 1, \ldots, 10 \\ (0.1)^n & \text{for } n \geq 11 \end{cases}$

(b) $h[n] = 1/n$ for $n > 0$

(c) $H(z) = \dfrac{(z-2)(z+1.2)}{z^3 - 0.3z^2 - 1.1755 + 0.75}$

(d) $H(z) = \dfrac{1 + 3z^{-1} + 10z^{-3}}{1 - 0.9z^{-1} - 0.275z^{-2} + 0.375z^{-3}}$

(e) $y[n] = 3x[n] + x[n-1] + 10x[n-2]$

(f) $2y[n+2] - y[n+1] - y[n] = 2x[n+2] - 2x[n]$

5.16 Consider a system with impulse response sequence $h[n]$, for $n \geq 0$. Is it true that if $h[n]$ is bounded and approaches 0 as $n \to \infty$, then the system is stable? If not, under what conditions will it be true?

5.17 Show that the necessary and sufficient conditions for

$$z^2 + a_1 z + a_2$$

to have all roots lying inside the unit circle are $|a_2| < 1$ and $|a_1| < 1 + a_2$. Use the Jury test to establish the condition. Can you prove it directly from the roots of the polynomial without using the Jury test?

5.18 What is the steady-state response of the DT system with transfer function $H(z) = (z+1)/(z^2 - 0.64)$ excited by $x[n] = 1 + \sin 0.1n$, for $n \geq 0$?

5.19 What is the steady-state response of the DT system with transfer function $H(z) = 0.6(z-1)/(z-0.16)$ excited by $x[n] = 10 \sin 0.01n + \cos 3.1n$, for $n \geq 0$? Is the system a high-pass or a low-pass filter?

5.20 Plot the magnitude and phase responses of the transfer function in Problem 5.19.

5.21 Consider

$$H(z) = \frac{3z}{2z^3 - 0.8z^2 - 1.1z + 1.7}$$

What are bn, an and bp, ap? Plot its step responses by using dstep(bn,an) and dstep(bp,ap). Which yields the correct step response? Is it true that they differ only by a delay or advance of one sample? Give your reasons.

5.22   Plot the magnitude and phase responses of the $H(z)$ in Problem 5.21 by using freqz(bn,an) and freqz(bp,ap). Which yields the correct frequency response? Is it true that their magnitude responses are the same but their phase responses are different? Give your reasons.

5.23   For the transfer function in Problem 5.21, what are the results of

[z,p,k]=tf2zp(bp,ap),[Bp,Ap]=zp2tf(z,p,k)

where tf2zp stands for transfer function to zero pole, and

[z,p,k]=tf2zp(bn,an),[Bn,An]=zp2tf(z,p,k)?

Can you conclude that the two MATLAB function tf2zp and zp2tf are defined for positive-power coefficients?

5.24   Let $x[n]$ be positive time and periodic for $n \geq 0$. Then we have

$$\mathcal{F}_d[x[n]] = Z[x[n]]|_{z=e^{j\omega}} + 0.5\mathcal{F}_d[x_p[n]]$$

where $\mathcal{F}_d$ denotes DT Fourier transform and $x_p[n]$ is the periodic extension of $x[n]$ to all n (Ref. 28). Use this formula to show that the spectrum of $x[n] = \cos \omega_0 n$, for $n \geq 0$, is

$$X_d(\omega) = \frac{1 - (\cos \omega_0)e^{-j\omega}}{1 - 2(\cos \omega_0)e^{-j\omega} + e^{-j2\omega}} + 0.5\pi\delta(\omega - \omega_0) + 0.5\pi\delta(\omega + \omega_0)$$

5.25   Use the formula in Problem 5.24 to show that the DTFT or spectrum of $x[n] = 1$, for $n \geq 0$, is

$$X_d(\omega) = \frac{1}{1 - e^{-j\omega}} + \pi\delta(\omega)$$

5.26   Discuss the applicability of the DTFT in computing the unit step responses of the systems in Problems 5.11 and 5.12?

5.27   Consider a DT system with proper and stable transfer function $H(z)$. The inverse of $H(z)$ is called its *inverse system*. Under what conditions will the inverse system be proper and stable?

Ideal and Some Practical Digital Filters

6.1 Introduction

With the background in the preceding chapter, we are ready to discuss the design of digital filters. Before plunging into specific design techniques, we discuss in this chapter a number of general issues in filter design. We then discuss the design of first- and second-order digital filters. For these simple filters, we discuss how to use poles and zeros to shape their magnitude responses. We also show how to use them to process CT signals. We then introduce digital sinusoidal generators, comb filters, all-pass filters, and some other related topics. Finally, we discuss ideal analog low-pass filters, their relationship with ideal D/A converters, and the reason for introducing antialiasing analog filters before A/D conversion.

As discussed in Section 5.3.2, the sampling period T does not play any role in non real-time processing and is implicitly assumed to be larger than the time required to compute each output data in real-time processing. Once output data are computed, they are displayed immediately or stored in registers. Delivery to the output is timed by a separate control logic. Therefore, we may assume, without loss of generality, $T = 1$ in designing digital filters. Under this assumption, frequency responses of DT systems are specified only over $(-\pi, \pi]$ or $[0, \pi]$.

6.2 Ideal Digital Filters

Consider the digital filter with frequency response specified by

$$H(e^{j\omega}) = \begin{cases} 1 \times e^{-j\omega n_0} & \text{for } |\omega| \leq \omega_c \\ 0 & \text{for } \omega_c < |\omega| \leq \pi \end{cases} \tag{6.1}$$

and plotted in Fig. 6.1, where n_0 is a positve integer. Note that the plot can be extended periodically with period 2π to all ω. The filter has magnitude 1 and a linear phase $-\omega n_0$ for $|\omega| \leq \omega_c$ and zero magnitude and unspecified phase for $\omega_c < |\omega| \leq \pi$.

Let us apply the sinusoidal sequence $x[n] = \sin n\omega_0$ to the digial filter. Then the output of the filter is given by, using Theorem 5.2,

$$y_{\text{ss}}[n] := \lim_{n \to \infty} y[n] = |H(e^{j\omega_0})| \sin \left[n\omega_0 + \not{\star} \, H(e^{j\omega_0}) \right] \tag{6.2}$$

If $|\omega_0| \leq \omega_c$, then

$$y_{\text{ss}}[n] = \sin(n\omega_0 - n_0\omega_0) = \sin[\omega_0(n - n_0)] \tag{6.3}$$

It means that, after the transient, the sinusoid can pass through the filter without any attenuation except a delay of n_0 samples. If $|\omega_0| > \omega_c$, then $y_{\text{ss}} = 0$ or the sinusoid is eventually blocked by the filter. More generally, consider $x[n] = x_s[n] + x_n[n]$, where $x_s[n]$ and $x_n[n]$ denote, respectively, a desired signal and an unwanted noise. Suppose their frequency spectra $X_s(\omega)$ and $X_n(\omega)$ are as shown in Fig. 6.2(a). Because the output frequency spectrum, as shown in (5.75), is the product of the filter frequency response and the input frequency spectrum, we have

$$Y(\omega) = H(e^{j\omega})X(\omega) = X_s(\omega)e^{-jn_0\omega} \tag{6.4}$$

which implies, using (3.51),

$$y[n] = x_s[n - n_0] \tag{6.5}$$

Thus the filter rejects completely the noise and passes without any attenuation the desired signal except a delay of n_0 samples. This is called a distortionless transmission of $x_s[n]$, and the filter is called a *digital ideal low-pass filter*.

Filters to be designed in the real world are required to have real coefficients. Such filters have even magnitude responses and odd phase responses. Thus we often plot frequency responses only in the positive frequency range $[0, \pi]$, as shown in Fig. 6.3(a). The frequency range $[0, \omega_c]$ is called the *passband* and the range $[\omega_c, \pi]$, the *stopband*. The frequency ω_c is called the *cutoff* frequency. Figures 6.3(b)–(d) show the magnitude responses of digital ideal high-pass, bandpass, and bandstop filters. These filters are called collectively *frequency-selective filters*.

6.3 Realizability

Can an ideal filter be built in the real world? As discussed in (5.10), a necessary condition for physical realizability is that the filter be causal or, equivalently,

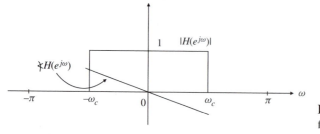

Figure 6.1 Ideal digital low-pass filter.

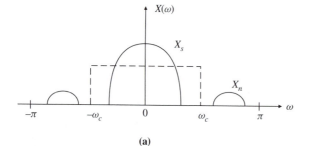

(a)

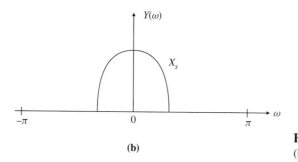

(b)

Figure 6.2 (a) Input frequency spectrum. (b) Output frequency spectrum.

$$h[n] = 0 \qquad \text{for } n < 0 \tag{6.6}$$

We shall check this condition for the ideal low-pass filter defined in (6.1). As discussed in (5.61), the inverse z-transform of (6.1) can be computed from the inverse DT Fourier transform of $H(e^{j\omega})$ as

$$h[n] = \frac{1}{2\pi} \int_{\omega=-\pi}^{\pi} H(e^{j\omega}) e^{jn\omega} d\omega \tag{6.7}$$

Substituing (6.1) into (6.7) yields

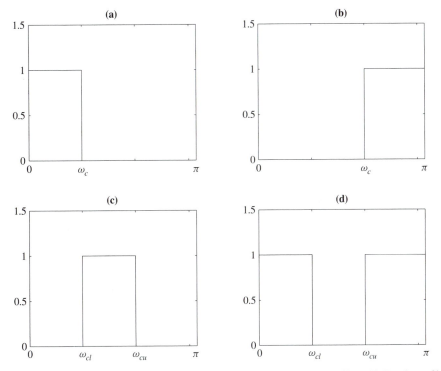

Figure 6.3 (a) Ideal digital low-pass filter. (b) High-pass filter. (c) Bandpass filter. (d) Bandstop filter.

$$h[n] = \frac{1}{2\pi} \int_{\omega=-\omega_c}^{\omega_c} e^{-jn_0\omega} e^{jn\omega} d\omega = \frac{\sin[(n-n_0)\omega_c]}{(n-n_0)\pi} \tag{6.8}$$

with $h[n_0] = \omega_c/\pi$. It is plotted in Fig. 6.4(a) for $n_0 = 5$ and $\omega_c = 1$. Clearly, $h[n] \neq 0$ for $n < 0$. Thus the digital ideal low-pass filter is not causal and, consequently, not physically realizable.

A question naturally arises at this point: How do we select a frequency response so that there exists a causal filter to have the selected response? Let $H(e^{j\omega}) = A(\omega)e^{-j\theta(\omega)}$. Generally, if $A(\omega)$ and $\theta(\omega)$ are specified independently, then no causal filter exists to meet the specifications. If only one of them, say, $A(\omega)$, is specified, then it is possible to find a causal filter whose magnitude response is as close as desired to the given $A(\omega)$. For example, let us specify only the magnitude response in Fig. 6.1, with the phase response left unspecified. We discuss in the following two causal filters to approximate the magnitude response.[1]

---

[1] This is consistent with the so-called Paley-Wiener condition. If $A(\omega)$ is identically zero over a nonzero frequency interval, then $\log|H(e^{j\omega})| = \log A(\omega) = -\infty$, and $A(\omega)$ violates the Paley-Wiener condition

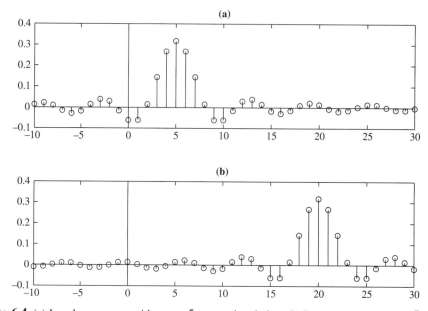

Figure 6.4 (a) Impulse response with $n_0 = 5$ versus time index. (b) Impulse response with $\bar{n}_0 = 20$ versus time index.

Consider the noncausal impulse response in (6.8). Let us select a large integer n_0, say, $n_0 = 20$, such that $h[n] \approx 0$ for $n < 0$ as shown in Fig. 6.4(b). By simply truncating $h[n]$ for $n < 0$, we will obtain a causal filter. The magnitude response of this causal filter will approximate the ideal magnitude response. Clearly, the larger n_0, the better the approximation. However, a large n_0 will introduce a large phase shift and a long time delay. Nevertheless, this does establish the assertion that a causal filter exists to approximate as closely as desired the magnitude response.

Next we discuss a Butterworth filter to approximate the magnitude response in Fig. 6.1 with $\omega_c = 1$. We type the following MATLAB functions

Program 6.1
```
[b,a]=butter(40,1/pi);
[H,w]=freqz(b,a,256);
subplot(1,2,1)
plot(w,abs(H)),title('(a)')
subplot(1,2,2)
plot(w,angle(H)*180/pi),title('(b)')
```

---

and has no causal filter realization. However, if $A(\omega)$ is nonzero but less than, for example, 10^{-10}, then it meets the condition and there exists a causal filter whose magnitude response equals $A(\omega)$. In engineering, $A(\omega)$ clearly can be considered to be zero if $A(\omega) < 10^{-10}$. Thus the Paley-Wiener condition may be more of theoretical interest than of practical importance.

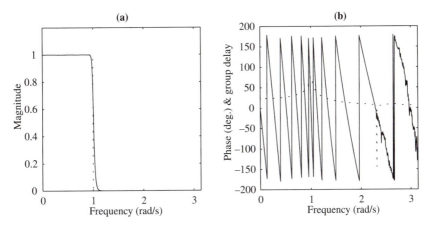

Figure 6.5 (a) Magnitude responses of digital Butterworth (solid line) and digital ideal (dotted line) filters. (b) Phase response (solid line) and group delay (dotted line) of digital Butterworth filter.

The first line generates a fortieth-order Butterworth filter, which we will discuss in Chapter 8. The second line computes 256 points of its frequency response, equally spaced in $[0, \pi)$. Its magnitude response is plotted in Fig. 6.5(a) with a solid line. Figure 6.5(a) also plots the ideal magnitude response with a dotted line. Figure 6.5(b) plots the phase response with a solid line, in degrees, of the Butterworth filter. The phase plot is said to be *wrapped* in the sense that the phase is limited to $\pm 180°$ by subtracting or adding repeatedly $360°$. We see that the magnitude response of the Butterworth filter approximates well the ideal magnitude response. This again establishes the assertion.

6.3.1 Filter Specifications

Even though we can find a causal filter to *approximate* any given magnitude response, the order of the filter may become very large. It is clear that the larger the order, the more computation the filter requires. This may impose a restriction in real-time computation. It will also be more costly if it is implemented with dedicated hardware. Thus in practice, we like to design a filter with an order as small as possible. If specifications are tight, then the required filter will be complex. Thus in practical design, the specifications are often given as shown in Fig. 6.6. The magnitude response is divided into three regions: passband, stopband, and transition bands. The frequency ω_p is called the *passband edge frequency*, and ω_s the *stopband edge frequency*. Instead of requiring passband and stopband to equal, respectively, 1 and 0, they are permitted to lie in the ranges shown, where ϵ_p and ϵ_s are positive numbers and specify allowable tolerances. The magnitude in the transition band is not specified. We see that if $\epsilon_p = 0$, $\epsilon_s = 0$, and $\omega_p = \omega_s =: \omega_c$, then the specification reduces to the one of an ideal low-pass filter. Clearly, the smaller ϵ_p and ϵ_s, and the narrower the transition band, the more complex the required filter.

The phase response $\theta(\omega)$ of a digital filter can be similarly specified. However, such a specification will be inconvenient because phases are often wrapped between $\pm 180°$ as shown in Fig. 6.5(b). Thus the phase response is often specified in terms of the *group delay* $\tau(\omega)$ defined as

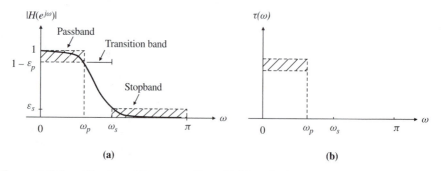

Figure 6.6 Specification of low-pass filter: (a) Magnitude response. (b) Group delay.

$$\tau(\omega) = -\frac{d\theta(\omega)}{d\omega}$$

The group delay of the ideal low-pass filter in (6.1) equals n_0 because $\theta(\omega) = -n_0\omega$. This is the time delay in (6.3). If $\tau(\omega)$ is not a constant, distortion will occur in (6.5). Thus it is desirable to keep $\tau(\omega)$ constant or, if not possible, in a narrow range, as shown in Fig. 6.6(b). The group delays in the transition and stopbands are not important and are not specified. For example, the fortieth-order Butterworth filter generated by Program 6.1 has the group delay shown in Fig. 6.5(b) with dotted line. It is obtained by typing at the end of Program 6.1

```
[gd,w]=grpdelay(b,a,256);
plot(w,gd,':')
```

For ω small, $\tau(\omega)$ roughly equals 20. Thus in practice, magnitude and phase responses are often specified as shown in Fig. 6.6.

We will discuss in the next two chapters the design of FIR and IIR filters. Under some symmetry or antisymmetry property, FIR filters will automatically have linear phases. We then search filters from this class of FIR filters to meet specifications on magnitude responses. In designing IIR filters, the situation is more complex. If we specify both the magnitude and phase responses, then it is difficult to find a discrete-time transfer function to meet both specifications. Thus in the design of IIR filters, we often specify only the magnitude response. If the resulting filter does not have a satisfactory phase response, we may then design an allpass filter to improve its phase response, as we will discuss later in this chapter.

In conclusion, we like to design a simplest possible causal digital filter to have a desired magnitude response. Simplicity can be achieved by requiring the transfer function to be a rational function of z with a degree as small as possible; causality can easily be achieved by requiring the rational function to be proper. In addition, the transfer function must be stable. Thus the design of digital filters reduces to the search of a discrete-time stable proper transfer function of a smallest possible degree to meet a given magnitude-response specification such as the one shown in Fig. 6.6(a) or 6.7. Figure 6.7 shows the specification

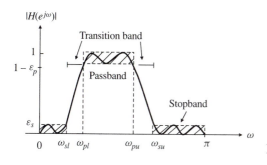

Figure 6.7 Specification of bandpass filter.

of a bandpass filter. It has one passband in $[\omega_{pl}, \ \omega_{pu}]$ with tolerance ϵ_p and two sections of stopband in $[0, \ \omega_{sl}]$ and $[\omega_{su}, \ \pi]$, with tolerance ϵ_s. The frequencies ω_{pu} and ω_{su} are called, respectively, the *upper* passband and stopband edge frequencies; ω_{pl} and ω_{sl} are called *lower* passband and stopband edge frequencies. There are two sections of transition bands as shown, where the magnitudes are not specified. The magnitude response is required to lie inside the shaded areas. The filter is said to have *bandwidth* $(\omega_{pu} - \omega_{pl})$. The low-pass filter specified in Fig. 6.6(a) is said to have bandwidth ω_p. The bandwidth is defined only in the positive frequency range $[0, \ \pi]$. High-pass and bandstop filters can be similarly specified.

6.3.2 Digital Processing of Analog Signals

What has been discussed for digital filters can be applied to analog filters with only minor modification. The frequency range of DT systems is $(-\pi, \ \pi]$ with $T = 1$; the frequency range of CT systems is $(-\infty, \ \infty)$. Thus if $\pm\pi$ in Figs. 6.1, 6.2, 6.3, 6.6, and 6.7 are replaced by $\pm\infty$, then the specifications apply to analog filters.

Consider the processing of CT signals using an analog filter shown in Fig. 6.8(a). The processing can be replaced by the setup shown in Fig. 6.8(b). In digital processing, the CT signal $x(t)$ must first pass through an A/D converter to yield a digital signal $x(nT)$. We then design a digital filter to process $x(nT)$ to yield $\bar{y}(nT)$. The signal $\bar{y}(nT)$ is then converted back to an analog signal $\bar{y}(t)$. Hopefully, $\bar{y}(t)$ will be close to the $y(t)$ in Fig. 6.8(a). In this process, the first step is to select a sampling period T. It must be chosen to avoid frequency aliasing due to time sampling or, equivalently, T must be chosen so that the frequency spectrum of $x(nT)$ roughly equals the spectrum of $x(t)$ divided by T in the frequency range $(-\pi/T, \ \pi/T)$. See (3.68). Because the frequency spectrum of $x(nT)$ is reduced by the factor $1/T$, one may wonder whether the magnitude response of the digital filter to be designed must be increased by the factor T. The answer is negative because the processed $\bar{y}(nT)$ will be converted to $\bar{y}(t)$ and the frequency spectrum of $\bar{y}(t)$ is the spectrum of $\bar{y}(nT)$ multiplied by T. Thus the magnitude responses of analog and digital filters should be the same. If the specification for the analog filter is as shown in Fig. 6.8(a), then the specification of the digital filter should be as shown in Fig. 6.8(c) with frequency range $[0, \ \pi/T]$.

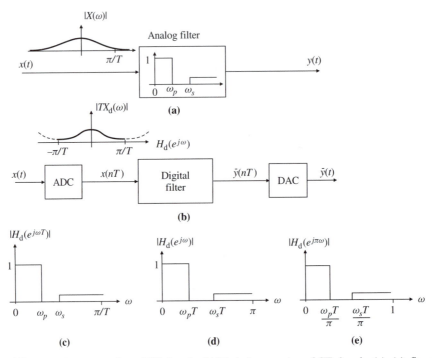

Figure 6.8 (a) Analog processing of CT signals. (b) Digital processing of CT signals. (c)–(e): Specifications of digital filter with different frequency coordinates.

As discussed in Section 5.3.2, actual computation of $\bar{y}(nT)$ is dictated by the hardware used and the complexity of the filter. The sampling period T is used to deliver $\bar{y}(nT)$ to the output at the right instant. Thus in the design of digital filters, we may assume T to be 1 or any other value. Selecting $T = 1$ is equivalent to normalize the Nyquist frequency range $(-\pi/T, \pi/T]$ to $(-\pi, \pi]$. In this case, the specification becomes the one in Fig. 6.8(d). In MATLAB, all filters are designed with frequency range normalized to $[0, 1]$. This is equivalent to select T as π if we use rad/s as the unit of frequency or as 1/2 if we use Hz as the unit of frequency. In terms of $[0, 1]$, the specification becomes the one in Fig. 6.8(e). The three specifications in Figs. 6.8(c)–(e) all lead to the same digital filter, and any one can be used. Once a digital filter $H(z) = B(z)/A(z)$ is obtained, typing

[H,w]=freqz(bn,an);plot(w,abs(H))

where bn and an are negative-power coefficients of $B(z)$ and $A(z)$, will yield the plot in Fig. 6.8(d). Recall that the function freqz(bn,an) automatically select 512 equally spaced frequencies in $[0, \pi)$. If we want to generate the plots in Figs. 6.8(c) and (e), we replace plot(w,abs(H)), respectively, by plot(w./T,abs(H)) and plot(w./pi,abs(H)). In this text, we use mostly the specification in Fig. 6.8(d).

6.4 First-Order Digital Filters

A first-order digital filter has a transfer function of degree one. Such a filter has one pole and none or one zero. We discuss the roles of pole and zero in shaping its magnitude response. Consider

$$H(z) = \frac{1}{z - a} = \frac{z^{-1}}{1 - az^{-1}} \tag{6.9}$$

with $|a| < 1$. If $|a| \geq 1$, the system is not stable and cannot be used as a filter. The transfer function has no finite zero; it has one pole at $z = a$ as shown in Fig. 6.9(a). The pole actually is a vector as shown. The point $e^{j\omega}$ is on the unit circle; it is a vector with magnitude 1 and angle ω. Thus $\alpha := e^{j\omega} - a$ is a vector emitting from a and ending at $e^{j\omega}$. If the vector has magnitude p and angle ϕ as shown in Fig. 6.9(a), then we have

$$\alpha := e^{j\omega} - a = pe^{j\phi}$$

Note that the angle ϕ is positive if it is measured counterclockwise; it is negative if measured clockwise. The frequency response of (6.9) is

$$H(e^{j\omega}) = \frac{1}{e^{j\omega} - a} = \frac{1}{pe^{j\phi}} = \frac{1}{p}e^{-j\phi} \tag{6.10}$$

From this equation, we see that the magnitude response is the inverse of the distance from the pole to a point on the unit circle as the point moves from $\omega = 0$ to π; the phase response is the angle of the vector α with the sign reversed. For example, if $a = 0$ or the pole is at the origin on the z plane, then the vector α has magnitude 1 for all ω and has angle ranging from 0 at $\omega = 0$ to π at $\omega = \pi$. Thus the magnitude response equals 1 for all ω, as shown in Fig. 6.9(b) with a solid line, and the phase response decreases from 0 at $\omega = 0$ to $-\pi$ at $\omega = \pi$, as shown in Fig. 6.9(b) with a dotted line. Such a filter will pass all signals without any attenuation and is called an *all-pass* filter. This filter happens to have a linear phase and its group delay equals 1 for all ω. Actually it is a unit delay element. This is a trivial filter.

If $a = 0.8$, then the vector α has magnitude 0.2 and angle 0 at $\omega = 0$ and magnitude 1.8 and angle π at $\omega = \pi$. Thus the magnitude response of (6.9) will decrease from $1/0.2 = 5$ at $\omega = 0$ to $1/1.8 = 0.56$ at $\omega = \pi$, and the phase response of (6.10) will decrease from 0 to $-\pi$, as shown in Fig. 6.9(c). Using the same argument, we can obtain the frequency response in Fig. 6.9(d) for $a = -0.8$. Clearly, the filter with $a = 0.8$ is a low-pass filter, whereas the filter with $a = -0.8$ is a high-pass filter. Recall that the highest frequency of discrete-time signal with $T = 1$ is π. In conclusion, because the magnitude response is inversely proportional to a pole's distance from the unit circle, the closer the pole to $e^{j\omega}$, the larger the magnitude. Before proceeding, we mention that in graphical interpretation, it is more convenient to use positive-power transfer functions. If we use the negative-power form, the interpretation will not be transparent.

Next we discuss the transfer function

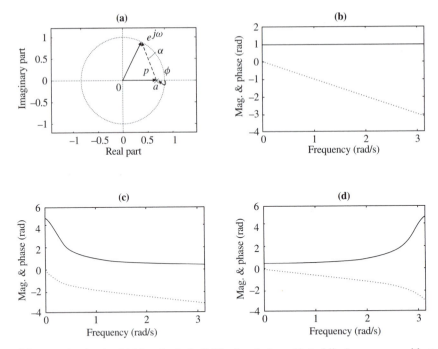

Figure 6.9 (a) Role of pole. (b) Magnitude (solid line) and phase (dotted line) responses with $a = 0$; (c) $a = 0.8$; (d) $a = -0.8$.

$$H(z) = \frac{z - b}{z - a} = \frac{1 - bz^{-1}}{1 - az^{-1}} \qquad (6.11)$$

It has one pole at a and one zero at b. For stability, we require the pole to lie inside the unit circle or $|a| < 1$. The zero, however, can lie inside, on, or outside the unit circle. In this section we consider only $|b| \leq 1$ or zeros lying inside or on the unit circle for reasons to be given later. Let

$$\alpha = e^{j\omega} - a = pe^{\phi} \quad \text{and} \quad \beta = e^{j\omega} - b = qe^{j\psi}$$

The vector α, which emits from the pole a to the point $e^{j\omega}$, has magnitude p and angle ϕ as shown in Fig. 6.10(a). The vector β, which emits from the zero b to $e^{j\omega}$, has magnitude q and angle ψ as shown. Thus the frequency response of (6.11) is

$$H(e^{j\omega}) = \frac{e^{j\omega} - b}{e^{j\omega} - a} = \frac{qe^{j\psi}}{pe^{j\phi}} = \frac{q}{p} e^{j(\psi - \phi)} \qquad (6.12)$$

To see the role of the zero, we assume $a = 0$ or $p = 1$ for all ω. Then the magnitude response of (6.11) equals q, the distance from the zero to the unit circle. Thus farther away the zero from $e^{j\omega}$, the larger the magnitude. For example, if $a = 0$ and $b = -1$, then the magnitude response

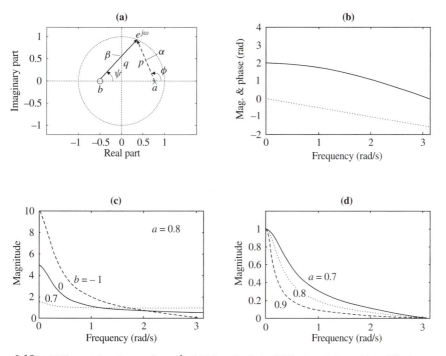

Figure 6.10 (a) Filter with pole a and zero b. (b) Magnitude (solid line) and phase (dotted line) responses with $a = 0$ and $b = -1$. (c) Magnitude responses of (6.11) with $a = 0.8$ and $b = 0, 0.7, -1$. (d) Magnitude responses of (6.13) with $a = 0.7, 0.8, 0.9$.

decreases from 2 at $\omega = 0$ to 0 at $\omega = \pi$, as shown in Fig. 6.10(b) with a solid line. Its phase response is also shown with a dotted line; it equals $\psi - \phi$. It is an FIR filter of order 1 and length 2. In conclusion, the farther away a zero from $e^{j\omega}$, the larger the magnitude of $H(e^{j\omega})$. This is in contrast to the fact that the closer a pole to $e^{j\omega}$, the larger the magnitude of $H(e^{j\omega})$. Thus poles and zeros play opposite roles in magnitude responses.

Now we can combine the effects of pole and zero to design a better first-order low-pass filter. Suppose the pole is located at $a = 0.8$. If the zero is located at the origin or $b = 0$, then the magnitude response of (6.11), as shown in Fig. 6.10(c) with a solid line, is identical to the one in Fig. 6.9(c), which has no zero. Their phase responses, however, will be different. Thus the zero is not utilized in shaping the magnitude response. If the zero is close to the pole, say, $b = 0.7$, then the zero will cancel the effect of the pole, and the resulting magnitude response will be poor. This is indeed the case, as shown in Fig. 6.10(c) with a dotted line. Thus, in order to design a good low-pass filter, the zero should be placed as far away as possible from the pole. Because we consider only zeros lying on or inside the unit circle, the farthest possible zero is $b = -1$. The magnitude response of (6.11) with $a = 0.8$ and $b = -1$ is shown in Fig. 6.10(c) with a dashed line. It is the best low-pass filter among the three shown. Thus a good first-order low-pass digital filter should have the zero at -1 and a pole close to 1.

Next we discuss the bandwidth of the digital low-pass filter

$$H(z) = \frac{b_0(z+1)}{z-a} = \frac{1-a}{2}\frac{z+1}{z-a} \tag{6.13}$$

where $0 \leq a < 1$ and the constant b_0 is chosen so that $H(e^{j\times 0}) = H(1) = 1$. Figure 6.10(d) shows its magnitude responses for $a = 0.7$ (solid line), $a = 0.8$ (dotted line), and $a = 0.9$ (dashed line).

For different passband tolerances, we can define different bandwidths. The 3-dB bandwidth is often used. The decibel (dB) is the unit of logarithmic magnitude defined as

$$20\log_{10}|H(e^{j\omega})| = 10\log_{10}|H(e^{j\omega})|^2 \quad \text{in dB} \tag{6.14}$$

For example, the magnitude of (6.13) at $\omega = 0$ is $2(1-a)/2(1-a) = 1$; thus it is 0 dB. If the maximum magnitude is 1 or 0 dB, the 3-dB passband is defined as the frequency range in $[0, \pi]$ in which the magnitude lies between 0 and -3 dB. The width of the range is called the 3-dB bandwidth. Note that the bandwidth is defined in the positive frequency range $[0, \pi]$, not in $(-\pi, \pi]$.

As shown in Fig. 6.10(d), the magnitude response of (6.13) decreases monotonically from 1 to 0 in $[0, \pi]$. Thus the 3-dB passband edge frequency can be obtained by solving

$$-3 = 20\log_{10}|H(e^{j\omega_p})| = 10\log_{10}|H(e^{j\omega_p})|^2$$

which implies

$$|H(e^{j\omega_p})|^2 = \log_{10}^{-1}\left(\frac{-3}{10}\right) = \frac{1}{2} = 0.5$$

and

$$|H(e^{j\omega_p})| = \log_{10}^{-1}\left(\frac{-3}{20}\right) = \frac{1}{\sqrt{2}} = 0.707$$

For $H(z)$ with real coefficients, we have

$$|H(e^{j\omega})|^2 = H(e^{j\omega})H^*(e^{j\omega}) = H(e^{j\omega})H(e^{-j\omega}) \tag{6.15}$$

where the asterisk denotes complex conjugate. Thus ω_p can be solved from

$$\frac{1-a}{2}\frac{e^{j\omega_p}+1}{e^{j\omega_p}-a}\frac{1-a}{2}\frac{e^{-j\omega_p}+1}{e^{-j\omega_p}-a} = \frac{(1-a)^2}{4}\frac{2+2\cos\omega_p}{1+a^2-2a\cos\omega_p} = \frac{1}{2}$$

or

$$(1-a)^2(1+\cos\omega_p) = 1+a^2-2a\cos\omega_p$$

Its solution is

$$\omega_p = \cos^{-1}\left(\frac{2a}{1+a^2}\right) \tag{6.16}$$

Thus the 3-dB bandwidth of the first-order digital filter in (6.13) can be computed from (6.16). Note that power is proportional to $|H(e^{j\omega})|^2$. Because $|H(0)|^2 = 1$ and $|H(e^{j\omega_p})|^2 = 0.5$, the 3-dB bandwidth is also called the half-power bandwidth.

Let $\beta =: 1 - a$. It is the distance from the pole $a > 0$ to the unit circle. If β is small, the bandwidth can be approximated by

$$\omega_p \approx \beta := 1 - a \tag{6.17}$$

Developing (6.17) from (6.16) is not simple. However, (6.17) can be easily established by using Fig. 6.11 on which the pole and zero of (6.13) are plotted. The magnitude of (6.13) at $z = e^{j0} = 1$ is $\beta \times 2/(2\beta) = 1$ or 0 dB. For β small, the point $e^{j\beta}$ on the unit circle can be approximated by the point A shown in Fig. 6.11, where A is on the vertical line emitting from $z = 1$ and has distance β from $z = 1$. The distance from the zero $b = -1$ to $e^{j\beta}$ roughly equals 2; the distance from the pole a to $e^{j\beta}$ roughly equals $\sqrt{2}\beta$. Thus the magnitude of (6.13) at $z = e^{j\beta}$ roughly equals $\beta \times 2/(2\sqrt{2}\beta) = 1/\sqrt{2} = 0.707$ or -3 dB. This establishes (6.17).

We list in Table 6.1 the exact bandwidth computed from (6.16) and the approximate bandwidth computed from (6.17). We see that if β is less than 0.2, the difference between the exact and approximate bandwidths is less than 10%. In this case, the bandwidth equals simply the distance from the pole to the unit circle. We use (6.17) in the next example to design a digital filter to process analog signals.

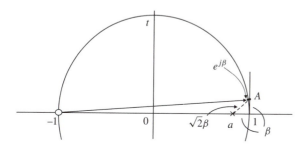

Figure 6.11 Derivation of (6.17).

Table 6.1 Exact and Approximate Bandwidths

a	Exact bandwidth	Approx. bandwidth β
0.6	0.49	0.4
0.7	0.35	0.3
0.8	0.22	0.2
0.85	0.16	0.15
0.9	0.10	0.1
0.95	0.05	0.05

◆ **Example 6.1**

Consider the CT signal

$$x(t) = \sin 7t + \sin 200t$$

It consists of one low-frequency sinusoid and one high-frequency sinusoid. Design a first-order digital filter to eliminate the high-frequency sinusoid.

To design a digital filter to process the CT signal, we must first select a sampling period T. Clearly T must be smaller than $\pi/200 = 0.0157$ to avoid frequency aliasing. Arbitrarily, we select $T = 0.015$. Then the magnitude spectra of $x(t)$ and $Tx(nT)$ are as shown in Figs. 6.12(a) and (b). The frequency range of $x(t)$ is $(-\infty, \infty)$ and the frequency range of $x(nT)$ is $(-\pi/T, \pi/T] = (-209.3 \ 209.3]$. Note that the spectrum of $x(nT)$ can be extended periodically with period $2\pi/T$ to all ω. Next we normalize the frequency range of $x(nT)$ to $(-\pi, \pi]$. This can be achieved by multiplying 7 and 200 by $T = 0.015$ to yield 0.105 and 3, as shown in Fig. 6.12(c). Because the spectra of the low- and high-frequency sinusoids are

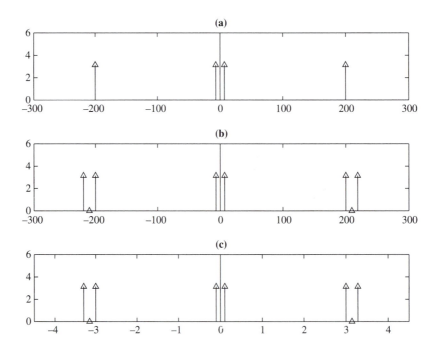

Figure 6.12 (a) Magnitude spectrum of $x(t)$. (b) Magnitude spectrum of $Tx(nT)$ with $T = 0.015$. (c) Magnitude spectrum of $Tx(nT)$ with Nyquist frequency range normalized to $(-\pi, \pi]$. All horizontal coordinates are frequency in rad/s. The Nyquist frequency range is indicated in (b) and (c) by a pair of small triangles.

widely apart, it is easy to design a digital low-pass filter to eliminate the latter. Clearly, the bandwidth of the digital low-pass filter must be larger than 0.105. Arbitrarily, we select the bandwidth as 0.2. Using (6.17), we have $a = 0.8$ and the filter in (6.13) becomes

$$H(z) = \frac{1 - 0.8}{2}\frac{z + 1}{z - 0.8} = \frac{z + 1}{10z - 8} \tag{6.18}$$

We plot in Fig. 6.13(a) the signal $x(t)$ and in Fig. 6.13(b) the output of the filter with a solid line and $\sin 7t$ with a dotted line. It is generated by the program that follows:

Program 6.2
```
t=0:0.015:10;
x=sin(7*t)+sin(200*t);
x1=sin(7*t);
b=[1 1];a=[10 -8];
y=filter(b,a,x);
subplot(2,1,1)
plot(t,x),title('(a)')
subplot(2,1,2)
plot(t,y,t,x1,':'),title('(b)')
```

Note that the MATLAB function **plot** automatically connects two neighboring points by a straight line. This is one way of D/A conversion and will be called the *linear interpolation* (Problem 6.15). We see that the high-frequency sinusoid is

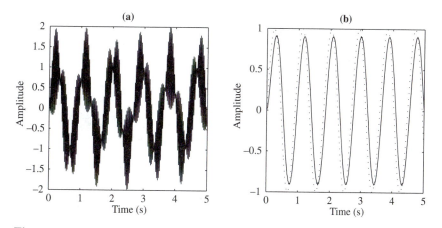

Figure 6.13 (a) Input signal. (b) Filter output (solid line) and $\sin 7t$ (dotted line).

completely blocked by the filter. If we use Table 5.3, then the output takes roughly $21 \times 0.015 = 0.3$ s to reach steady state; thus the output in Fig. 6.13(b) shows hardly any transient. The steady-state output is the same as $\sin 7t$ except for a small attenuation and a small time delay. If we increase the gain of the filter in (6.18), the attenuation can be avoided. There is however no way to eliminate the time delay.

We next discuss first-order high-pass filters. If we place the pole close to $-1 = e^{j\pi}$ and the zero at $1 = e^{j \times 0}$, then we will obtain the following high-pass filter

$$H(z) = \frac{b_0(z - 1)}{z - a} = \frac{1 + a}{2} \frac{z - 1}{z - a} \tag{6.19}$$

where $-1 < a \leq 0$ and b_0 is chosen so that $H(-1) = 1$. Figure 6.14(a) shows the magnitude responses of (6.19) for $a = -0.7$ (solid line), $a = -0.8$ (dotted line), and $a = -0.9$ (dashed line). Figure 6.14(b) shows the corresponding phase responses. If the pole is close to the unit circle, then the 3-dB bandwidth of the high-pass filter equals roughly $1 - |a|$.

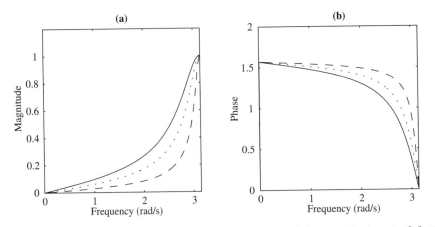

Figure 6.14 (a) Magnitude responses for $a = -0.7$ (solid line), -0.8 (dotted line), and -0.9 (dashed line). (b) Corresponding phase responses.

◆ **Example 6.2**

Design a first-order high-pass digital filter to eliminate the low-frequency sinusoid of the analog signal $x(t) = \sin 7t + \sin 200t$. Let us select $T = 0.015$ as in Example 6.1 and a high-pass filter with bandwidth 0.2. Then we have $a = -0.8$, and the high-pass digital filter in (6.19) becomes

$$H(z) = \frac{1 - 0.8}{2} \frac{z - 1}{z + 0.8} = \frac{z - 1}{10z + 8}$$

We plot in Fig. 6.15 with a solid line the straight-line interpolation of the output of the filter, with a dash-and-dotted line the straight-line interpolation of $\sin 200nT$, and with a dotted line $\sin 200t$. We see that the solid line roughly follows $\sin 200nT$ and has the same period as $\sin 200t$. Thus the digital high-pass filter blocks the low-frequency frequency component and passes the high-frequency component.

In digital processing of CT signals, outputs of digital filters must be converted back to CT signals. If we use the MATLAB function **plot**, the conversion is carried out by straight-line interpolation. For $T = 0.015$, we take only slightly more than two time samples per period of $\sin 200t$. Thus the result is poor as shown.

There are two possible ways to improve the result. One way is to use a more sophisticated interpolation scheme. For example, if we use the ideal interpolation formula in (3.69) to carry out the D/A conversion, then the result may be greatly improved. However, (3.69) is, as we will discuss later, essentially an analog ideal low-pass filter and cannot be realized in real time. The second way is to select a smaller T. For example, if we select $T = 0.0015$, then the linear interpolation may yield a satisfactory result. However, for $T = 0.0015$, the normalized low and high frequencies becomes $7 \times 0.0015 = 0.0103$ and $200 \times 0.0015 = 0.3$. They are

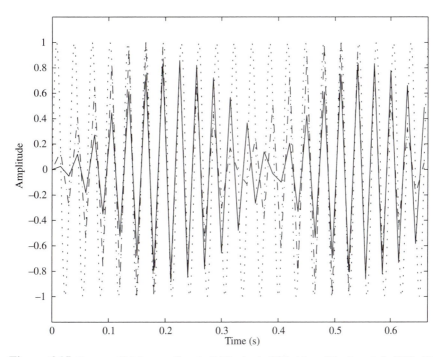

Figure 6.15 Output of high-pass filter (solid line), $\sin 200t$ (dotted line), and $\sin(200nT)$ (dash-dotted line).

close in the frequency range $[0, \pi]$, and it will be very difficult to design a high-pass filter that passes frequency 0.3 and stops frequency 0.0105. Therefore, it is generally difficult to design a digital high-pass filter to eliminate the low-frequency component of $x(t) = \sin 7t + \sin 200t$.

6.4.1 Second-Order Digital Filters

We next discuss second-order digital filters. Such a filter has two poles and none, one, or two zeros. Suppose it has two poles and two zeros as shown in Fig. 6.16 or has the transfer function

$$H(z) = \frac{(z - b_1 e^{jb_2})(z - b_1 e^{-jb_2})}{(z - a_1 e^{ja_2})(z - a_1 e^{-ja_2})} \tag{6.20}$$

where a_i and b_i are implicitly assumed to be real and positive. To find the magnitude of $H(z)$ at $z = e^{j\omega}$, we draw vectors from the two zeros and two poles to the point $e^{j\omega}$. Let the magnitudes and angles of the four vectors be denoted as

$$e^{j\omega} - b_1 e^{jb_2} = q_1 e^{j\psi_1} \quad e^{j\omega} - b_1 e^{-jb_2} = q_2 e^{j\psi_2}$$

$$e^{j\omega} - a_1 e^{ja_2} = p_1 e^{j\phi_1} \quad e^{j\omega} - a_1 e^{-ja_2} = p_2 e^{j\phi_2}$$

Then the magnitude and phase of $H(e^{j\omega})$ are, respectively,

$$|H(e^{j\omega})| = \frac{q_1 q_2}{p_1 p_2} \qquad \measuredangle\, H(e^{j\omega}) = \psi_1 + \psi_2 - \phi_1 - \phi_2$$

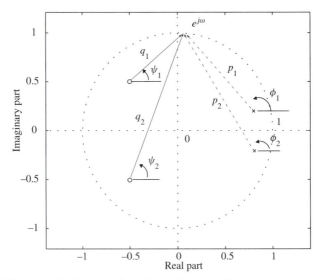

Figure 6.16 Transfer function with two poles and two zeros.

Clearly the larger the distances from the zeros to $e^{j\omega}$ and the smaller the distances from the poles to $e^{j\omega}$, the larger the magnitude of $H(e^{j\omega})$. Therefore, by placing them properly, we can obtain various magnitude responses.

Digital Resonators A digital resonator is a second-order filter that has a pair of complex conjugate poles $ae^{\pm j\omega_0}$, with a close to 1, as shown in Fig. 6.17(a). Clearly the magnitude response in the neighborhood of ω_0 is large. The magnitude response of a good resonator should drop off sharply to zero as ω moves away from ω_0. This will be achieved by utilizing zeros. Let the two zeros be complex conjugate and be located as shown in Fig. 6.17(a). Because the zeros are close to high frequencies, the magnitude response will fall off sharply as ω approaches π, as shown in Fig. 6.17(b) with a solid line. However, because the zeros are far away from low frequencies, the magnitude response will not fall off sharply as ω approaches $\omega = 0$. For comparison, we plot in Fig. 6.17(b) with a dotted line the magnitude response of the filter without any zeros and with a gain so that its peak magnitude equals that of the solid line. From the plot, we see that if the two zeros are complex, the magnitude response is improved for $\omega > \omega_0$ but worsen for $\omega < \omega_0$ or vise versa. Thus the two zeros of a resonantor should not be complex.

If the two zeros are real and located at $z = 0$, then they have no effect on the magnitude response. Now if we place one real zero at 1 and the other at -1, as shown in Fig. 6.17(c), then

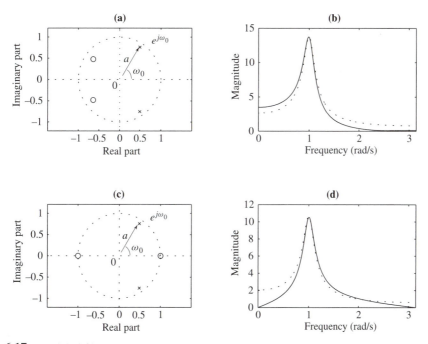

Figure 6.17 (a) Digital filter. (b) Magnitude responses. (c) Digital resonator. (d) Magnitude responses.

its magnitude response is as shown in Fig. 6.17(d) with a solid line. Comparing with the dotted line, which is the magnitude response without any zeros, we conclude that the solid line is a better resonator. Thus the transfer function of a good digital resonator should be of the form

$$H(z) = \frac{b_0(z+1)(z-1)}{(z - ae^{j\omega_0})(z - ae^{-j\omega_0})} \tag{6.21}$$

where b_0 can be chosen so that the peak magnitude is 1. If a is close to 1, the maximum magnitude will locate roughly at $\omega = \omega_0$, called the *resonance frequency*. Using the argument in Fig. 6.11, we can readily show that the magnitudes at $\omega_0 \pm \beta$, where $\beta = 1 - a$, roughly equals 70.7% of the peak magnitude. Thus *the resonator has roughly 3-dB bandwidth* $2(1 - a) = 2\beta$, twice the bandwidth in (6.17) for low-pass filters.

◆ **Example 6.3**

Design a digital filter to eliminate the lower-frequency sinusoid of $x(t) = \sin 7t + \sin 200t$. We will design a resonator with resonance frequency 200 to achieve the task. We select $T = 0.002$, roughly taking 15 samples of $\sin 200t$ in one period. Then the normalized resonance frequency is $200 \times 0.002 = 0.4$. The normalized frequency of $\sin 7t$ is $7 \times 0.002 = 0.014$. If we select the bandwidth as 0.02, then 0.014 will be way outside the passband [0.39, 0.41]. Because the bandwidth of resonators is $2(1 - a)$, we solve

$$2(1 - a) = 0.02$$

which yields $a = 0.99$. Thus the transfer function of the resonator is

$$H(z) = \frac{b_0(z+1)(z-1)}{(z - 0.99e^{j0.4})(z - 0.99e^{-j0.4})} = \frac{b_0(z^2 - 1)}{z^2 - 2 \cdot 0.99 \cos 0.4 \, z + (0.99)^2}$$

$$= \frac{b_0(z^2 - 1)}{z^2 - 1.8237z + 0.9801}$$

Using MATLAB, we can readily compute

$$H(e^{j0.4}) = (100.49 + j1.2)b_0 = 100.49e^{j0.012}b_0$$

Thus if $b_0 = 1/100.49$, then $|H(e^{j0.4})| = 1$. The output of the resonator is shown in Fig. 6.18 with a solid line. We also plot in Fig. 6.18 $\sin 200t$ with a dotted line. We see that after transient, the output is identical to the high-frequency sinusoid. Thus the low-frequency sinusoid $\sin 7t$ is completely eliminated. If we use Table 5.3, then the filter output takes roughly $448 \times 0.002 = 0.896$ s to reach the steady-state response.

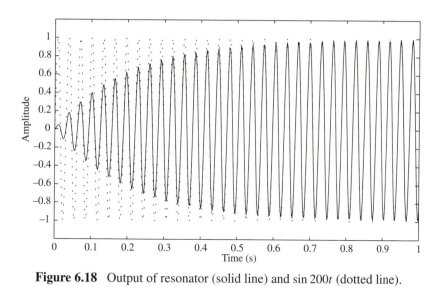

Figure 6.18 Output of resonator (solid line) and $\sin 200t$ (dotted line).

Notch Filters When instruments are powered by 60-Hz power supplies, unwanted 60-Hz noise often arises. Thus it is useful to have a filter that will eliminate 60-Hz noise but will have no effect on the rest of the signal. We introduce here such a filter.

A *notch filer* is a second-order filter whose magnitude response is 0 or null at $\omega = \pm\omega_0$ and roughly equals 1 for all other frequencies in $(-\pi, \pi]$. In order to have null at $\pm\omega_0$, the transfer function must have zeros at $e^{\pm j\omega_0}$. We then use two poles to make the magnitude response roughly constant for all ω. This can be achieved by placing the poles very close to the two zeros as shown in Fig. 6.19(a). Then the transfer function is

$$H(z) = \frac{(z - e^{j\omega_0})(z - e^{-j\omega_0})}{(z - ae^{j\omega_0})(z - ae^{-j\omega_0})} \tag{6.22}$$

with $0 \leq a < 1$. If $a = 0$, $H(z)$ is an FIR filter and the two poles are not utilized. For a small, the notch will be wide. The magnitude responses of $H(z)$ are shown in Fig. 6.19(b) for $a = 0.9$ (solid line) and $a = 0.95$ (dotted line). This can be easily explained from the pole-zero plot. Let $\beta := 1 - a$. For ω far away from $\omega = \omega_0$, the distances from the zeros $e^{\pm j\omega_0}$ and the poles $ae^{\pm j\omega_0}$ to $e^{j\omega}$ are roughly the same. Thus we have $|H(e^{j\omega})| \approx 1$. If $\omega = \omega_0 \pm \beta$ and if β is small, then using the similar argument as in Fig. 6.11, we have

$$H(e^{j(\omega_0 \pm \beta)}) \approx \frac{\beta}{\sqrt{2}\beta} = \frac{1}{\sqrt{2}} = 0.707$$

Note that the distances from the zero $e^{-j\omega_0}$ and the pole $ae^{-j\omega_0}$ to $e^{j(\omega_0 \pm \beta)}$ are roughly the same.

Because of the zero at $e^{j\omega_0}$, we have $H(e^{j\omega_0}) = 0$. The filter is said to have notches or nulls at $\omega = \pm\omega_0$. The smaller β is, the narrower the notches.

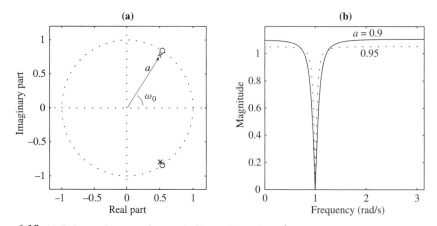

Figure 6.19 (a) Poles and zeros of a notch filter with null at $\pm\omega_0$. (b) Its magnitude responses for $a = 0.9$ (solid line) and $a = 0.95$ (dotted line).

◆ **Example 6.4**

Design a digital notch filter to eliminate the 60-Hz sinusoid in

$$x(t) = \sin(60 \times 2\pi t) + x_1(t)$$

where $x_1(t)$ is to be discussed shortly. The period of the 60-Hz sinusoid is $1/60 = 0.0166$; thus the sampling period must be less than $0.01666/2 = 0.00833$ to avoid frequency aliasing in sampling the sinusoid. We select arbitrarily $T = 0.002$ and consider

$$x(nT) = \sin(120\pi nT) + x_1(nT)$$

with frequency range $(-\pi/T, \pi/T] = (-500\pi, 500\pi]$. The sinusoid has frequency 120π rad/s in $(-\pi/T, \pi/T]$. If we normalize the frequency range to $(-\pi, \pi]$, then the sinusoid has frequency

$$60 \times 2\pi \times T = 120\pi \times 0.002 = 0.754$$

Thus the notch filter can be chosen as

$$H(z) = \frac{(z - e^{j0.754})(z - e^{-j0.754})}{(z - 0.95e^{j0.754})(z - 0.95e^{-j0.754})}$$

$$= \frac{z^2 - 2z \cos 0.754 + 1}{z^2 - 2 \cdot 0.95z \cos 0.754 + 0.95^2}$$

$$= \frac{z^2 - 1.4579z + 1}{z^2 - 1.3850z + 0.9025}$$

$$= \frac{1 - 1.4579z^{-1} + z^{-2}}{1 - 1.3850z^{-1} + 0.9025z^{-2}} \tag{6.23}$$

To test the effectiveness of this notch filter, we shall select a $x_1(nT)$ that has a constant frequency spectrum over all frequencies. Arbitrarily, we select $x_1[n] = 2\delta[n - 200]$, an impulse sequence at $n = 200$ or at $nT = 0.4$ s with magnitude 2. Its magnitude spectrum is 2 for all ω. Figure 6.20(a) shows the input $x(nT)$, and Fig. 6.20(b) shows the filter output. After transient, the notch filter eliminates the sinusoid and has no effect on the signal $x_1[n]$. It takes roughly $88 \times 0.002 = 0.176$ s for the transient to die out, where 88 is taken from Table 5.3.

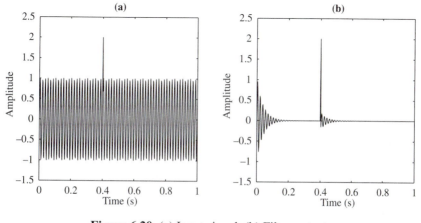

Figure 6.20 (a) Input signal. (b) Filter output.

6.5 Reciprocal Roots and All-Pass Filters

Consider the two transfer functions

$$H_1(z) = \frac{z - b}{z - a} = \frac{1 - bz^{-1}}{1 - az^{-1}} \quad \text{and} \quad H_2(z) = \frac{b(z - 1/b)}{z - a} = \frac{b - z^{-1}}{1 - az^{-1}} \tag{6.24}$$

The transfer function $H_1(z)$ has pole a and zero b; the transfer function $H_2(z)$ has pole a and zero $1/b$, as shown in Figs. 6.21(a) and (b). The two zeros b and $1/b$ are said to be *reciprocal*. More general, $be^{j\theta}$ and $(1/b)e^{j\theta}$ are said to be mutually reciprocal. If one root lies inside the unit circle, then its reciprocal root lies outside the unit circle. Note that a root on the unit circle is reciprocal to itself. The physical significance of reciprocal roots is that the ratio of their distances to the unit circle is a constant or

$$\frac{|e^{j\omega} - be^{j\theta}|}{|e^{j\omega} - (1/b)e^{j\theta}|} = |b| \tag{6.25}$$

for all ω. To show this, we show

$$|e^{j\omega} - be^{j\theta}|^2 = |be^{j\omega} - e^{j\theta}|^2$$

Indeed, we have

$$
\begin{aligned}
|e^{j\omega} - be^{j\theta}|^2 &= |(\cos\omega - b\cos\theta) + j(\sin\omega - b\sin\theta)|^2 \\
&= (\cos\omega - b\cos\theta)^2 + (\sin\omega - b\sin\theta)^2 \\
&= (\cos^2\omega + \sin^2\omega) - 2b(\cos\omega\cos\theta + \sin\omega\sin\theta) + b^2(\cos^2\theta + \sin^2\theta) \\
&= 1 + b^2 - 2b\cos(\omega - \theta)
\end{aligned}
$$

Using the same procedure, we can show

$$|be^{j\omega} - e^{j\theta}|^2 = 1 + b^2 - 2b\cos(\omega - \theta)$$

This establishes (6.25).

Consider again $H_1(z)$. If $|b| > 1$ or the zero is outside the unit circle, we may replace the zero by its reciprocal zero and introduce b as in $H_2(z)$. Then $H_2(z)$ has its zero inside the unit circle and has the same magnitude response as $H_1(z)$. More generally, consider

$$H_1(z) = \frac{z^2 + b_1 z + b_2}{A(z)} = \frac{(z - be^{j\theta})(z - be^{-j\theta})}{A(z)}$$

with $b_1 = -2b\cos\theta$ and $b_2 = b^2$. If the two zeros are outside the unit circle, we can replace them by their reciprocal zeros and introduce the gain b as

$$
\begin{aligned}
H_2(z) &= \frac{b(z - b^{-1}e^{j\theta})b(z - b^{-1}e^{-j\theta})}{A(z)} \\
&= \frac{b^2(z^2 - 2b^{-1}(\cos\theta)z + b^{-2})}{A(z)} \\
&= \frac{b_2 z^2 + b_1 z + 1}{A(z)}
\end{aligned}
$$

Then we have $|H_1(e^{j\omega})| = |H_2(e^{j\omega})|$ for all ω and $H_2(z)$ has all its zeros inside the unit circle. Thus we conclude that *in designing a digital filter to meet a given magnitude response, there is no loss of generality to place all zeros on or inside the unit circle* as we did in the preceding sections.

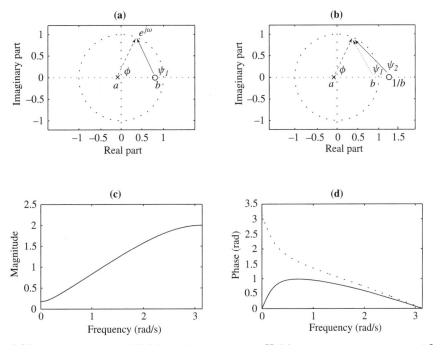

Figure 6.21 (a) Pole and zero of $H_1(z)$. (b) Pole and zero of $H_2(z)$. (c) Magnitude responses of $H_1(z)$ and $H_2(z)$. (d) Phase responses of $H_1(z)$ (solid line) and $H_2(z)$ (dotted line).

Minimum-Phase Transfer Functions Even though $H_1(z)$ and $H_2(z)$ in (6.24) have the same magnitude responses, their phase responses are different. The phase of $H_i(z)$ at $e^{j\omega}$ equals $\psi_i - \phi$, where the angles are shown in Figs. 6.21(a) and (b). If $0 < b < 1$, we can see from Fig. 6.21(b) that, for $0 \leq \omega \leq \pi$,

$$\psi_1 \leq \psi_2$$

Consequently, we have

$$\angle\ H_1(e^{j\omega}) = \psi_1 - \phi \leq \angle\ H_2(e^{j\omega}) = \psi_2 - \phi \tag{6.26}$$

for $0 \leq \omega \leq \pi$. Indeed, this is the case as shown in Fig. 6.21(d). For this reason, a zero inside the unit circle is called a *minimum-phase* zero and a transfer function will all zeros inside the unit circle is called a *minimum-phase transfer function*. Note that (6.26) does not hold for $-\pi \leq \omega < 0$ and for $-1 < b < 0$.[2]

---

[2] The terminology *minimum phase* is widely used in the CT case. A zero in $H(s)$ is called a minimum-phase zero if it lies inside the left-half s plane. Because the left-half s plane is mapped into the interior of the unit circle on the z plane, zeros of $H(z)$ lying inside the unit circle are called minimum-phase zeros.

All-Pass Filter Consider the transfer function

$$H(z) = \frac{a(z - 1/a)}{z - a} = \frac{az - 1}{z - a} = \frac{a - z^{-1}}{1 - az^{-1}}$$

with $|a| < 1$. This is a stable transfer function. Because the zero is the reciprocal of the pole, the magnitude response of $H(z)$ is 1 for all ω. Thus the filter is called an *all-pass filter*. This can be extended to the general case as

$$\begin{aligned} H(z) &= \pm\frac{a_{N+1}z^N + a_N z^{N-1} + \cdots + a_1}{a_1 z^N + a_2 z^{N-1} + \cdots + a_{N+1}} \\ &= \pm\frac{a_{N+1} + a_N z^{-1} + \cdots + a_1 z^{-N}}{a_1 + a_2 z^{-1} + \cdots + a_{N+1} z^{-N}} \end{aligned} \tag{6.27}$$

for any real a_i and $a_1 \neq 0$ so long as $H(z)$ is stable. In this equation, if $a_1 = 0$ and $a_{N+1} \neq 0$, then the transfer function becomes improper and the filter is not causal. A slightly different form of all-pass filters is, by assuming $a_1 \neq 0$ and $a_{N+1} \neq 0$,

$$H(z) = \pm\frac{a_{N+1}z^N + a_N z^{N-1} + \cdots + a_1}{z^k(a_1 z^N + a_2 z^{N-1} + \cdots + a_{N+1})} \tag{6.28}$$

for any positive integer k.

To show (6.27) to be an all-pass filter or $|H(e^{j\omega})| = 1$ for all ω, we define

$$\begin{aligned} A(z) &:= a_1 z^N + a_2 z^{N-1} + \cdots + a_{N+1} \\ &= z^N(a_1 + a_2 z^{-1} + \cdots + a_{N+1} z^{-N}) \end{aligned} \tag{6.29}$$

Then we have

$$B(z) := a_{N+1}z^N + a_N z^{N-1} + \cdots + a_1 = z^N A(z^{-1}) \tag{6.30}$$

Thus we can express (6.27) as

$$H(z) = \pm\frac{B(z)}{A(z)} = \pm\frac{z^N A(z^{-1})}{A(z)}$$

Let us compute

$$\begin{aligned} |H(e^{j\omega})|^2 &= H(e^{j\omega})H^*(e^{j\omega}) = H(e^{j\omega})H(e^{-j\omega}) = \frac{\pm B(e^{j\omega})}{A(e^{j\omega})}\frac{\pm B(e^{-j\omega})}{A(e^{-j\omega})} \\ &= \frac{e^{jN\omega}A(e^{-j\omega})}{A(e^{j\omega})}\frac{e^{-jN\omega}A(e^{j\omega})}{A(e^{-j\omega})} = 1 \end{aligned}$$

This shows that the magnitude response of $H(z)$ equals 1 for all ω. Thus $H(z)$ is an all-pass filter.

Next we show that the poles and zeros of $H(z)$ are reciprocal. If $\lambda e^{j\theta}$, with $\lambda \neq 0$, is a pole or if $A(\lambda e^{j\theta}) = 0$, then

$$B(\lambda^{-1} e^{-j\theta}) = \lambda^{-N} e^{-jN\theta} A(\lambda e^{j\theta}) = 0$$

which implies that $\lambda^{-1} e^{-j\theta}$ is a zero. Note that $\lambda^{-1} e^{-j\theta}$ is not a reciprocal root of $\lambda e^{j\theta}$. However, because complex-conjugate roots must appear in pairs, $H(z)$ contains the reciprocal zero $\lambda^{-1} e^{j\theta}$. If $\lambda = 0$ is a pole, then its reciprocal zero is at ∞, which will show up in (6.28) with $k \neq 0$. Thus we conclude that all poles and all zeros of any all-pass filter are mutually reciprocal. For example, the all-pass filter

$$H(z) = \frac{0.5(z-2)0.8(z-1.25e^{j3\pi/4})0.8(z-1.25e^{-j3\pi/4})}{z(z-0.5)(z-0.8e^{j3\pi/4})(z-0.8e^{-j3\pi/4})}$$

$$= \frac{0.32z^3 + 0.075z^2 + 0.63z + 1}{z(z^3 + 0.63z^2 + 0.075z^1 - 0.32)}$$

has its poles and zeros shown in Fig. 6.22. The reciprocal of the pole at $z = 0$ is an infinite zero and is not shown. All other poles and zeros are indeed reciprocal.

All-pass filters can be used as *phase* or *delay equalizers*. Consider a filter $H(z)$ with a desired magnitude response but an unsatisfactory phase response. In this case, we may connect to it an all-pass filter $H_a(z)$. Then we have

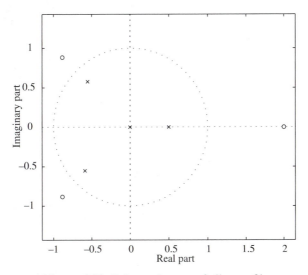

Figure 6.22 Poles and zeros of all-pass filter.

$$|H(e^{j\omega})H_a(e^{j\omega})| = |H(e^{j\omega})||H_a(e^{j\omega})| = |H(e^{j\omega})|$$

and

$$\measuredangle \left(H(e^{j\omega})H_a(e^{j\omega}) \right) = \measuredangle H(e^{j\omega}) + \measuredangle H_a(e^{j\omega})$$

Because the magnitude response of $H(z)$ is not affected by introducing an all-pass filter, we can use an all-pass filter to improve the phase response of $H(z)H_a(z)$. There is, however, no simple method of designing $H_a(z)$.

6.6 Miscellaneous Topics

This section introduces comb filters, digital sinusoidal generators, and the Goertzel algorithm.

6.6.1 Comb Filters

The notch filter discussed in Section 6.4.1 can be extended to have more notches or nulls. For example, if we extend (6.22) as

$$H(z) = \frac{z-1}{z-a} \frac{(z-e^{j\omega_0})(z-e^{-j\omega_0})}{(z-ae^{j\omega_0})(z-ae^{-j\omega_0})} \frac{z+1}{z+a} \tag{6.31}$$

where $0 \le a < 1$, then the filter has four nulls at frequency $\omega = 0$, $\pm\omega_0$ and π as shown in Fig. 6.23. Note that the frequency range is $(-\pi, \pi]$; thus $-\pi$ is not a null in the range.

A comb filter is a notch filter with a number of equally spaced nulls. It can be easily obtained from a notch filter $H(z)$ by replacing z by z^M, where M is a positive integer. This can be developed from the fact that the roots of $z^M - a = 0$ or $z^M = ae^{jk2\pi}$, for all integer k, are

$$z = a^{1/M} e^{j2k\pi/M} \qquad \text{for } k = 0, \pm 1, \pm 2, \ldots$$

There are M equally spaced distinct roots on the circle with radius $a^{1/M}$. For example, consider the notch filter

$$H(z) = \frac{z-1}{z-a} = \frac{1-z^{-1}}{1-az^{-1}} \tag{6.32}$$

with $0 \le a < 1$. It has one null at $\omega = 0$ as shown in Fig. 6.24(a) with $a = 0.8$. Let us replace z by z^4. Then we have

$$
\begin{aligned}
H(z^4) &= \frac{z^4 - 1}{z^4 - a} \\
&= \frac{z-1}{z-a^{1/4}} \frac{z - 1e^{j\pi/2}}{z - a^{1/4}e^{j\pi/2}} \frac{z - 1e^{-j\pi/2}}{z - a^{1/4}e^{-j\pi/2}} \frac{z - 1e^{j\pi}}{z - a^{1/4}e^{j\pi}}
\end{aligned}
\tag{6.33}
$$

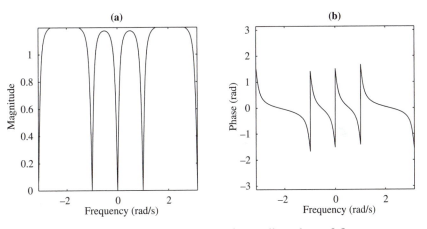

Figure 6.23 Notch filter with four nulls and $a = 0.8$.

Its magnitude response is shown in Fig. 6.24(c) with $a^{1/4} = 0.8^{1/4} = 0.946$ and has the shape of a comb. It does show four nulls in $(-\pi, \pi]$. Because $1 > a^{1/4} > a$, the notches in Fig. 6.24(c) are narrower than the one in Fig. 6.24(a).

If we use the notch filter

$$H(z) = \frac{z+1}{z+a}$$

which has null at $z = -1$ or $\omega = \pi$, as shown in Fig. 6.24(b), then the comb filter

$$H(z^4) = \frac{z^4+1}{z^4+a}$$

has the magnitude response shown in Fig. 6.24(d) with $a = 0.8$. It has nulls at $\pm\pi/4$ and $\pm 3\pi/4$, but not at π.

6.6.2 Sinusoid Generators

Filters are designed to process signals. Their poles are required to lie inside the unit circle on the z plane and their time responses will all die out eventually. Thus outputs of filters are essentially part of the applied inputs. A signal generator is a device that will maintain a sustained output once it is excited by an input. Once it is excited, no more input will be applied. Thus the response must be sustained by the device itself. If the transfer function of the device has all poles lying inside the unit circle, its response will eventually die out. If it has one or more poles outside the unit circle or has repeated poles on the unit circle, its response will grow unbounded. Thus *the transfer function of a signal generator must have simple poles on the unit circle.*

Consider the two systems with transfer functions

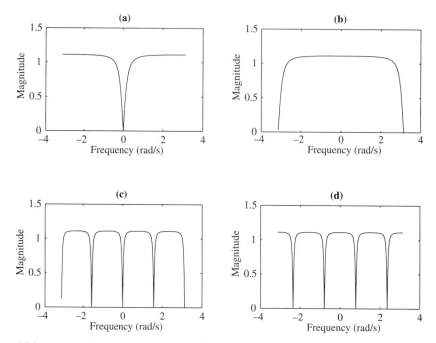

Figure 6.24 (a) Notch filter with null at $\omega = 0$. (b) Notch filter with null at $\omega = \pi$. (c) Comb filter developed from (a). (d) Comb filter developed from (b).

$$H_1(z) = \frac{Y_1(z)}{X(z)} = \frac{(\sin \omega_0)z^{-1}}{1 - 2(\cos \omega_0)z^{-1} + z^{-2}} \qquad (6.34)$$

and

$$H_2(z) = \frac{Y_2(z)}{X(z)} = \frac{1 - (\cos \omega_0)z^{-1}}{1 - 2(\cos \omega_0)z^{-1} + z^{-2}} \qquad (6.35)$$

They are the z-transforms of $b^n \sin \omega_0 n$ and $b^n \cos \omega_0 n$ with $b = 1$ as listed in Table 5.1. Both transfer functions have poles at $e^{\pm j\omega_0}$; they are simple poles on the unit circle. Now if $x[n] = A\delta[n]$, then $X(z) = A$ and (6.34) and (6.35) imply

$$Y_1(z) = H_1(z)X(z) = \frac{A(\sin \omega_0)z^{-1}}{1 - 2(\cos \omega_0)z^{-1} + z^{-2}}$$

and

$$Y_2(z) = H_2(z)X(z) = \frac{A(1 - (\cos \omega_0)z^{-1})}{1 - 2(\cos \omega_0)z^{-1} + z^{-2}}$$

Their time responses can be obtained from Table 5.1 as

$$y_1[n] = A \sin \omega_0 n \quad \text{and} \quad y_2[n] = A \cos \omega_0 n$$

Thus if we design two systems with transfer functions $H_1(z)$ and $H_2(z)$, then their outputs generate sine and cosine sequences once they are excited by $A\delta[n]$.

Next we develop a block diagram for $H_1(z)$ and $H_2(z)$. Let us introduce a new sequence $v[n]$. It is defined through its z-transform as

$$\frac{X(z)}{V(z)} = 1 - 2(\cos \omega_0)z^{-1} + z^{-2} \tag{6.36}$$

Then we have

$$[1 - 2(\cos \omega_0)z^{-1} + z^{-2}]V(z) = X(z)$$

which implies, in the time domain,

$$v[n] = 2(\cos \omega_0)v[n-1] - v[n-2] + x[n] \tag{6.37}$$

This recursive difference equation has order two and needs two unit-delay elements as shown in Fig. 6.25. Let us assign the input of the upper unit-delay element as $v[n]$. Then its output is $v[n-1]$, which, in turn, is the input of the lower unit-delay element. Thus the output of the lower unit-delay element is $v[n-2]$. Using (6.37), we can readily complete the left-hand half of Fig. 6.25.

The multiplication of (6.34) and (6.36) yields

$$\frac{Y_1(z)}{V(z)} = (\sin \omega_0)z^{-1}$$

which implies

$$y_1[n] = (\sin \omega_0)v[n-1] \tag{6.38}$$

The multiplication of (6.35) and (6.36) yields

$$\frac{Y_2(z)}{V(z)} = 1 - (\cos \omega_0)z^{-1}$$

which implies

$$y_2[n] = v[n] - (\cos \omega_0)v[n-1] \tag{6.39}$$

Using (6.38) and (6.39), we can complete the diagram in Fig. 6.25. Once the system in Fig. 6.25 is excited by $x[0] = A$, its two outputs will generate $A \sin \omega_0 n$ and $A \cos \omega_0 n$ even if no further input is applied.

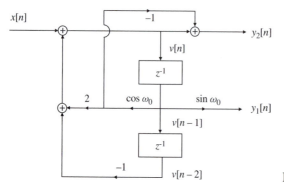

Figure 6.25 Digital sinusoidal generator.

Instead of applying an impulse sequence as an input, sinusoidal sequences can also be excited by setting initial conditions. If $v[-1] = v[-2] = 0$ and if $x[0] = A$, then (6.37) implies $v[0] = A$. Thus instead of applying $x[n] = A\delta[n]$, we may simply set $v[0] = A$. Note that unit-delay elements are actually registers. By setting the register values as $v[-1] = 0$ and $v[0] = A$, we can verify, using (6.37) through (6.39), that

$$y_1[n] = A \sin \omega_0 n \quad \text{and} \quad y_2[n] = A \cos \omega_0 n$$

for $n \geq 0$ (Problem 6.14). This is a simpler way of starting the oscillation. More will be said regarding sinusoidal generators in Chapter 9. To conclude this part, we mention that sinusoidal signals can also be generated from a table, as discussed in Problems 2.24, 3.20, and 3.21.

6.6.3 Goertzel Algorithm[3]

This part discusses a system interpretation of the DFT. Let $x[n]$ for $n = 0, 1, \ldots, N - 1$ be a sequence of length N. Then its DFT is

$$X[m] = \sum_{n=0}^{N-1} x[n] e^{-j2\pi nm/N} = \sum_{n=0}^{N-1} x[n] W^{nm}$$

where $W := e^{-j2\pi/N}$. Because $W^{-mN} = 1$ for all integer m, we can write the DFT as

$$X[m] = W^{-mN} \sum_{n=0}^{N-1} x[n] W^{nm} = \sum_{n=0}^{N-1} x[n] W^{-m(N-n)} \tag{6.40}$$

This is of the form of convolution. Let us define a new sequence

---

[3] This subsection may be skipped without loss of continuity.

$$y_m[n] = \sum_{l=0}^{n} x[l]W^{-m(n-l)} \tag{6.41}$$

This is the convolution of $x[n]$ and W^{-mn}. Thus we can consider $y_m[n]$ as the output of a causal system with impulse response

$$h_m[n] = \begin{cases} W^{-mn} & \text{for } n \geq 0 \\ 0 & \text{for } n < 0 \end{cases} \tag{6.42}$$

excited by the finite-sequence input $x[n]$. The input $x[n]$ is implicitly assumed to be 0 for $n < 0$ and $n > N - 1$. Because $x[N] = 0$, we have

$$y_m[N] = \sum_{l=0}^{N} x[l]W^{-m(N-l)} = \sum_{l=0}^{N-1} x[l]W^{-m(N-l)}$$

and

$$X[m] = y_m[n]|_{n=N} = y_m[N]$$

for $m = 0, 1, \ldots, N - 1$. Thus the DFT can be computed from a bank of N systems with complex-valued impulse responses in (6.42).

Taking the z-transform of (6.41) yields the following transfer function

$$H_m(z) = \frac{1}{1 - W^{-m}z^{-1}} = \frac{Y_m(z)}{X(z)} \tag{6.43}$$

which implies

$$y_m[n] = W^{-m}y_m[n-1] + x[n]$$

For each m, we compute $y_m[n]$, for $n = 0, 1, \ldots, N$. Then we have $X[m] = y_m[N]$. This is called the *Goertzel algorithm* for computing DFT.

The algorithm can be modified as follows. Multiplying the numerator and denominator of (6.43) by $1 - W^m z^{-1}$ yields

$$H_m(z) = \frac{1 - W^m z^{-1}}{(1 - W^m z^{-1})(1 - W^{-m}z^{-1})} = \frac{1 - W^m z^{-1}}{1 - 2(\cos 2m\pi/N)z^{-1} + z^{-2}} \tag{6.44}$$

This is of the form shown in (6.35). As in (6.36), we define

$$X(z)/V_m(z) = 1 - 2(\cos 2m\pi/N)z^{-1} + z^{-2}$$

Then we can readily obtain the difference equation

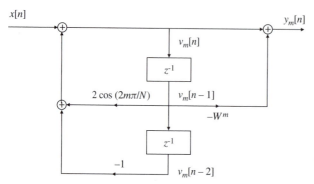

Figure 6.26 Block diagram of (6.45) and (6.46).

$$v_m[n] = 2\left(\cos\frac{2m\pi}{N}\right)v_m[n-1] - v_m[n-2] + x[n] \tag{6.45}$$

$$y_m[n] = v_m[n] - W^m v_m[n-1] \tag{6.46}$$

and the block diagram in Fig. 6.26. Note that (6.45) involves only real coefficients and (6.46) is to be computed only once at $n = N$ after $v_m[N]$ and $v_m[N-1]$ are computed. This is a system interpretation of DFT.

6.7 Analog Ideal Low-Pass Filters

In this section, we discuss CT ideal low-pass filters and develop the interpolation formula in (3.69) from the filtering point of view. Other related issues will also be discussed. Before proceeding, we mention that the z-transform, DT transfer function $H(z)$, and the frequency response $H(e^{j\omega})$ in the DT case must be replaced by the Laplace transform, CT transfer function $H(s)$, and the frequency response $H(j\omega)$ in the CT case.

The frequency range of DT systems is $(-\pi, \pi]$ with $T = 1$ and the frequency range of CT systems is $(-\infty, \infty)$. Thus the specifications discussed for DT filters are applicable to CT filters if $\pm\pi$ in Figs. 6.1, 6.2, 6.3, and 6.6 are replaced by $\pm\infty$. For example, an ideal analog low-pass filter is specified by, as in (6.1),

$$H(j\omega) = \begin{cases} 1 \times e^{-jwt_0} & \text{for } |\omega| \le w_c \\ 0 & \text{for } \omega_c < |\omega| < \infty \end{cases} \tag{6.47}$$

This filter will pass, without any distortion except a delay of t_0 seconds, any CT signal whose frequency spectrum lies entirely inside the range $[-\omega_c, \omega_c]$, and stop completely any signal whose spectrum lies outside the range.

The inverse Laplace transform of (6.47) can be computed from the inverse CT Fourier transform as

$$h(t) = \frac{\sin \omega_c (t - t_0)}{\pi (t - t_0)} \tag{6.48}$$

See (5.98) and Example 3.3. This impulse response is nonzero for $t < 0$, as shown in Fig. 3.3(b) for $t_0 = 0$; thus the low-pass filter is not causal and cannot be realized in the real world. For this reason, it is called an ideal filter. Even though it cannot be implemented, it is useful in explaining a number of issues, as we will do next.

Ideal D/A Converter In Section 3.8, we showed that if a CT signal $x(t)$ is bandlimited to W, then $x(t)$ can be recovered from its sampled sequence $x(nT)$ if the sampling period T is less than π / W. Furthermore, $x(t)$ can be computed from $x(nT)$ as

$$x(t) = \sum_{n=-\infty}^{\infty} x(nT) \frac{\sin[\pi(t - nT)/T]}{\pi(t - nT)/T} \tag{6.49}$$

See (3.69). This ideal interpolation formula was obtained there by relating the DT and CT Fourier transforms.

Now we will develop (6.49) from a different viewpoint. A D/A converter can be modeled as shown in Fig. 6.27. It consists of a converter that transforms the input sequence $x(nT)$ into

$$x_s(t) = \sum_{n=-\infty}^{\infty} x(nT) \delta(t - nT) \tag{6.50}$$

See (5.89). It consists of a sequence of impulses with weight $x(nT)$ at $t = nT$. The signal $x_s(t)$ then drives an analog ideal low-pass filter with gain T, cutoff frequency $\omega_c = \pi / T$, and zero phase. We shall show that the output of this ideal low-pass filter equals (6.49). Before proceeding, we discuss the reason for converting $x(nT)$ into $x_s(t)$. If we apply the sequence $x(nT)$ directly to the analog ideal low-pass filer, the output $y(t)$ will be identically zero. In order to drive a CT system, the input must have nonzero area or power. A sequence that consists of only a stream of numbers has zero width; therefore, it cannot drive any CT system. Although the impulse has zero width, it has area 1. Thus once $x(nT)$ is transformed into $x_s(t)$, it can drive the analog ideal low-pass filter.

The impulse response of the analog ideal low-pass filter with gain T, cutoff frequency $\omega_c = \pi / T$, and $t_0 = 0$ is, as in (6.48),

$$h(t) = T \frac{\sin(\pi t / T)}{\pi t} \tag{6.51}$$

The input and output of any CT system are related by the convolution shown in (5.81), where the input is applied from $t = 0$. If the input is applied from $t = -\infty$, then the lower integration limit must be replaced by $-\infty$. See also Section 5.7.3. Using (6.50), (6.51), and (A.6), we have

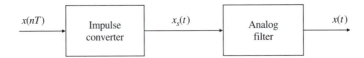

Figure 6.27 Ideal D/A converter.

$$y(t) = \int_{\tau=-\infty}^{\infty} h(t - \tau)x_s(\tau)d\tau$$

$$= \int_{\tau=-\infty}^{\infty} T\frac{\sin[\pi(t-\tau)/T]}{\pi(t-\tau)}\left(\sum_{n=-\infty}^{\infty} x(nT)\delta(\tau - nT)\right)d\tau$$

$$= T\sum_{n=-\infty}^{\infty} x(nT)\left(\int_{\tau=-\infty}^{\infty}\frac{\sin[\pi(t-\tau)/T]}{\pi(t-\tau)}\delta(\tau - nT)d\tau\right)$$

$$= T\sum_{n=-\infty}^{\infty} x(nT)\frac{\sin[\pi(t-\tau)/T]}{\pi(t-\tau)}\bigg|_{\tau=nT}$$

$$= \sum_{n=-\infty}^{\infty} x(nT)\frac{\sin[\pi(t-nT)/T]}{\pi(t-nT)/T}$$

This is the same as (6.49). This shows that the ideal interpolation formula developed in (3.69) can be viewed as a filtering problem.

It is of interest to compare the frequency response of a practical D/A converter with the frequency response of an ideal low-pass filter. Most practical D/A converters use a zero-order hold, which holds the current output constant until the arrival of the next output. Thus if the input of a zero-order hold is 1, then its output is 1 until the next input arrives. Thus the practical D/A converter can be modeled as shown in Fig. 6.27, in which the analog filter has the impulse response

$$h(t) = \begin{cases} 1 & \text{for } 0 \le t \le T \\ 0 & \text{for } t < 0 \text{ and } t > T \end{cases} \tag{6.52}$$

Its Laplace transform is, using (5.84),

$$H(s) = \mathcal{L}[h(t)] = \int_{t=0}^{T} e^{-st}dt = \frac{e^{-sT} - 1}{-s} = \frac{1 - e^{-sT}}{s} \tag{6.53}$$

Its magnitude and phase responses are plotted with solid lines in Figs. 6.28(a) and (b) for $T = 1$. They are quite different from those of the ideal low-pass filter, which are plotted with dotted lines. If we compute the frequency response of a first-order hold, then it will be closer to the ideal low-pass filter. In conclusion, practical D/A conversion is far from ideal. However, as T becomes smaller, so does the conversion error.

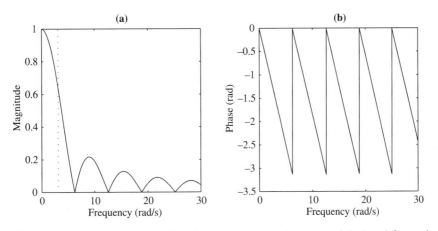

Figure 6.28 (a) Magnitude responses of analog ideal low-pass filter (dotted line) and first-order hold (solid line). (b) Phase response of first-order hold (solid line). The ideal filter has zero phase.

6.7.1 Why Antialiasing Filters?

In digital processing, analog signals are ofter prefiltered as shown in Fig. 1.11. Prefiltering can reduce or eliminate the effect of frequency aliasing; therefore, it is reasonable to expect that prefiltering will yield a better result than the one without prefiltering. In this subsection, we use analog ideal low-pass filters to establish this conjecture formally.

Consider the arrangement shown in Fig. 6.29(a). The CT signal $x(t)$ is prefiltered to become $\bar{x}(t)$. It is then applied to the A/D converter to yield $\bar{x}(nT)$. We assume that the sampled signal $\bar{x}(nT)$ is converted back immediately to the analog signal $\hat{x}(t)$ using a D/A converter. The low-pass prefilter and the D/A converter are assumed to be analog ideal low-pass filters. In Fig. 6.29(b), the CT signal $x(t)$ is not prefiltered and is converted into the DT signal $x(nT)$ and then back to the analog signal $\tilde{x}(t)$.

Now we will compare the outputs $\hat{x}(t)$ (with prefiltering) and $\tilde{x}(t)$ (without prefiltering) with the original signal $x(t)$. If $x(t)$ is bandlimited to W and if the prefilter has a cutoff frequency equal to W or larger, then $\bar{x}(t) = x(t)$. If the sampling period T is smaller than π/W, then the A/D conversion will not introduce frequency aliasing, and we have $\bar{x}(nT) = x(nT)$ and $\bar{X}_d(\omega) = X_d(\omega) = X(\omega)$, for $|\omega| \leq \pi/T$. Thus we have $x(t) = \hat{x}(t) = \tilde{x}(t)$. In conclusion, if $x(t)$ is bandlimited, there is no difference whether $x(t)$ is prefiltered or not in its digital processing.

If $x(t)$ is not bandlimited, digital processing of $x(t)$ will always introduce errors. Thus the two outputs $\hat{x}(t)$ and $\tilde{x}(t)$ in Fig. 6.29 cannot equal $x(t)$. In order to compare the two processes, we introduce a criterion. Let us define

$$E_1 := \int_{-\infty}^{\infty} |x(t) - \hat{x}(t)|^2 dt \tag{6.54}$$

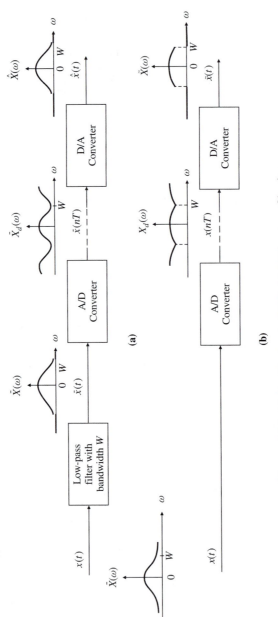

Figure 6.29 (a) Prefiltering in DSP. (b) Without prefiltering.

and

$$E_2 := \int_{-\infty}^{\infty} |x(t) - \tilde{x}(t)|^2 dt \tag{6.55}$$

They are called integral squared errors. Applying the Parseval formula in (3.27) to (6.54) yields

$$E_1 = \frac{1}{2\pi} \int_{-\infty}^{\infty} |X(\omega) - \hat{X}(\omega)|^2 \, d\omega \tag{6.56}$$

If the low-pass prefilter is ideal, we have

$$\bar{X}(\omega) = \begin{cases} X(\omega) & \text{for } |\omega| \le \pi/T \\ 0 & \text{for } |\omega| > \pi/T \end{cases}$$

Because $\bar{x}(t)$ is bandlimited, its sampling will not introduce frequency aliasing. Thus we have

$$\bar{X}_d(\omega) = \bar{X}(\omega)/T \qquad \text{for } |\omega| \le \pi/T$$

Note that $\bar{X}_d(\omega)$ can be extended periodically with period $2\pi/T$ to all ω. The output $\hat{x}(t)$ of the ideal D/A converter has the spectrum

$$\hat{X}(\omega) = \begin{cases} T\bar{X}_d(\omega) = \bar{X}(\omega) = X(\omega) & \text{for } |\omega| \le \pi/T \\ 0 & \text{for } |\omega| > \pi/T \end{cases}$$

Using this equation, the integral squared error between the input $x(t)$ and the output $\hat{x}(t)$ in Fig. 6.29(a) can be reduced as

$$E_1 = \frac{1}{2\pi} \left(\int_{-\infty}^{-\pi/T} |X(\omega)|^2 d\omega + \int_{\pi/T}^{\infty} |X(\omega)|^2 d\omega \right) \tag{6.57}$$

Note that if $x(t)$ is bandlimited, then $E_1 = 0$.

Next we compute the integral squared error in Fig. 6.29(b). If $x(t)$ is not prefiltered, frequency aliasing will occur and $X_d(\omega) \ne X(\omega)$ for $|\omega| \le \pi/T$. Thus we have

$$\tilde{X}(\omega) \begin{cases} \ne X(\omega) & \text{for } |\omega| \le \pi/T \\ = 0 & \text{for } |\omega| > \pi/T \end{cases}$$

and

$$E_2 = \frac{1}{2\pi} \int_{-\infty}^{\infty} |X(\omega) - \tilde{X}(\omega)|^2 d\omega$$

$$= \frac{1}{2\pi} \left(\int_{-\infty}^{-\pi/T} |X(\omega)|^2 d\omega + \int_{-\pi/T}^{\pi/T} |X(\omega) - \tilde{X}(\omega)|^2 d\omega + \int_{\pi/T}^{\infty} |X(\omega)|^2 d\omega \right)$$

Comparing this equation with (6.57) and because

$$\int_{-\pi/T}^{\pi/T} |X(\omega) - \tilde{X}(\omega)|^2 d\omega > 0$$

we conclude $E_1 < E_2$. This shows that the integral squared error with prefiltering is less than the integral squared error without prefiltering.

The preceding discussion is carried out using ideal low-pass filters. In practice, no such filters can be built. However, the same statement probably still holds in using practical filters and zero-order holds. Thus in practice, CT signals are often prefiltered before they are processed digitally. The low-pass prefilter is also called an antialiasing filter.

PROBLEMS

6.1   Compute the impulse response of the digital ideal high-pass filter specified by

$$H_{hp}(e^{j\omega}) = \begin{cases} 0 & \text{for } |\omega| < \omega_c \\ 1 \times e^{-j\omega n_0} & \text{for } \omega_c \le |\omega| \le \pi \end{cases}$$

Use $(-\pi, \ \pi]$ or $[0, \ 2\pi)$, whichever is simpler, to carry out the computation.

6.2   Let $H_{ap}(e^{j\omega}) := 1 \times e^{-j\omega n_0}$ for $|\omega| \le \pi$. It is a digital ideal all-pass filter. Let $H_{lp}(e^{j\omega})$ be defined as in (6.1). Show

$$H_{ap}(e^{j\omega}) = H_{hp}(e^{j\omega}) + H_{lp}(e^{j\omega})$$

and use this equation to find the impulse response in Problem 6.1.

6.3   Find the impulse response of the digital ideal bandpass filter shown in Fig. 6.3(c).

6.4   Find the impulse response of the digital ideal bandstop filter shown in Fig. 6.3(d). Can you obtain it from the impulse response in Problem 6.3.

6.5   Given

$$H(z) = \frac{z + 0.707}{z - 0.707}$$

Use measurement on the pole-zero plot to obtain $H(e^{j\omega})$ at $\omega = 0, \pi/4, \pi/2, 3\pi/4$, and π. Sketch its magnitude and phase responses for ω in $[0, \ \pi]$.

6.6   Design a digital low-pass filter to eliminate the high-frequency component of $x(t) = \cos 5t + \sin 176t$. Roughly how long will the response take to reach steady state?

6.7   Design a digital high-pass filter to eliminate the low-frequency component of $x(t) = \cos 5t + \sin 176t$. Is the result satisfactory?

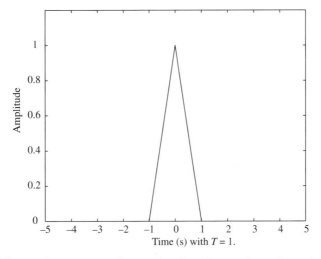

Figure 6.30 Impulse response of an analog filter that performs linear interpolation.

6.8 Design a digital resonator to eliminate the low-frequency component and pass the high-frequency component of $x(t) = \cos 5t + \sin 176t$.

6.9 Design a digital notch filter to eliminate the low-frequency component of $x(t) = \cos 5t + \sin 176t$.

6.10 Design a notch filer with nulls at $\omega = 0$ and ± 2.1 rad/s.

6.11 Design a comb filter with five equally spaced nulls, including $\omega = 0$, in $(-\pi, \pi]$.

6.12 Design a comb filter with five equally spaced nulls, including $\omega = \pi/5$, in $(-\pi, \pi]$.

6.13 Develop a block diagram whose outputs will generate $\sin 0.5n$ and $2 \cos 0.5n$ when it is excited by an impulse sequence.

6.14 Verify that if $v[n] = 0$, for $n \leq -1$, $v[0] = A$, and $x[n] = 0$, for all n, then the outputs of Fig. 6.25 yield $y_1[n] = A \sin \omega_0 n$ and $y_2[n] = A \cos \omega_0 n$, for $n \geq 0$.

6.15 The MATLAB function plot connects two neighboring points by a straight line. It is a type of D/A conversion and is called the *linear interpolation*. Show that if the process is modeled as in Fig. 6.27, then the analog filter has the impulse response shown in Fig. 6.30. Is it a causal system? Can it be used in real-time processing without any delay? Can it be used in real-time processing if the delay of one sample is permitted?

6.16 Suppose the DT transfer function $H(z)$ has two unstable poles at $ae^{\pm j\omega_0}$ with a real, positive, and larger than 1. Find $H_1(z)$ that has two corresponding poles inside the unit circle and has the same magnitude response as $H(z)$. This is called the *DT stabilization*.

6.17 Suppose the CT transfer function $H(s)$ has two unstable poles at $\alpha \pm j\beta$ with α real and positive. Find $H_1(s)$ that has two corresponding poles inside the left-half s plane and has the same magnitude response as $H(s)$. This is called the *CT stabilization*.

CHAPTER 7

Design of FIR Filters

7.1 Introduction

A digital filter is called a finite-impulse-response (FIR) filter if its impulse response has a finite number of nonzero entries such as $h[n]$, for $n = 0, 1, \ldots, N$. Generally we assume implicitly $h[0] \neq 0$ and $h[N] \neq 0$. The filter has at most $N + 1$ nonzero entries and is said to have length $N + 1$. Let $x[n]$ and $y[n]$ be the input and output of the filter. Then they can be described by the convolution

$$y[n] = h[0]x[n] + h[1]x[n-1] + \cdots + h[N]x[n-N] \tag{7.1}$$

This is, as discussed in Section 5.3.1, also an Nth-order nonrecursive difference equation. Thus the filter is said to have order N. An Nth-order FIR filter clearly has length $N + 1$.

Applying the z-transform to (7.1) and assuming zero initial conditions, we obtain the transfer function of the filter as

$$\frac{Y(z)}{X(z)} =: H(z) = h[0] + h[1]z^{-1} + \cdots + h[N]z^{-N} \tag{7.2}$$

$$= \frac{h[0]z^N + h[1]z^{N-1} + \cdots + h[N]}{z^N} \tag{7.3}$$

It has N poles and N zeros. All its poles are located at $z = 0$. Thus every FIR filter is stable. This also follows from the fact that its impulse response is always absolutely summable.

We study in this text only FIR filters that have linear phase. They will be designed to approximate a desired frequency response of the form

$$H_d(e^{j\omega}) = D(\omega)e^{-jM\omega}$$

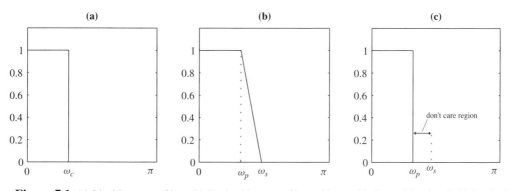

Figure 7.1 (a) Ideal low-pass filter. (b) Desired lowpass filter with specified transition band. (c) Desired lowpass filter with "don't care" transition band.

where M is a positive constant and $D(\omega)$ denotes a desired magnitude response such as the ones shown in Fig. 7.1 for low-pass filters. The one in Fig. 7.1(a) has a passband and a stopband with cutoff frequency ω_c; it has no transition band. The ones in Fig. 7.1(b) and (c) have a passband, a transition band, and a stopband. The transition band in Fig. 7.1(b) is specified such as the straight line as shown. The transition band in Fig. 7.1(c), however, is not specified or is a "don't care" region. We will introduce various methods to design linear-phase FIR filters to meet each of the specifications in Fig. 7.1.

Before proceeding, we mention that the MATLAB function [H,w]= freqz(bn,an) computes the frequency response, at 512 equally spaced frequencies in [0, π), of a transfer function expressed in negative-power form. For the FIR filter in (7.2), we have bn=h;an=1;; thus its magnitude response can be easily obtained as [H,w]=freqz(h,1);plot(w,abs(H)).

7.2 Classification of Linear-Phase FIR Filters

FIR filters to be designed in this chapter are required to have a linear phase. Depending on whether filter orders are even or odd and whether filter coefficients are symmetric or antisymmetric, linear-phase FIR filters are classified into four types.

Type I FIR Filters [Symmetric and Even Order (Odd Length)] An FIR filter is called type I if its order N is even and its impulse response $h[n]$ has the symmetric property

$$h[n] = h[N - n] \quad \text{for } 0 \leq n \leq N \tag{7.4}$$

as shown in Fig. 7.2(a) for $N = 8$. Let $M = N/2$. Then the symmetric condition can be reduced to

$$h[n] = h[N - n] \quad \text{for } 0 \leq n \leq M \tag{7.5}$$

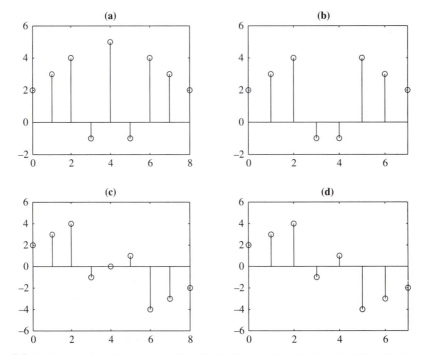

Figure 7.2 (a) Symmetric and even order (Type I). (b) Symmetric and odd order (Type II). (c) Antisymmetric and even order (Type III). (d) Antisymmetric and odd order (Type IV).

Thus there are only $M + 1$ independent parameters. Because $h[M] = h[N - M] = h[M]$, no restriction is imposed on $h[M]$. We compute its frequency response:

$$
\begin{aligned}
H(e^{j\omega}) = \sum_{n=0}^{N} h[n]e^{-jn\omega} &= e^{-jM\omega} \sum_{n=0}^{N} h[n]e^{j(M-n)\omega} \\
&= e^{-jM\omega} \big[h[0](e^{jM\omega} + e^{-jM\omega}) + h[1](e^{j(M-1)\omega} + e^{-j(M-1)\omega}) \\
&\quad + \cdots + h[M-1](e^{j\omega} + e^{-j\omega}) + h[M] \big] \\
&= e^{-jM\omega} \big[h[M] + 2h[M-1]\cos\omega + \cdots \\
&\quad + 2h[1]\cos(M-1)\omega + 2h[0]\cos M\omega \big]
\end{aligned}
\tag{7.6}
$$

Define $d[0] := h[M]$ and

$$
d[n] := 2h[M - n] \ \text{ for } 1 \le n \le M
\tag{7.7}
$$

We then use $d[n]$ to define

$$A(\omega) := \sum_{n=0}^{M} d[n] \cos n\omega \tag{7.8}$$

Clearly, $A(\omega)$ is a real-valued function of ω, and will be called an *amplitude response*. In contrast to the magnitude response, which is always zero or positive, the amplitude response can assume negative numbers. Using (7.8), we write (7.6) as

$$H(e^{j\omega}) = A(\omega)e^{-jM\omega} = |A(\omega)|e^{-j(M\omega - \alpha(\omega))} \tag{7.9}$$

where $\alpha(\omega) = 0$ if $A(\omega) \geq 0$ and $\alpha(\omega) = \pi$ if $A(\omega) < 0$. Thus $\alpha(\omega)$ is piecewise constant. Excluding $\alpha(\omega)$, $H(z)$ has a linear phase. The group delay of (7.9) is, for all ω,

$$\tau(\omega) = M \pm \pi \delta(\omega - \omega_i)$$

where ω_i are frequecies at which $A(\omega)$ changes signs. If $A(\omega)$ is all positive in passbands, as is often the case, then such a filter has a constant group delay in the passband and will introduce a time delay of M samples.

We next discuss zeros of type I FIR filters. Substituting (7.4) into (7.2) and defining $\bar{n} = N - n$, we obtain

$$H(z) = \sum_{n=0}^{N} h[n]z^{-n} = \sum_{n=0}^{N} h[N-n]z^{N-n}z^{-N} = z^{-N} \sum_{\bar{n}=N}^{0} h[\bar{n}]z^{\bar{n}}$$

$$= z^{-N} \sum_{n=0}^{N} h[n]z^{n} = z^{-N} H(z^{-1}) \tag{7.10}$$

Thus if $ae^{j\theta}$ is a zero, so is $a^{-1}e^{-j\theta}$. Note that we assume implicitly $h[N] \neq 0$, which implies $a \neq 0$. Because complex-conjugate zeros must appear in pairs due to real coefficients, we conclude that zeros of type I FIR filters are mutually reciprocal. Note that a zero on the unit circle is reciprocal to itself and can be simple or repeated. We show in Fig. 7.3(a) the zeros of the type I FIR filter in Fig. 7.2(a). There are four zeros on the unit circle; they are reciprocal to themselves. There are two pairs of complex-conjugate zeros. They are mutually reciprocal. Note that all poles are located at $z = 0$ and are not shown.

Type II FIR Filters [Symmetric and Odd Order (Even Length)] An FIR filter is type II if its order N is odd and its impulse sequence $h[n]$ has the symmetric property $h[n] = h[N - n]$ for $0 \leq n \leq N$. Figure 7.2(b) shows such a sequence for $N = 7$ or length 8. Let $M = N/2$ and $\bar{M} = (N - 1)/2$. Note that M is not an integer. Then the symmetric condition can be reduced to

$$h[n] = h[N - n] \quad \text{for } 0 \leq n \leq \bar{M} \tag{7.11}$$

Thus there are only $\bar{M} + 1$ independent coefficients. As in (7.6), we have

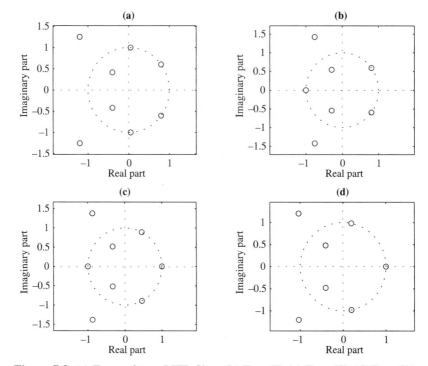

Figure 7.3 (a) Zeros of type I FIR filter. (b) Type II. (c) Type III. (d) Type IV.

$$H(e^{j\omega}) = \sum_{n=0}^{N} h[n]e^{-jn\omega} = e^{-jM\omega}\sum_{n=0}^{N} h[n]e^{j(M-n)\omega}$$

$$= e^{-jM\omega}\big[h[0](e^{jM\omega} + e^{-jM\omega}) + h[1](e^{j(M-1)\omega} + e^{-j(M-1)\omega})$$

$$+ \cdots + h[\bar{M}](e^{j(M-\bar{M})\omega} + e^{-j(M-\bar{M})\omega})\big]$$

$$= e^{-jM\omega}\big[2h[\bar{M}]\cos 0.5\omega + 2h[M-1]\cos 1.5\omega + \cdots$$

$$+ 2h[1]\cos(M-1)\omega + 2h[0]\cos M\omega\big] \tag{7.12}$$

Define

$$d[n] := 2h[\bar{M} - n] \quad \text{for } 0 \le n \le \bar{M} \tag{7.13}$$

and

$$A(\omega) := \sum_{n=0}^{\bar{M}} d[n]\cos(n + 0.5)\omega \tag{7.14}$$

Clearly, $A(\omega)$ is a real-valued function of ω, and is called an amplitude response. Then we can write (7.12) as

$$H(e^{j\omega}) = A(\omega)e^{-jM\omega} = |A(\omega)|e^{-j(M\omega-\alpha(\omega))} \tag{7.15}$$

where $\alpha(\omega) = 0$ if $A(\omega) \geq 0$ and $\alpha(\omega) = \pi$ if $A(\omega) < 0$. Thus $\alpha(\omega)$ is piecewise constant. Excluding $\alpha(\omega)$, $H(z)$ has a linear phase. In the passband, the filter $H(z)$ has a constant group delay M if $A(\omega)$ is always positive. Thus the filter will introduce a time delay of M samples. Note that M is not an integer, and the delay of M samples in DT signals does not have a simple physical interpretation. However, if the digital filter is used to process CT signals as shown in Fig. 6.8, then it can be interpreted as introducing a delay of MT seconds in CT signals.

As in type I filters, (7.10) holds for type II FIR filters. Thus all zeros of type II FIR filters are mutually reciprocal. In addition, (7.10) and N odd imply, at $z = -1$,

$$H(-1) = (-1)^N H((-1)^{-1}) = -H(-1)$$

This holds only if $H(-1) = 0$. Thus every type II FIR filters has at least one zero at $z = -1$, as shown in Fig. 7.3(b). Such a filter clearly cannot be used to design a high-pass filter.

Type III FIR Filters [Antisymmetric and Even Order (Odd Length)] An FIR filter is type III if its order N is even and $h[n]$ has the antisymmetric property

$$h[n] = -h[N - n] \quad \text{for } 0 \leq n \leq N \tag{7.16}$$

Let $M = N/2$. Then the antisymmetric condition can be reduced to

$$h[n] = -h[N - n] \quad \text{for } 0 \leq n \leq M \tag{7.17}$$

as shown in Fig. 7.2(c) for $N = 8$. The condition $h[M] = -h[M]$ implies $h[M] = 0$; thus there are only M independent coefficients.

We compute its frequency response:

$$
\begin{aligned}
H(e^{j\omega}) &= \sum_{n=0}^{N} h[n]e^{-jn\omega} = e^{-jM\omega} \sum_{n=0}^{N} h[n]e^{j(M-n)\omega} \\
&= e^{-jM\omega} \big[h[0](e^{jM\omega} - e^{-jM\omega}) + h[1](e^{j(M-1)\omega} - e^{-j(M-1)\omega}) \\
&\quad + \cdots + h[M+1](e^{j\omega} - e^{-j\omega}) + h[M] \big] \\
&= je^{-jM\omega} [2h[M-1]\sin\omega + 2h[M-2]\sin 2\omega + \cdots \\
&\quad + 2h[1]\sin(M-1)\omega + 2h[0]\sin M\omega]
\end{aligned}
\tag{7.18}
$$

where we have used $h[M] = 0$. Define

$$d[n] := 2h[M - n] \text{ for } 1 \leq n \leq M \tag{7.19}$$

and

$$A(\omega) := \sum_{n=1}^{M} d[n] \sin n\omega \tag{7.20}$$

Clearly, $A(\omega)$ is a real-valued function of ω, and is called an amplitude response. Then we can write (7.18) as

$$H(e^{j\omega}) = A(\omega) j e^{-jM\omega} = |A(\omega)| e^{-j[M\omega - \alpha(\omega)]} \tag{7.21}$$

where $\alpha(\omega) = \pi/2$ if $A(\omega) \geq 0$ and $\alpha(\omega) = -\pi/2$ if $A(\omega) < 0$. Thus $\alpha(\omega)$ is piecewise constant. Excluding $\alpha(\omega)$, $H(z)$ has a linear phase.

We discuss zeros of type III FIR filters. Substituting (7.16) into (7.2) and defining $\bar{n} = N - n$, we obtain

$$H(z) = \sum_{n=0}^{N} h[n] z^{-n} = -\sum_{n=0}^{N} h[N - n] z^{N-n} z^{-N} = -z^{-N} \sum_{\bar{n}=N}^{0} h[\bar{n}] z^{\bar{n}}$$

$$= -z^{-N} \sum_{n=0}^{N} h[n] z^{n} = -z^{-N} H(z^{-1}) \tag{7.22}$$

Thus, as in types I and II, zeros of type III FIR filters are mutually reciprocal. Furthermore, (7.22) and N even imply $H(1) = 0$ and $H(-1) = 0$. Thus every type III FIR filter has zeros at $z = \pm 1$ as shown in Fig. 7.3(c).

Type IV FIR Filters [Antisymmetric and Odd Order (Even Length)] If N is odds and $h[n]$ has the antisymmetric property in (7.16), then the filter, as shown in Fig. 7.2(d) for $N = 7$, is called a type IV FIR filter. Let $M = N/2$ and $\bar{M} = (N - 1)/2$. Note that M is not an integer. Then the antisymmetric condition can be written as

$$h[n] = -h[N - n] \quad \text{for } 0 \leq n \leq \bar{M} \tag{7.23}$$

Thus there are $\bar{M} + 1$ independent coefficients. Define

$$d[n] := 2h[\bar{M} - n] \text{ for } 0 \leq n \leq \bar{M} \tag{7.24}$$

and

$$A(\omega) := \sum_{n=0}^{\bar{M}} d[n] \sin(n + 0.5)\omega \tag{7.25}$$

Then the frequency response of every type IV FIR filter can be expressed as

$$H(e^{j\omega}) = A(\omega)je^{-jM\omega} = |A(\omega)|e^{-j[M\omega-\alpha(\omega)]} \tag{7.26}$$

where $\alpha(\omega) = \pi/2$ if $A(\omega) \geq 0$ and $\alpha(\omega) = -\pi/2$ if $A(\omega) < 0$. Thus $\alpha(\omega)$ is piecewise constant. Excluding $\alpha(\omega)$, $H(z)$ has a linear phase.

As in type III, (7.22) holds for type IV FIR filters. Thus all zeros of type IV FIR filters are mutually reciprocal. In addition, (7.22) implies, at $z = 1$,

$$H(1) = -(1)^{-N}H(1^{-1}) = -H(1)$$

This holds only if $H(1) = 0$. Thus every type IV FIR filter has at least one zero at $z = 1$ as shown in Fig. 7.3(d). Note that (7.22) does not imply $H(-1) = -H(-1)$ for N odd. Thus a type IV filter has no zero at $z = -1$. The preceding four types of FIR filters are summerized in Table 7.1.

One may wonder why we introduce these four types of FIR filters. To see the reason, we discuss the design of digital low-pass filters. Their magnitude responses should be close to 1 at low frequencies. However, if we use type III or IV filters that have zeros at $\omega = 0$, then the design will always be poor. Thus we must use type I or II to design digital low-pass filters. Using similar argument, we can establish the following.

Table 7.1 Nth-Order Linear-Phase FIR Filters

$$H(e^{j\omega}) = \sum_{n=0}^{N} h[n]e^{jn\omega} = A(\omega)e^{-jM\omega}$$

$$M = N/2 \qquad \bar{M} = (N-1)/2$$

Type	N		$A(\omega)$		Zeros
I	even	$h[n] = h[N-n]$	$\displaystyle\sum_{n=0}^{M} d[n]\cos n\omega$	$h[n] = d[M-n]/2, h[M] = d[0]$ $n = 0,\ldots,M-1$	
II	odd	$h[n] = h[N-n]$	$\displaystyle\sum_{n=0}^{\bar{M}} d[n]\cos(n+0.5)\omega$	$h[n] = d[\bar{M}-n]/2$ $n = 0,\ldots,\bar{M}$	$z = -1$
III	even	$h[n] = -h[N-n]$	$\displaystyle\sum_{n=0}^{M} d[n]\sin n\omega$	$h[n] = d[M-n]/2, h[M] = 0$ $n = 0,\ldots,M-1$	$z = \pm 1$
IV	odd	$h[n] = -h[N-n]$	$\displaystyle\sum_{n=0}^{\bar{M}} d[n]\sin(n+0.5)\omega$	$h[n] = d[\bar{M}-n]/2$ $n = 0,\ldots,\bar{M}$	$z = 1$

- Low-pass filter: Use type I or II
- High-pass filter: Use type I or IV
- Bandpass filter: Use any type
- Bandstop filter: Use only type I

To conclude this section, we compare even and odd sequences defined in Chapter 4 and in this section. Consider the sequence $x[n]$ for $n = 0, 1, \ldots, N$. The finite sequence is defined to be even (odd) in Chapter 4 if its periodic extension with period $N + 1$ is even (odd) or $x[n] = x[N + 1 - n]$ ($x[n] = -x[N + 1 - n]$) for all n, in particular, for $n = 0, 1, \ldots, N$. It is defined in this section to be even (odd) if $x[n] = x[N - n]$ ($x[n] = -x[N - n]$) for $n = 0, 1, \ldots, N$. These two definitions are clearly different. Compare also Figs. 7.2 and 4.5.

7.3 Least-Squares Optimal Filters—Direct Truncation

This section discusses a general property of the inverse DT Fourier transform and its truncation. Consider a desired frequency response $H_d(e^{j\omega})$ specified for all ω in $(-\pi, \pi]$. We compute its inverse DT Fourier transform

$$h_d[n] = \frac{1}{2\pi} \int_{\omega=-\pi}^{\pi} H_d(e^{j\omega})e^{jn\omega}d\omega \tag{7.27}$$

for all integers n in $(-\infty, \infty)$. Clearly we have

$$H_d(e^{j\omega}) = \sum_{n=-\infty}^{\infty} h_d[n]e^{-jn\omega} \tag{7.28}$$

In general, $h_d[n]$ is two sided and of infinite length. The problem is to design an FIR causal filter whose frequency response approximates $H_d(e^{j\omega})$.

A simple way of obtaining an Nth-order FIR causal filter is to truncate $h_d[n]$ as

$$h[n] = \begin{cases} h_d[n] & \text{for } 0 \leq n \leq N \\ 0 & \text{for } n < 0 \text{ and } n > N \end{cases} \tag{7.29}$$

This is a finite sequence of length $N+1$. We first show that this truncated sequence approximates optimally the desired frequency response in some sense. Let $H(e^{j\omega})$ be the frequency response of $h[n]$. We compare $H_d(e^{j\omega})$ and $H(e^{j\omega})$. Define

$$e(\omega) = H_d(e^{j\omega}) - H(e^{j\omega}) \tag{7.30}$$

It is the error at frequency ω between the desired and the actual frequency responses. Because errors may exist at all frequencies, the total squared error in $(-\pi, \pi]$ is defined as[1]

---

[1] Note that the total error

$$E = \int_{\omega=-\pi}^{\pi} e^*(\omega)e(\omega)d\omega$$

$$= \int_{\omega=-\pi}^{\pi} [H_d(e^{j\omega}) - H(e^{j\omega})]^*[H_d(e^{j\omega}) - H(e^{j\omega})]d\omega$$

$$= \int_{\omega=-\pi}^{\pi} |H_d(e^{j\omega}) - H(e^{j\omega})|^2 d\omega \tag{7.31}$$

where the asterisk denotes complex conjugate. Using the Parseval formula in (3.53) with $T = 1$, we can write (7.31) as

$$E = 2\pi \left(\sum_{n=-\infty}^{\infty} |h_d[n] - h[n]|^2 \right)$$

$$= 2\pi \left(\sum_{n=-\infty}^{-1} |h_d[n]|^2 + \sum_{n=0}^{N} |h_d[n] - h[n]|^2 + \sum_{n=N+1}^{\infty} |h_d[n]|^2 \right) \tag{7.32}$$

If $h[n] = h_d[n]$, for $0 \le n \le N$, then (7.32) reduces to

$$E = 2\pi \left(\sum_{n=-\infty}^{-1} h_d^2[n] + \sum_{n=N+1}^{\infty} h_d^2[n] \right) \tag{7.33}$$

If $h[n] \ne h_d[n]$ for some n in $[0, N]$, then the error in (7.32) will be larger than the one in (7.33). Thus we conclude that for the given desired frequency response, the causal FIR filter obtained in (7.29) by direct truncation has the least integral squared (LS) error and will be called the LS optimal filter.

The preceding discussion is applicable to any $H_d(e^{j\omega})$, in particular, to

$$H_d(e^{j\omega}) = D(\omega)e^{-jM\omega} \tag{7.34}$$

where M is a positive constant and $D(\omega)$ denotes a desired magnitude response such as the one in Fig. 7.1(a) or (b). We first discuss the total error in (7.31) due to different group delay M and filter order N. For a fixed M, it is clear that the total error decreases as N increases. Thus the more interesting question is: For a fixed N, what is the total error as a function of M?

Suppose the inverse DT Fourier transform of $D(\omega)$ and $M = 0$ is as shown in Fig. 7.4(a). Retaining $h[n]$ for $0 \le n \le 10$ (solid dots), we will obtain a tenthth-order FIR filter. This filter

---

$$E = \int_{\omega=-\pi}^{\pi} e(\omega)d\omega$$

is meaningless because positive and negative errors may cancel out and small E may not imply small $|e(\omega)|$. The total error $E = \int_{-\pi}^{\pi} |e(\omega)|d\omega$ is acceptable, but it does not lead to a simple analytical solution.

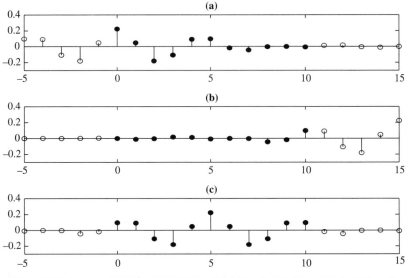

Figure 7.4 Inverse DTFT of (7.34) for (a) $M = 0$, (b) $M = 15$, (c) $M = 5$.

is optimal in the sense of (7.31), and its total error is the sum of squared $h[n]$ for $n < 0$ and $n > 10$ (hollow dots). From Fig. 7.4(a), we see that $h[n]$ for $n < 0$ is not small; thus the error is quite large. However, the FIR filter still has the smallest total error among all tenth-order FIR filters to approximate $D(\omega)e^{-j0\cdot\omega}$ and is LS optimal.

Next we consider $M = 15$ or find an FIR filter to approximate $D(\omega)e^{-j15\omega}$. Its inverse DTFT is shown in Fig. 7.4(b); it is the delay of the one in Fig. 7.4(a) by 15 samples. Retaining $h[n]$ for $n \leq n \leq 10$ (solid dots), we will obtain a tenth-order LS optimal FIR filter. It is optimal but its total error is very large as we can see from Fig. 7.4(b). Thus even though it has the smallest possible error among all tenth-order FIR filters, the filter is not satisfactory. Next we consider $M = 5$ or find an FIR filter to approximate $D(\omega)e^{-j5\omega}$. Its inverse DTFT is shown in Fig. 7.4(c); it is the delay of the one in Fig. 7.4(a) by five samples. Retaining $h[n]$ for $0 \leq n \leq 10$ (solid dots), we obtain an optimal FIR filter. From Fig. 7.4, we see that this optimal filter has the smallest total error among the three optimal filters and should be the one used in practice. In conclusion, optimality may not yield a satisfactory filter unless it is properly formulated.

For ideal low-pass filters with $D(\omega)$ shown in Fig. 7.1(a) and specified as in (7.34), it is possible to show that the error defined in (7.31) is smallest when $M = N/2$. See Problem 7.4. We conjecture that this is also the case in general. Thus from now on, we design linear-phase FIR filters to approximate

$$H_d(e^{j\omega}) = D(\omega)e^{-jM\omega}$$

with $M = N/2$, where N is the order of the filter to be designed and can be even or odd.

7.4 Window Method

With the preceding background, we are ready to discuss specific design methods. Let us design a causal Nth-order FIR filter to approximate the ideal low-pass filter specified by

$$H_d(e^{j\omega}) = \begin{cases} 1 \times e^{-jM\omega} & \text{for } |\omega| \leq \omega_c \\ 0 & \text{for } \omega_c < |\omega| \leq \pi \end{cases} \tag{7.35}$$

with $M = N/2$. Its inverse DT Fourier transform can be computed as, for all integer n,

$$h_d[n] = \frac{1}{2\pi} \int_{\omega=-\omega_c}^{\omega_c} e^{-jM\omega} e^{jn\omega} d\omega$$

$$= \frac{e^{j(n-M)\omega_c} - e^{-j(n-M)\omega_c}}{2j\pi(n-M)} = \frac{\sin(n-M)\omega_c}{\pi(n-M)}$$

and $h_d[M] = \omega_c/\pi$. This ideal filter is infinitely long and is noncausal. We truncate it as

$$h[n] = \begin{cases} \dfrac{\sin(n-M)\omega_c}{\pi(n-M)} & \text{for } 0 \leq n \leq N \\ 0 & \text{otherwise} \end{cases} \tag{7.36}$$

Because

$$h[N-n] = \frac{\sin(N-n-M)\omega_c}{\pi(N-n-M)} = \frac{\sin(M-n)\omega_c}{\pi(M-n)} = \frac{\sin(n-M)\omega_c}{\pi(n-M)} = h[n]$$

the impulse response in (7.36) is symmetric. Thus the FIR filter is type I if N is even and type II if N is odd. This filter is least-squares optimal or has the smallest error defined in (7.31) among all Nth-order causal FIR filters.

Figure 7.5 shows the magnitude responses of the causal filters in (7.36) with $\omega_c = 1$, $N = 10$ ($M = 5$) (solid line), and $N = 101$ ($M = 50.5$) (dashed line). They are obtained by using the program that follows

Program 7.1

```
N= ;M=N/2;
if N/2-floor(N/2)<0.1 (% N even)
  n1=0:M-1;h1=sin(n1-M)./(pi*(n1-M));
  h=[h1 1/pi fliplr(h1)];
else
  n=0:N; (% N odd)
  h=sin(n-M)./(pi*(n-M));
end
[H,w]=freqz(h,1);
plot(w,abs(H))
```

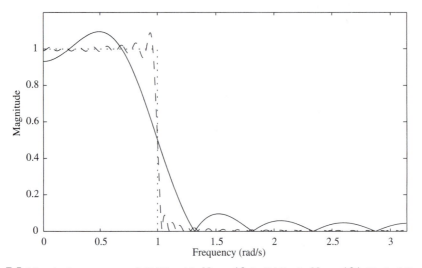

Figure 7.5 Magnitude responses of (7.36) with $N = 10$ (solid line), $N = 101$ (dashed line), and $N = \infty$ (dotted line).

We first explain the program. If N is even, then $N/2 - \text{floor}(N/2) = 0$, where floor rounds a number downward to an integer. If N is odd, then $N/2 - \text{floor}(N/2) = 0.5$. Thus the if-else-end statement can be used to check whether N is even or odd. Note that the value 0.1 in the statement can be replaced by any number between 0 and 0.5. For N even, the filter has $N + 1 = 2M + 1$ coefficients. The first M coefficients are generated in h1. The coefficient $h[M]$ must be typed in because a computer will generate NaN (not a number) instead of $1/\pi$. The last M coefficients, because of symmetric, equal the left right flip of the first M coefficients. This can be achieved in MATLAB as fliplr(h1). For N odd, all coefficients can be generated from (7.36) because we will not encounter "zero divided by zero." For FIR filters, we have bn=h, an=1. Thus freqz(h,1) generates its frequency response at 512 equally spaced frequencies in $[0, \pi)$. From Fig. 7.5, we see that the magnitude responses have leakage and ripples. As discussed in Chapter 3, as N increases, the leakage will decrease and the ripples will move closer to ω_c, the discontinuity point. But the peak of the ripples remains the same, roughly 9% of the amount of the discontinuity. This is called the Gibbs phenomenon and is not desirable. Thus even though direct truncation yields an LS optimal filter, the filter exhibits the Gibbs phenomenon and is not satisfactory.

There are ways to reduce or eliminate the ripples. In this section, we discuss the window method. We first list a number of sequences of length $N + 1$:[2]

$$\text{Rectangular: } w_1[n] = 1 \tag{7.37}$$

---

[2] The length can be even or odd. If we define a window as $w[n]$ for $0 \leq |n| \leq M$, then its length is limited to an odd integer.

$$\text{Triangular: } w_2[n] = 1 - \frac{2|n - N/2|}{N} \tag{7.38}$$

$$\text{Hamming: } w_3[n] = 0.54 - 0.46 \cos\left(\frac{2\pi n}{N}\right) \tag{7.39}$$

$$\text{Blackman: } w_4[n] = 0.42 - 0.5 \cos\left(\frac{2\pi n}{N}\right)$$

$$+ 0.08 \cos\left(\frac{4\pi n}{N}\right) \tag{7.40}$$

for $0 \leq n \leq N$. They all equal zero for $0 < n$ and $n > N$ and have the symmetric property $w_i[n] = w_i[N - n]$. Their *envelopes* are plotted in Fig. 7.6 for $N = 40$. If we plot them using the MATLAB function stem, then the results will be poor. Thus we use the function plot to plot their interpolations. They are called, respectively, the rectangular or boxcar window (solid line), triangular or Bartlett window (dotted line), Hamming window (dashed line), and Blackman window (dash-dotted line). Their magnitude spectra are shown in Fig. 7.7 for $N = 20$. They can be obtained using 1024-point FFT by padding zeros as discussed in Section 4.5.1 or using the MATLAB function freqz.

Direct truncation of $h_d[n]$ is the same as multiplying $h_d[n]$ by the rectangular window in (7.37). The frequency spectrum of the truncated sequence in Fig. 7.5 is, as shown in (3.56), the convolution of the desired frequency response and the spectrum of the rectangular window shown in Fig. 7.7 with a solid line. The spectrum of the rectangular window has one main lobe and a number of side lobes. As discussed in Sections 3.5 and 3.7, the leakage or smearing in the neighborhood of $\omega_c = 1$ in Fig. 7.5 is due to the main lobe, and the ripples are due to the side lobes. As shown in Fig. 7.7 with a solid line, the rectangular window has fairly large side lobes; thus it causes significant ripples in Fig. 7.5.

Now, instead of direct truncation, $h[n]$ in (7.29) will be obtained by multiplying $h_d[n]$ with one of the windows as

$$\bar{h}[n] = h_d[n]w_i[n] \tag{7.41}$$

for $i = 2, 3, 4$. For example, if we use a Hamming window, inserting

```
n=0:N;
win=0.54-0.46*cos(2*pi*n/N);
h=h.*win;
```

after h=[h1 1/pi fliplr(h1)] in Program 7.1, will yield

-0.0060 -0.0144 0.0071 0.1058 0.2483 0.3183 0.2483 0.1058
0.0071 -0.0144 -0.0060

Thus the transfer function of the resulting filter is

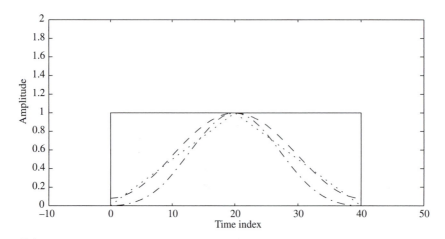

Figure 7.6 Envelopes or interpolations of rectangular (solid line), triangular (dotted line), Hamming (dashed line), and Blackman (dash-dotted line) windows with $N = 40$ or length 41.

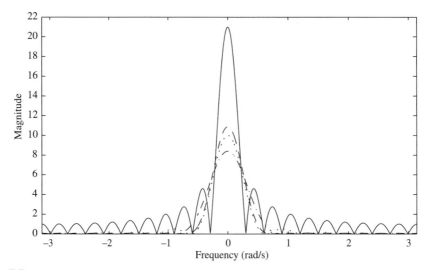

Figure 7.7 Magnitude spectra of rectangular (solid line), triangular (dotted line), Hamming (dashed line), and Blackman (dash-dotted line) windows with length 21.

$$H(z) = -0.0060(1 + z^{-10}) - 0.0144(z^{-1} + z^{-9}) + 0.0071(z^{-2} + z^{-8})$$
$$+ 0.1058(z^{-3} + z^{-7}) + 0.2483(z^{-4} + z^{-6}) + 0.3183z^{-5} \qquad (7.42)$$

It is a type I FIR filter of order 10 and length 11. Thus the design of FIR filters using windows is simple and straightforward.

We now compare the FIR filters obtained by using the four windows in (7.37) through (7.40). Figure 7.8(a) shows the magnitude responses of the tenth-order FIR filters obtained by using a rectangular window (solid line), triangular window (dotted line), Hamming window (dashed lined), and Blackman window (dash-dot line). Figure 7.8(b) shows the corresponding magnitude responses expressed in decibels (dB). The unit of decibels stretches the vertical linear scale $[0, 1]$ to the vertical logarithmic scale $(-\infty, 0]$; thus it is easier to compare in Fig. 7.8(b) magnitudes close to 0. However, it is easier to compare in Fig. 7.8(a) magnitudes close to 1 as shown. From Fig. 7.8, we see that the filters obtained by using the triangular, Hamming, and Blackman windows have negligible ripples but wider leakages. This is due to the fact that the three windows have negligible sides lobes but wider main lobes as shown in Fig. 7.7. From Fig. 7.8, we also see that the Hamming window introduces least leakage. Thus the Hamming window is the most widely used among the four windows. We mention that FIR filters obtained by windows other than the rectangular window are not optimal in the least-squares sense.

In addition to the aforementioned windows, the following Kaiser window

$$w[n] = \frac{I_0\left(0.5N\beta\sqrt{(0.5N)^2 - (n - 0.5N)^2}\right)}{I_0(0.5N\beta)} \tag{7.43}$$

for $0 \leq n \leq N$, is also widely cited. The function I_0 is the modified zeroth-order Bessel function and can be expressed as

$$I_0(x) = 1 + \sum_{k=1}^{\infty} \left(\frac{(0.5x)^2}{k!}\right)^2$$

Because the infinite power series converges rapidly, we may use a finite number of terms such as 20 terms in its computation. The constant β in (7.43) is a parameter that can be used to adjust the tradeoff between the main lobe width and side lobe levels. Side lobe levels govern the stopband attenuation; increasing β will increase the attenuation. If the attenuation is required to be less than $R_s := -20 \log b$ dB, where b is the largest magnitude in the stopband, the following values are listed in Ref. 21:

$$\beta = \begin{cases} 0.1102(R_s - 8.7) & 50 < R_s \\ 0.5842(R_s - 21)^{0.4} + 0.07886(R_s - 21) & 21 \leq R_s \leq 50 \\ 0 & R_s < 21 \end{cases}$$

Note that the Kaiser window reduces to the rectangular window if $\beta = 0$.

The MATLAB function fir1(n,f,window) generates an nth-order lowpass FIR filter with cutoff frequency f using the specified window. The cutoff frequency f must be normalized to lie in $[0, 1]$. If no window is specified, the default is the Hamming window. For example, typing h=fir1(10,1/pi) will yield the type I FIR filter in (7.42) using a Hamming window. Note that the cutoff frequency $\omega_c = 1$ in the frequency range $[0, \pi]$ must be normalized as $1/\pi$ in $[0, 1]$. If we use a Kaiser window, then we first generate a Kaiser window of length 11 as ka=kaiser(11,b) with a selected β. Then h=fir1(10,1/pi,ka) returns a tenth-order type I FIR filter using a Kaiser window.

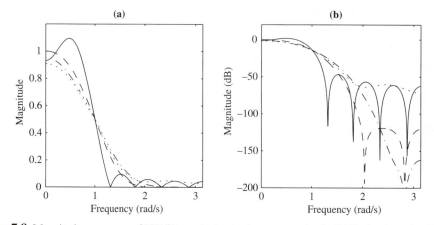

Figure 7.8 Magnitude responses of FIR filters obtained using rectangular (solid line), triangular (dotted line), Hamming (dashed line), and Blackman (dash-dot line) windows.

If f is a two-element vector such as f=[f1 f2], then fir1(n,f,window) returns a bandpass filter with passband edge frequencies f_1 and f_2 in [0, 1] or $f_1\pi$ and $f_2\pi$ in [0, π]. The functions fir1(n,f,window,'high') and fir1(n,[f1 f2],window,'stop') return, respectively, high-pass and bandstop FIR filters using the specified window or a Hamming window if no window is specified. As shown in Fig. 7.3, every type II FIR filter has at least one zero at $\omega = \pi$. Thus type II (odd-order) FIR filters cannot be used to design high-pass and bandstop filters. If the filter order n in fir1(n,f,window,'high') or fir1(n,[f1 f2],window,'stop') is odd, then fir1 automatically increases n by 1. In designing low-pass and bandpass filters, we can select n to be even or odd.

7.5 Desired Filters with Specified Transition Bands

The Gibbs phenomenon due to direct truncation of ideal filters was alleviated in the preceding section by using various windows. In this section, we discuss a different method to deal with the problem.

The ideal low-pass frequency response used in the preceding section contains no transition band. Thus there is a vertical rolloff from the passband to the stopband. This discontinuity is the main reason for causing the Gibbs phenomenon. Now we will insert a transition band between the passband and stopband. One way to specify the transition band is to connect a smooth curve between the edge frequencies. If we connect them with a straight line, then the desired frequency response becomes

$$D(\omega) = \begin{cases} 1 & \text{for } 0 \le \omega \le \omega_p \\ (\omega_s - \omega)/(\omega_s - \omega_p) & \text{for } \omega_p < \omega < \omega_s \\ 0 & \text{for } \omega_s \le \omega \le \pi \end{cases} \tag{7.44}$$

as shown in Fig. 7.1(b). To compute the inverse DTFT of $D(\omega)e^{-jM\omega}$ with $M \neq 0$ is fairly complex, To simplify the computation, we assume $M = 0$ in this section. After the inverse $h_d[n]$ of $D(\omega)$ is computed, we will truncate $h_d[n]$ for $|n| > M$ and delay it M samples to obtain a causal filter. Thus all filters in this section have odd length or, equivalently, even order.

First we extend Fig. 7.1(b) to Fig. 7.9(a) for $-\pi < \omega \leq \pi$. We can show that the frequency response in Fig. 7.9(a) is the convolution of the frequency responses in Figs. 7.9(b) and (c) with

$$\omega_c = \frac{\omega_p + \omega_s}{2} \qquad \Delta = \omega_s - \omega_p$$

(Problem 7.5). Note that the three responses are all periodic with period 2π. The inverse DT Fourier transforms of Figs. 7.9(b) and (c) can be readily computed as

$$\frac{\sin \omega_c n}{\pi n} \quad \text{and} \quad \frac{\sin 0.5\Delta n}{\Delta \pi n}$$

Then the inverse DTFT of Fig. 7.9(a), as shown in (3.54) with $T = 1$, is the product of 2π and the preceding two inverse DTFT; that is,

$$\begin{aligned} h_d[n] &= 2\pi \times \frac{\sin \omega_c n}{\pi n} \times \frac{\sin 0.5\Delta n}{\Delta \pi n} \\ &= \frac{2(\sin 0.5\Delta n)(\sin \omega_c n)}{\pi \Delta n^2} \end{aligned} \tag{7.45}$$

with $h[0] = \omega_c/\pi$.[3] Truncating $h_d[n]$, for $|n| > M$, and then delaying it M samples, we obtain an Nth-order type I FIR filter with $N = 2M$. For example, if $\omega_p = 1$ and $\omega_s = 1.5$, then $\omega_c = 1.25$ and $\Delta = 0.5$, and the following

```
n=1:4;
hp=2.0.*sin(0.5*0.5.*n).*sin(1.25.*n)./(pi*0.5.*n.*n);
h=[fliplr(hp) 1.25/pi hp]
```

will generate the eighth-order FIR filter that follows

$$\begin{aligned} H(z) &= -0.0642(1 + z^{-8}) - 0.0551(z^{-1} + z^{-7}) + 0.0913(z^{-2} + z^{-6}) \\ &\quad + 0.2989(z^{-3} + z^{-5}) + 0.3979z^{-4} \end{aligned} \tag{7.46}$$

This filter is least-squares optimal.

Figure 7.10 shows the magnitude responses of the eighth-order (solid line), twentieth-order (dotted line), and sixtieth-order (dashed line) filters with $\omega_p = 1$ and $\omega_s = 1.5$. As the order

---

[3] This can be computed from (7.45) by using l'Hôpital's rule or, more easily, directly from the inverse DTFT of Fig. 7.9(a) with $n = 0$.

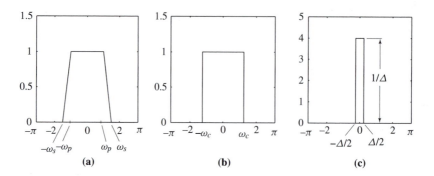

Figure 7.9 (a) Magnitude response, which equals the convolution of (b) and (c).

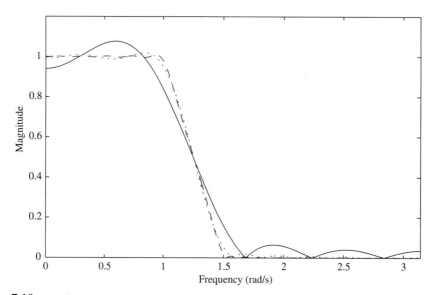

Figure 7.10 Magnitude responses of eighth- (solid line), twentieth- (dotted line), and sixtieth- (dashed line) order FIR filters obtained by direct truncation.

increases, the leakage and ripples decrease. It does not show any Gibbs phenomenon as in Fig. 7.5. This is expected because the desired frequency response has no discontinuity.

The transition band in Fig. 7.9(a) is described by a polynomial of ω and of degree one as in (7.44). Convolving the inverse DTFT of Figs. 7.9(a) and (c), we will obtain a transition band that is describable by a second-order spline function that consists of two polynomials of degree 2. Proceeding forward, we can develop a transition band that is describable by a higher-order spline function. Another possibility is to use the raised cosine function as

$$0.5 + 0.5 \cos\left(\frac{(\omega - \omega_p)\pi}{\omega_s - \omega_p}\right)$$

The LS optimal filters with these transition bands are extensively discussed in Ref. 3. However, the straight-line connection appears to be satisfactory in eliminating the Gibbs phenomenon.

7.5.1 Design by Frequency Sampling

This subsection discusses a different method of designing FIR filters. Consider the desired frequency response shown in Fig. 7.9(a). It is plotted in Fig. 7.11 with ω in $[0, 2\pi)$ instead of $(-\pi, \pi]$. Recall that DT frequency responses are periodic with period 2π. Let us take its $N + 1$ equally spaced frequency samples and compute its inverse DFT. Then we will obtain an impulse sequence of length $N + 1$. This will be the coefficients of an Nth-order FIR filter. The frequency response of this filter matches exactly the $N + 1$ frequency samples. This is called *design by frequency sampling*.

We use an example to illustrate the procedure. Let us design an FIR filter to match nine frequency samples of Fig. 7.11 with $\omega_p = 1$ and $\omega_s = 1.5$. Then the frequency resolution is $D = 2\pi/9 = 0.698$, and the nine frequencies will be at kD for $k = 0, 1, \ldots, 8$. The nine frequency samples are

$$H = [1\ 1\ 0.2074\ 0\ 0\ 0\ 0\ 0.2074\ 1]$$

as shown in Fig. 7.11 with crosses where the value 0.2074 is computed from (7.44). This is the input of inverse FFT.

Strictly speaking, from one inverse FFT computation, we cannot determine uniquely, as discussed in Section 4.7, the location of the time sequence corresponding the frequency samples. However, because the frequency spectrum given in Fig. 7.9(a) is real and even, its inverse DTFT should also be real and even. Thus we will take the inverse DFT of H as the value of $h[n]$ for $n = -4:4$. Let $\tilde{h}[n]$ be the periodic extension of $h[n]$ with period $N = 9$. Then the output of ifft(H) yields $\tilde{h}[n]$, for $n = 0, 1, \ldots, 8$, as

$$0.3794,\ 0.2893,\ 0.1064,\ -0.0230,\ -0.0624,$$
$$-0.0624,\ -0.0230,\ 0.1064,\ 0.2893$$

By shifting the last four numbers to the front, we obtain

$$h[-4] = -0.0624,\ h[-3] = -0.023,\ h[-2] = 0.1064,\ h[-1] = 0.2893,$$
$$h[0] = 0.3794,\ h[n] = h[-n] \quad \text{for } n = 1, 2, 3, 4$$

This is the inverse DFT of H and is a noncausal filter. Delaying it by four samples, we obtain the following eighth-order type I FIR filter

$$H(z) = -0.0624(1 + z^{-8}) - 0.023(z^{-1} + z^{-7}) + 0.1064(z^{-2} + z^{-6})$$
$$+ 0.2893(z^{-3} + z^{-5}) + 0.2893z^{-4} \tag{7.47}$$

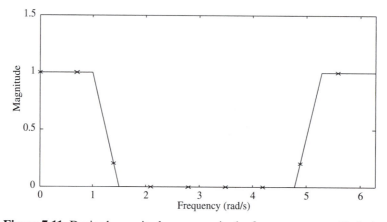

Figure 7.11 Desired magnitude response in the frequency range $[0, 2\pi)$.

Its magnitude response is plotted in Fig. 7.12(a) with solid line for ω in $[0, \pi)$. It is obtained by typing

h=shift(ifft(H));[H,w]=freqz(h,1);plot(w,abs(H))

It matches exactly the five frequency samples in $[0, \pi)$. The other four matchings are in $(-\pi, 0)$ or $[\pi, 2\pi)$ and are not shown.

Figure 7.12(b) shows the magnitude response of the twentieth-order FIR filter obtained by matching 21 frequency samples. The result is better than the one in Fig. 7.12(a), but the filter order is higher.

Let us discuss the relationship between the FIR filters obtained in this and preceding sections. The impulse response $h_d[n]$ in (7.45) is the inverse DTFT of the desired frequency response in Fig. 7.9(a). The $h[n]$, for $n = 0, 1, \ldots, N$, obtained in this section is the inverse DFT of the $N + 1$ frequency samples of Fig. 7.9(a). As discussed in Section 4.2.1, we have

$$h[n] = \sum_{k=-\infty}^{\infty} h_d[n + k(N + 1)]$$

Because $h_d[n]$ is of infinite length, time aliasing will always occur in $h[n]$. For $N = 8$, time aliasing is not small, and the coefficients in (7.46) and (7.47) differ appreciably. We plot in Fig. 7.13(a) the magnitude responses of (7.46) (solid line) and (7.47) (dotted line), and in Fig. 7.13(b) the corresponding plots with units in dB. They are clearly different. However, as N increases, the effect of time aliasing will decrease, and the two FIR filters will become closer.

We compare in the following the three methods of designing FIR filters introduced so far.

1. Find the inverse DTFT of an ideal filter with no transition band and then truncate it. Its direct truncation will result in an LS optimal filter; but the filter will exhibit the Gibbs

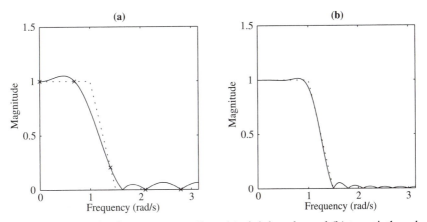

Figure 7.12 Design by frequency sampling: (a) eighth-order and (b) twentieth-order.

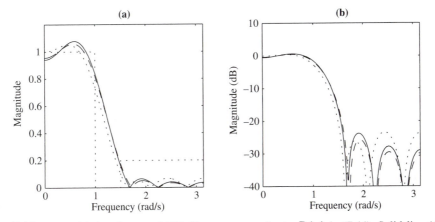

Figure 7.13 Three eighth-order type I FIR filters to approximate $D(\omega)$ in (7.44): Solid line (direct truncation or LS optimal), dotted line (frequency sampling or matching 9 frequency samples), and dashed line (LS optimal with "don't care" transition band, which we will introduce in Section 7.6.1). (a) Magnitude responses. (b) Corresponding plots with unts of dB.

phenomenon. If we truncate it using a window (other than a rectangular window), the resulting filter will not exhibit the Gibbs phenomenon but is no longer LS optimal.

2. Find the inverse DTFT of a desired filter with a specified transition band and then truncate it. Generally we assume the desired filter to have zero phase to simplify the computation of its inverse DTFT. Thus we must delay the truncated sequence to obtain a causal filter. There is no need to use windows in this method. The resulting causal FIR filter is LS optimal with respect to the selected desired filter. But computing the inverse DTFT will be more complex than the computation in Method 1.

3. Let us take $N + 1$ equally-spaced frequency samples of the desired filter in Method 2. Its inverse FFT yields an Nth order FIR filter whose magnitude response matches the desired one at the $N + 1$ frequencies. This filter will approach the LS optimal filter in Method 2 as N increases because the effect of time aliasing due to frequency sampling will decrease. This design method is called design by frequency sampling and can be carried out using inverse FFT.

From the preceding discussion, we conclude that the third method is easiest to employ among the three methods.

The preceding design methods are applicable only if desired frequency responses are specified for all ω. That is, in addition to the passband and stopband, we must also specify the transition band. In the remainder of this chapter, we discuss methods to design filters that are best, in some sense, only in the passband and stopband. What happens in the transition band is completely disregarded or, equivalently, we don't care what happens in the transition band. The methods to be introduced will be flagged with "wdcr" (with don't care region) if they may confuse with the preceding methods.

7.6 Discrete Least-Squares Optimal FIR Filters

The design by frequency sampling has three limitations. First, the samples must be equally spaced. Second, the filter order is dictated by the number of frequency samples to be matched, and, finally, the transition band must be specified. In this section, these restrictions will be removed. Before introducing the design method, we use an example to illustrate the idea involved.

◆ Example 7.1

Design a second-order type I digital filter to match the desired filter specified by $H_d(\omega) = D(\omega)e^{-j\omega}$ with

$$D(\omega) = \begin{cases} 1 & 0 \leq \omega \leq 1 \\ 0 & 1.5 \leq \omega \leq \pi \end{cases}$$

at a number of frequencies. The frequency range $(1, 1.5)$ is not specified and is the don't care region.

For $N = 2$, we have $M = N/2 = 1$ and the type I FIR filter is of the form

$$H(e^{j\omega}) = h[0] + h[1]e^{-j\omega} + h[2]e^{-j2\omega} = A(\omega)e^{j\omega}$$

with

$$A(\omega) = d[0] + d[1]\cos\omega$$

where $h[0] = h[2] = d[1]/2$ and $h[1] = d[0]$. Because the filter of length 3 is required to have a linear phase, it has only two independent unknowns $d[0]$ and $d[1]$. If $A(\omega)$ is required to match only $D(\pi/4) = 1$, then we have

$$A(\pi/4) = d[0] + d[1]\cos\pi/4 = d[0] + 0.707d[1] = D(\pi/4) = 1$$

There are infinitely many solutions such as $d[0] = 1, d[1] = 0$ or $d[0] = 0, d[1] = 1/0.707$.

If $A(\omega)$ is required to match $D(\pi/4) = 1$ and $D(3\pi/4) = 0$, then we have

$$A(\pi/4) = d[0] + d[1]\cos\pi/4 = d[0] + 0.707d[1] = D(\pi/4) = 1$$

$$A(3\pi/4) = d[0] + d[1]\cos 3\pi/4 = d[0] - 0.707d[1] = D(3\pi/4) = 0$$

which can be expressed in matrix form as

$$\begin{bmatrix} 1 & 0.707 \\ 1 & -0.707 \end{bmatrix} \begin{bmatrix} d[0] \\ d[1] \end{bmatrix} = \begin{bmatrix} 1 \\ 0 \end{bmatrix}$$

Its solution can be obtained in MATLAB by typing a=[1 0.707;0 -0.707]; D=[1;0]; d=a\D, which yields $d = [0.5 \ 0.707]'$, where the prime denotes the transpose. This solution is unique.

If $A(\omega)$ is required to match $D(\pi/4) = 1$, $D(\pi/2) = 0$, and $D(3\pi/4) = 0$, then we have

$$A(\pi/4) = d[0] + d[1]\cos\pi/4 = d[0] + 0.707d[1] = D(\pi/4) = 1$$

$$A(\pi/2) = d[0] + d[1]\cos\pi/2 = d[0] + 0 \cdot d[1] = D(\pi/2) = 0$$

$$A(3\pi/4) = d[0] + d[1]\cos 3\pi/4 = d[0] - 0.707d[1] = D(3\pi/4) = 0$$

which can be expressed in matrix form as

$$\begin{bmatrix} 1 & 0.707 \\ 1 & 0 \\ 1 & -0.707 \end{bmatrix} \begin{bmatrix} d[0] \\ d[1] \end{bmatrix} = \begin{bmatrix} 1 \\ 0 \\ 0 \end{bmatrix}$$

There are three equations and two unknowns and generally no solution exist in the set of equations. However, typing a=[1 0.707;1 0;0 -0.707]; D=[1;0;0];d=a\D in MATLAB, we obtain $d = [0.3333 \ 0.7072]'$. Substituting these into the preceding matrix equation yields

$$\begin{bmatrix} 1 & 0.707 \\ 1 & 0 \\ 1 & -0.707 \end{bmatrix} \begin{bmatrix} 0.3333 \\ 0.7072 \end{bmatrix} = \begin{bmatrix} 0.8333 \\ 0.3333 \\ -0.1667 \end{bmatrix} \neq \begin{bmatrix} 1 \\ 0 \\ 0 \end{bmatrix}$$

Clearly, $d = [0.3333 \ 0.7072]'$ is not a solution. However, it will be called the *least-squares (LS) solution* for reasons to be given shortly. In conclusion, if the number of equations is less than the number of unknowns, the set of equations has infinitely many solutions. If the number of equations equals the number of unknowns, the set of equations generally has a unique solution. If the number of equations is larger than the number of unknowns, the set of equations has an LS solution.

With the preceding background, we are ready to introduce the discrete least-squares optimal FIR filter design. Consider an Nth-order type I FIR filter expressed as

$$H(e^{j\omega}) = \sum_{n=0}^{N} h[n]e^{-jn\omega} = A(\omega)e^{-jM\omega} \tag{7.48}$$

with $M = N/2$ and

$$A(\omega) = \sum_{n=0}^{M} d[n] \cos n\omega \tag{7.49}$$

where $d[n]$ and $h[n]$ are related as shown in (7.7) or Table 7.1. The filter has $M+1$ independent coefficients $d[n]$. Suppose we are given a desired low-pass frequency response

$$H_d(e^{j\omega}) = D(\omega)e^{-jM\omega}$$

with

$$D(\omega) = \begin{cases} 1 & \text{for } 0 \leq \omega \leq \omega_p \\ 0 & \text{for } \omega_s \leq \omega \leq \pi \end{cases} \tag{7.50}$$

where the transition band (ω_p, ω_s) is not specified. The problem is to design (7.49) to match K samples of $D(\omega)$.

Let ω_i for $i = 1, 2, \ldots, K$ be the frequencies where $D(\omega)$ is to be matched. Then the design problem is to find $d[n]$ to meet

$$A(\omega_i) = D(\omega_i) =: D_i \tag{7.51}$$

for $i = 1, 2, \ldots, K$. These K equations will be arranged as a set of linear algebraic equations. For easier viewing, we develop the matrix equation for $N = 8$ and $K = 8$. Note that N and K can be any positive integers; they need not be the same. For $N = 8$, we have $M = 4$ and five unknowns $d[n]$, for $n = 0, 1, 2, 3, 4$. Substituting (7.49) into (7.51) yields

$$d[0] + d[1] \cos \omega_i + d[2] \cos 2\omega_i + d[3] \cos 3\omega_i + d[4] \cos 4\omega_i = D_i \tag{7.52}$$

These $K = 8$ equations can be expressed in matrix form as

$$
\begin{bmatrix}
1 & \cos \omega_1 & \cos 2\omega_1 & \cos 3\omega_1 & \cos 4\omega_1 \\
1 & \cos \omega_2 & \cos 2\omega_2 & \cos 3\omega_2 & \cos 4\omega_2 \\
1 & \cos \omega_3 & \cos 2\omega_3 & \cos 3\omega_3 & \cos 4\omega_3 \\
1 & \cos \omega_4 & \cos 2\omega_4 & \cos 3\omega_4 & \cos 4\omega_4 \\
1 & \cos \omega_5 & \cos 2\omega_5 & \cos 3\omega_5 & \cos 4\omega_5 \\
1 & \cos \omega_6 & \cos 2\omega_6 & \cos 3\omega_6 & \cos 4\omega_6 \\
1 & \cos \omega_7 & \cos 2\omega_7 & \cos 3\omega_7 & \cos 4\omega_7 \\
1 & \cos \omega_8 & \cos 2\omega_8 & \cos 3\omega_8 & \cos 4\omega_8
\end{bmatrix}
\begin{bmatrix}
d[0] \\
d[1] \\
d[2] \\
d[3] \\
d[4]
\end{bmatrix}
=
\begin{bmatrix}
D_1 \\
D_2 \\
D_3 \\
D_4 \\
D_5 \\
D_6 \\
D_7 \\
D_8
\end{bmatrix}
\tag{7.53}
$$

or

$$\mathbf{Pd} = \mathbf{D} \tag{7.54}$$

where $\mathbf{P}$ is a $K \times (N/2+1) = 8 \times 5$ matrix, $\mathbf{d}$ is a 5×1 column vector, and $\mathbf{D}$ is an 8×1 column vector.[4] The matrix equation in (7.53) consists of eight scalar equations and five unknowns. Let us discuss its solution.

If the number of unknowns is larger than the number of frequency samples or, equivalently, the number of equations to be matched, solutions always exist in (7.53), and exact matching is always possible. In this case, the filter is not fully utilized. If the number of unknowns equals the number of equations, then the $\mathbf{P}$ matrix in (7.53) is square. If the frequencies ω_i are selected to be distinct, then the square matrix is nonsingular and a unique solution can be obtained as $\mathbf{d} = \mathbf{P}^{-1}\mathbf{D}$. If the number of unknowns is less than the number of equations, generally no solution exists in (7.53). However, we can find an approximate "solution" that is best in some sense as we will discuss next.

Let us define

$$e_i := D_i - (d[0] + d[1]\cos \omega_i + d[2]\cos 2\omega_i + d[3]\cos 3\omega_i + d[4]\cos 4\omega_i) \tag{7.55}$$

for $i = 1, 2, \ldots, K = 8$. It is the error at frequency ω_i between the desired magnitude $D(\omega_i)$ and the actual amplitude $A(\omega_i)$. Thus the total squared error at the eight frequencies is

$$E = \sum_{i=1}^{8} e_i^2 = \sum_{i=1}^{8} (D(\omega_i) - A(\omega_i))^2 \tag{7.56}$$

The "solution" of (7.54) that has the smallest total error will be called the *discrete least-squares solution*. We will develop such a solution.

Let us define $\mathbf{e} := [e_1 \; e_2 \; \cdots \; e_8]'$. It is an 8×1 vector and can be expressed as

---

[4] We use boldface symbols to denote matrices or vectors and light-face symbols to denote scalars.

$$\mathbf{e} = \mathbf{D} - \mathbf{Pd} \tag{7.57}$$

The total squared error in (7.56) then can be written as

$$E = \sum_{i=1}^{8} e_i^2 = \mathbf{e}'\mathbf{e} = (\mathbf{D} - \mathbf{Pd})'(\mathbf{D} - \mathbf{Pd}) = (\mathbf{D}' - \mathbf{d}'\mathbf{P}')(\mathbf{D} - \mathbf{Pd})$$

$$= \mathbf{D}'\mathbf{D} - \mathbf{D}'\mathbf{Pd} - \mathbf{d}'\mathbf{P}'\mathbf{D} + \mathbf{d}'\mathbf{P}'\mathbf{Pd} \tag{7.58}$$

The problem is to find $d[n]$ to minimize the squared error E or to find the discrete LS solution.

One way to find the LS solution is to set to zero the differentiation of E with respect to each $d[n]$. This approach will become unwieldy in notation; thus we use a different approach. Suppose $\mathbf{d}$ is the LS solution and E in (7.58) is the smallest error. Let us perturb $\mathbf{d}$ to $\mathbf{d} + \bar{\mathbf{d}}$, where $\bar{\mathbf{d}}$ is also a 5×1 vector and its components can assume any positive or negative numbers. We compute the error $\bar{E}$ due to $\mathbf{d} + \bar{\mathbf{d}}$:

$$\bar{E} = \mathbf{D}'\mathbf{D} - \mathbf{D}'\mathbf{P}(\mathbf{d} + \bar{\mathbf{d}}) - (\mathbf{d}' + \bar{\mathbf{d}}')\mathbf{P}'\mathbf{D} + (\mathbf{d}' + \bar{\mathbf{d}}')\mathbf{P}'\mathbf{P}(\mathbf{d} + \bar{\mathbf{d}})$$

$$= \mathbf{D}'\mathbf{D} - \mathbf{D}'\mathbf{Pd} - \mathbf{d}'\mathbf{P}'\mathbf{D} + \mathbf{d}'\mathbf{P}'\mathbf{Pd} - \mathbf{D}'\mathbf{P}\bar{\mathbf{d}} - \bar{\mathbf{d}}'\mathbf{P}'\mathbf{D}$$

$$+ \bar{\mathbf{d}}'\mathbf{P}'\mathbf{Pd} + \mathbf{d}'\mathbf{P}'\mathbf{P}\bar{\mathbf{d}} + \bar{\mathbf{d}}'\mathbf{P}'\mathbf{P}\bar{\mathbf{d}}$$

$$= E + (\mathbf{d}'\mathbf{P}'\mathbf{P} - \mathbf{D}'\mathbf{P})\bar{\mathbf{d}} + \bar{\mathbf{d}}'(\mathbf{P}'\mathbf{Pd} - \mathbf{P}'\mathbf{D}) + \bar{\mathbf{d}}'\mathbf{P}'\mathbf{P}\bar{\mathbf{d}}$$

$$= E + 2\bar{\mathbf{d}}'(\mathbf{P}'\mathbf{Pd} - \mathbf{P}'\mathbf{D}) + \bar{\mathbf{d}}'\mathbf{P}'\mathbf{P}\bar{\mathbf{d}}$$

where we have used (7.58). Note that every term in the preceding equation is scalar. For example, $\bar{\mathbf{d}}'\mathbf{P}'\mathbf{P}\bar{\mathbf{d}}$ is the product of 1×5, 5×8, 8×5, and 5×1 matrices and is a 1×1 scalar. Because E is assumed to be the least-squared error, the error $\bar{E}$ must equal E or be larger. The term $\bar{\mathbf{d}}'\mathbf{P}'\mathbf{P}\bar{\mathbf{d}}$ for any nonzero $\bar{\mathbf{d}}$ is always positive because of its symmetric form. The term $\bar{\mathbf{d}}'(\mathbf{P}'\mathbf{Pd} - \mathbf{P}'\mathbf{D})$, however, can be negative for some $\bar{\mathbf{d}}$ and positive for other $\bar{\mathbf{d}}$. Thus the condition to ensure $\bar{E} \geq E$ for every $\bar{\mathbf{d}}$ is

$$\mathbf{P}'\mathbf{Pd} - \mathbf{P}'\mathbf{D} = 0 \quad \text{or} \quad \mathbf{P}'\mathbf{Pd} = \mathbf{P}'\mathbf{D} \tag{7.59}$$

This is called a *normal equation*. Note that $\mathbf{P}$ is 8×5 and its transpose $\mathbf{P}'$ is 5×8. Thus $\mathbf{P}'\mathbf{P}$ is a 5×5 square matrix. If all ω_i are selected to be distinct, then $\mathbf{P}'\mathbf{P}$ is nonsingular and the LS solution can be computed from (7.59) as

$$\mathbf{d} = (\mathbf{P}'\mathbf{P})^{-1}\mathbf{P}'\mathbf{D} \tag{7.60}$$

The MATLAB function P\D yields the exact solution of $\mathbf{Pd} = \mathbf{D}$, if it exists, or the LS solution, if an exact solution does not exist. Thus the discrete LS solution can easily be obtained using MATLAB.

We discuss how to implement the preceding procedure in MATLAB. The 8×5 matrix $\mathbf{P}$ in (7.53) can be formed as follows. We arrange the eight frequencies ω_i as $W = [\omega_1 \ \omega_2 \ \ldots \ \omega_8]'$, where the prime denotes the transpose. It is a 8×1 column. The notation [0:4] in MATLAB is a 1×5 vector with entries from 0 to 4. We form

$$
WW := W * [0 : M] = \begin{bmatrix} \omega_1 \\ \omega_2 \\ \omega_3 \\ \omega_4 \\ \omega_5 \\ \omega_6 \\ \omega_7 \\ \omega_8 \end{bmatrix} \; [0\;1\;2\;3\;4]
$$

$$
= \begin{bmatrix}
0 & \omega_1 & 2\omega_1 & 3\omega_1 & 4\omega_1 \\
0 & \omega_2 & 2\omega_2 & 3\omega_2 & 4\omega_2 \\
0 & \omega_3 & 2\omega_3 & 3\omega_3 & 4\omega_3 \\
0 & \omega_4 & 2\omega_4 & 3\omega_4 & 4\omega_4 \\
0 & \omega_5 & 2\omega_5 & 3\omega_5 & 4\omega_5 \\
0 & \omega_6 & 2\omega_6 & 3\omega_6 & 4\omega_6 \\
0 & \omega_7 & 2\omega_7 & 3\omega_7 & 4\omega_7 \\
0 & \omega_8 & 2\omega_8 & 3\omega_8 & 4\omega_8
\end{bmatrix}
$$

It is an 8×5 matrix. Then the matrix **P** in (7.53) can be generated as $\cos(WW)$. Note that $\cos 0 = 1$. The program that follows forms (7.53), solves its LS solution, forms a type I FIR filter, and plots its frequency response.

Program 7.2
```
N= ;M=N/2;K= ; (% N must be even)
W=[w1 w2 ⋯ wK]';
D=[D1 D2 ⋯ DK]';
WW=W*[0:M]; P = cos(WW);
d=P\D;
k=0:M-1;
h1=[d(M+1-k)/2]';
h=[h1 d(1) fliplr(h1)]
[H,w]=freqz(h,1);
plot(w,abs(H))
```

Once $d[n]$ is obtained, we use Table 7.1 to form $h[n]$. Because d generated in MATLAB is a column vector, we use its transpose to form the row vector h1. Recall that indices in MATLAB start from 1 not 0, and must be enclosed by parentheses. For FIR filters, we have bn=h and an=1. Thus freqz(h,1) generates the frequency response at 512 equally spaced frequencies in $[0, \pi)$.

We give an example by designing an eighth-order type I FIR filter to approximate a desired low-pass magnitude response with passband $[0, 1]$ and stopband $[1.5, \pi]$. An eighth-order FIR filter has five independent coefficients and can match exactly the desired magnitude at five frequencies. Let us select arbitrarily

W=[0 1 1.5 2.3 pi]';

The first two frequencies are in the passband and the rest in the stopband. Thus we have D=[1 1 0 0 0]';. Program 7.2 with $N = 8$ and $K = 5$ will yield the magnitude response in Fig. 7.14(a). We see that the magnitude response matches exactly the desired magnitude response denoted by crosses at the five selected frequencies. If we select the five frequencies as W=[0.35 0.9 1.7 2.2 2.8]', then we will obtain the magnitude response in Fig. 7.14(b). The response again matches the crosses and is a better filter.

From Fig. 7.14, we see that the resulting filter depends on where the frequencies are selected. Clearly one set of frequencies will yield a better result than the other. Thus how to select the set of frequencies to be matched becomes a problem. One way to resolve the problem is by trial and error. This problem, however, can be resolved if we increase the number of frequencies to be matched as we will discuss next.

An eighth-order type I FIR filter can match exactly a desired magnitude response at five frequencies. Now we will try to match at $K = 8$ frequencies. Let us select

W=[0 0.33 0.67 1 1.5 2 2.5 3.14]';D=[1 1 1 1 0 0 0 0]';

The first four frequencies are equally spaced in the passband, and the last four frequencies are in the stopband. Then Program 7.2 with $N = 8$ and $K = 8$ will yield the magnitude response in Fig. 7.15(a). We see that the magnitude response does not go through the points marked by crosses as in Fig. 7.14. But the sum of their squared errors will be the smallest.

We repeat the design by matching 28 frequencies (11 in the passband and 17 in the stopband with frequency interval 0.1). Typing $K = 28$ and

W=[0:0.1:1 1.5:0.1:3.1]';
D=[ones(1,11) zeros(1,17)]';

in Program 7.2 will yield the filter

$$H(z) = -0.0611 - 0.0523z^{-1} + 0.0923z^{-2} + 0.2986z^{-3} + 0.3975z^{-4}$$
$$+ 0.2986z^{-5} + 0.0923z^{-6} - 0.0523z^{-7} - 0.0611z^{-8} \qquad (7.61)$$

and its magnitude response in Fig. 7.15(b) with a solid line. This is the best design among the four in Figs. 7.14 and 7.15. Thus if we select a sufficiently large K and place K frequencies evenly in the passband and stopband, Program 7.2 will yield a good filter.

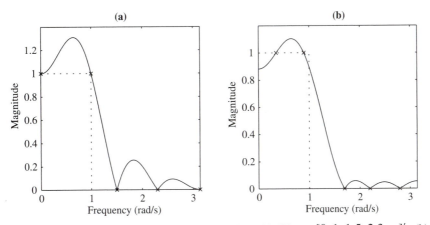

Figure 7.14 Exact matching of five frequency samples: (a) W $=$ $[0\ 1\ 1.5\ 2.3\ \pi]'$. (b) W$=$ $[0.35\ 0.9\ 1.7\ 2.2\ 2.8]'$.

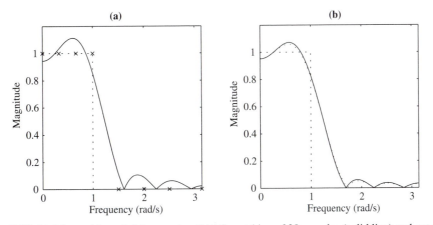

Figure 7.15 (a) LS matching of eight samples. (b) LS matching of 28 samples (solid line) and output of firls (dotted line).

The preceding method can be modified to include a weighting factor. Instead of minimizing (7.56), now we minimize

$$E = \sum W_t^2(\omega_i)[D(\omega_i) - A(\omega_i)]^2 = \sum [W_t(\omega_i)D(\omega_i) - W_t(\omega_i)A(\omega_i)]^2 \qquad (7.62)$$

where the weighting factor $W_t(\omega_i)$ must be positive for all ω_i. Let us define

$$\bar{\mathbf{e}} := \bar{\mathbf{D}} - \bar{\mathbf{P}}\mathbf{d} \qquad (7.63)$$

where the ith entry of $\bar{\mathbf{D}}$ equals the ith entry of $\mathbf{D}$ in (7.54) multiplied by $W_t(\omega_i)$ and the ith row of $\bar{\mathbf{P}}$ equals the ith row of $\mathbf{P}$ in (7.54) multiplied by $W_t(\omega_i)$. Using the procedure in (7.58)–(7.60), we can show that the weighted discrete LS solution is given by

$$\mathbf{d} = (\bar{\mathbf{P}}'\bar{\mathbf{P}})^{-1}\bar{\mathbf{P}}\bar{\mathbf{D}} \tag{7.64}$$

and can be obtained in MATLAB as $\mathsf{d}=\bar{P}\backslash\bar{D}$.

Although the preceding procedure is developed for type I FIR filters, it is applicable to type II, III, and IV filters with only minor modification.

7.6.1 Integral Least-Squares Optimal FIR Filters

The total error in the preceding section is the sum of squared $A(\omega_i) - D(\omega_i)$ at discrete frequencies in the passband and stopband. Now we modify it as

$$E = \int_{\omega=0}^{\omega_p} [A(\omega) - D(\omega)]^2 d\omega + \int_{\omega=\omega_s}^{\pi} [A(\omega) - D(\omega)]^2 d\omega \tag{7.65}$$

Note that the transition band is not included in the equation. This is different from the integral LS design in Sections 7.3 and 7.5 and will be called *integral squared error (wdcr)*. In this section, we design an FIR filter to minimize this error. The resulting filter is called the integral LS optimal filter (wdcr).

For convenience, we discuss the design of

$$A(\omega) = \sum_{n=0}^{4} d[n] \cos n\omega \tag{7.66}$$

to minimize the E in (7.65), where $D(\omega)$ equals 1 in $[0, \omega_p]$ and 0 in $[\omega_s, \pi]$. Substituting (7.66) and $D(\omega)$ into (7.65) yields

$$
\begin{aligned}
E &= \int_0^{\omega_p} [A(\omega) - 1]^2 d\omega + \int_{\omega_s}^{\pi} A^2(\omega) d\omega \\
&= \int_0^{\omega_p} A^2(\omega) d\omega - 2 \int_0^{\omega_p} A(\omega) d\omega + \int_0^{\omega_p} 1^2 d\omega + \int_{\omega_s}^{\pi} A^2(\omega) d\omega \\
&= \sum_{n=0}^{4} d^2[n] \int_{0,\omega_s}^{\omega_p,\pi} \cos^2 n\omega\, d\omega + 2 \sum_{n=0}^{4} \sum_{k>n}^{4} d[n]d[k] \int_0^{\omega_p} \cos n\omega\, \cos k\omega\, d\omega \\
&\quad - 2 \sum_{n=0}^{4} d[n] \int_0^{\omega_p} \cos n\omega\, d\omega + \omega_p
\end{aligned}
\tag{7.67}
$$

where we have used

$$\int_{0,\omega_s}^{\omega_p,\pi} \cos^2 n\omega d\omega := \int_0^{\omega_p} \cos^2 n\omega d\omega + \int_{\omega_s}^{\pi} \cos^2 n\omega d\omega$$

Differentiating E in (7.67) with respect to $d[n]$ and then setting it to zero, we obtain

$$d[n] \int_{0,\omega_s}^{\omega_p,\pi} \cos^2 n\omega d\omega + \sum_{m=1,k\neq n}^{4} d[k] \int_0^{\omega_p} \cos n\omega \cdot \cos k\omega dom = \int_0^{\omega_p} \cos n\omega d\omega$$

for $n = 0, 1, 2, 3, 4$. They can be arranged in matrix form as

$$\hat{\mathbf{P}}\mathbf{d} = \hat{\mathbf{D}} \tag{7.68}$$

where $\hat{\mathbf{P}} = P(i, j)$ is 5×5 with entries given by

$$P(i, i) = \int_0^{\omega_p} \cos^2[(i - 1)\omega]d\omega + \int_{\omega_s}^{\pi} \cos^2[(i - 1)\omega]d\omega$$

for $i = 1, 2, \ldots, 5$ and

$$P(i, j) = \int_0^{\omega_p} \cos[(i - 1)\omega]\cos[(j - 1)\omega]d\omega$$

for $i, j = 1, 2, \ldots, 5$ and $i \neq j$. The vector $\hat{\mathbf{D}}$ is 5×1 with its ith entry given by

$$D_i = \int_0^{\omega_p} \cos[(i - 1)\omega]d\omega$$

We see that computing $\hat{\mathbf{P}}$ and $\hat{\mathbf{D}}$ requires integrations. Fortunately, their integrations can be expressed in closed form. Once $\hat{\mathbf{P}}$ and $\hat{\mathbf{D}}$ are computed, the integral LS solution can be obtained as $\hat{\mathbf{P}}^{-1}\hat{\mathbf{D}}$ or $\hat{\mathbf{P}}\backslash\hat{\mathbf{D}}$.

The MATLAB function firls(n,f,d), an acronym for finite impulse response least squares, returns an nth-order linear phase FIR filters. The vector f consists of pairs of frequencies, in increasing order, in [0, 1]. Each pair denotes the frequency range in which its desired magnitude is specified in the corresponding pair in d. The regions between each pair are "don't care" transition bands. For example, f=[0 0.5 0.7 1] and d=[0 0.5*pi 0.2 0.2] specify the desired magnitude response shown in Fig. 7.16. The region between 0.5 and 0.7 is the don't care region. For the design problem in the preceding section, the following

f=[0 1/pi 1.5/pi 1];D=[1 1 0 0];h=firls(8,f,D)
[H,w]=freqz(h,1)

yields the filter

$$H(z) = -0.0568 - 0.0506z^{-1} + 0.0897z^{-2} + 0.2964z^{-3} + 0.3966z^{-4}$$

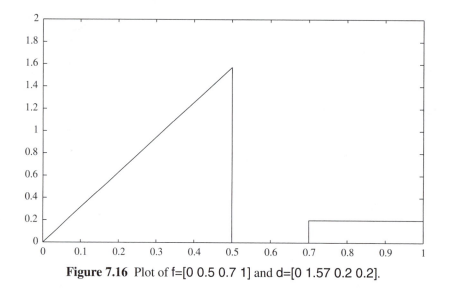

Figure 7.16 Plot of f=[0 0.5 0.7 1] and d=[0 1.57 0.2 0.2].

$$+ 0.2964z^{-5} + 0.0897z^{-6} - 0.0506z^{-7} - 0.0568z^{-8} \qquad (7.69)$$

and the magnitude response shown in Fig. 7.15(b) with a dotted line. This filter minimizes the integral squared error

$$E = \int_{\omega=0}^{1} [A(\omega) - 1]^2 d\omega + \int_{\omega=1.5}^{\pi} A^2(\omega)d\omega$$

We compare (7.69) with (7.61) which was obtained by minimizing the total squared errors at $K = 28$ discrete frequencies. Their coefficients are very close; so are their magnitude responses as shown in Fig. 7.15(b). This follows from the fact that the integral squared error can be approximated by the summed squared error as

$$E \approx 0.1 \left(\sum_{m=0}^{10} |A(0.1m) - 1|^2 + \sum_{m=0}^{16} |A(1.5 + 0.1m)|^2 \right)$$

Thus as K increases, the two designs will become identical.

The discrete LS design (wdcr) can be easily obtained by using Program 7.2, and is applicable to any desired magnitude response. In order for the integral LS design (wdcr) in (7.65) to become solving the linear algebraic equation in (7.68), we must carry out integrations. The MATLAB function firls is not developed from discrete LS, and is developed from direct integration, as discussed above.[5] It is limited to piecewise linear specifications; whereas the discrete LS method is applicable to any specification.

---

[5] Personal correspondence with Mr. T. Krauss, the author of firls. See also Ref. 3.

We compare the integral LS optimal filter (with a "don't care" transition band) with the ones (with a specified transition band) obtained in Sections 7.5 and 7.5.1 by plotting the magnitude response of (7.69) in Fig. 7.13 with a dashed line. It is probably better than the other two, but the improvement is not clearcut.

7.7 Minimax Optimal FIR Filters

The linear-phase FIR filters designed in the preceding sections are optimal in the least-squares sense. In this section, we discuss an optimization using a different criterion. The criterion is to minimize the largest absolute error, and the resulting filter is said to be optimal in the *minimax* sense.

We use an eighth-order type I FIR filter to discuss the basic idea and procedure. Consider

$$H(e^{j\omega}) = A(\omega)e^{-j4\omega}$$

with

$$A(\omega) = \sum_{n=0}^{4} d[n] \cos n\omega \qquad (7.70)$$

For $N = 8$, we have $M = N/2 = 4$ and $(M + 1)$ parameters in (7.70). The desired low-pass filter is given by $H_d(e^{j\omega}) = D(\omega)e^{-j4\omega}$ with

$$D(\omega) = \begin{cases} 1 & \text{for } \omega \text{ in } [0, \ \omega_p] \\ 0 & \text{for } \omega \text{ in } [\omega_s, \ \pi] \end{cases} \qquad (7.71)$$

Let us introduce the weighting function

$$W_t(\omega) = \begin{cases} e_s/e_p & \text{for } \omega \text{ in } [0, \ \omega_p] \\ 1 & \text{for } \omega \text{ in } [\omega_s, \ \pi] \end{cases} \qquad (7.72)$$

where e_s and e_p are positive constants and will be explained shortly. Then the weighted error at ω is

$$e(\omega) = W_t(\omega)[A(\omega) - D(\omega)] \qquad (7.73)$$

The design problem is to find $d[n]$ so that the largest absolute error $|e(\omega)|$ for ω in $[0, \ \omega_p]$ and $[\omega_s, \ \pi]$ is the smallest. This is called the minimization of the maximum error or a minimax problem.

It turns out that the posed problem has long been solved in the mathematical literature, and is called the *Chebyshev approximation*. The solution is that the optimal $e(\omega)$ must be equiripple in $[0, \ \omega_p]$ and $[\omega_s, \ \pi]$ as shown in Fig. 7.17(a), and has at least $M+2 = 6$ extremal frequencies. An extremal frequency is defined as a frequency ω_i that has the property $e(\omega_i) = e_m$ or $e(\omega_i) = -e_m$, where e_m is the maximum error. Furthermore, they must appear alternately. Excluding the band-edge frequencies, an extremal point can also be defined as a frequency at which $e(\omega)$ has zero slope. We state this formally as a theorem.

Alternation Theorem Let Ω be a closed subset of $[0, \pi]$ such as the union of $[0, \omega_p]$ and $[\omega_s, \pi]$. Then $A(\omega)$ in (7.70) is the unique best approximation of $D(\omega)$ in the minimax sense if and only if the error function $e(\omega)$ in (7.73) is *equiripple* and has at least $M + 2$ extremal ponts in W; that is, there exist in W

$$0 \leq \omega_1 < \omega_2 < \cdots < \omega_{M+2} \leq \pi$$

which include ω_p and ω_s, such that

$$e(\omega_i) = (-1)^i e_m \quad \text{for } i = 1, 2, \ldots, M + 2 \tag{7.74}$$

where

$$|e_m| = \max_{\{\omega \text{ in } \Omega\}} |e(\omega)| \tag{7.75}$$

and e_m can be a positive or negative number.

Before applying the theorem, we discuss the weighting factor in (7.72). If $D(\omega)$ equals 1 in the passband and 0 in the stopband, then (7.72), (7.73), and (7.75) imply

$$|A(\omega)| \leq |e_m| =: e_s \quad \text{for } \omega_s \leq \omega \leq \pi$$

$$|A(\omega) - 1| \leq \left| \frac{e_p}{e_s} e_m \right| = e_p \quad \text{for } 0 \leq \omega \leq \omega_p$$

Thus the best $A(\omega)$ must be equiripple in the passband bounded by $1 \pm e_p$, and in the stopband bounded by $\pm e_s$ as shown in Fig. 7.17(b). We call e_p the passband ripple and e_s the stopband ripple. If we require a smaller e_p than e_s, then the weighting factor e_s/e_p is larger than 1. In other words, if we put more weight in the passband, then the resulting filter has a smaller passband ripple. By adjusting the weighting factor e_s/e_p, we can control relative sizes of the passband and stopband ripples.

We now apply the Alternation Theorem to carry out the design. The theorem states that there are $M + 2$ or more extremal frequencies. However, the knowledge of $M + 2$ extreme frequencies is sufficient to solve our problem. Substituting (7.70) into (7.73) and then into (7.74) yields

$$\sum_{n=0}^{4} d[n] \cos n\omega_i - (-1)^i \frac{e_m}{W_t(\omega_i)} = D(\omega_i) =: D_i \tag{7.76}$$

for $i = 1, 2, \ldots, M + 2$ with $M = 4$. There are a total of $M + 2$ equations in (7.76). We arrange them in matrix form as

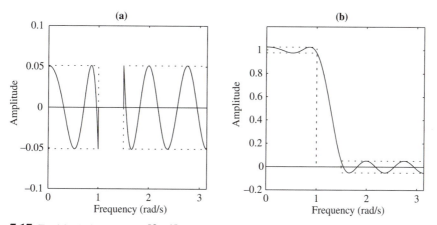

Figure 7.17 Equiripple in passband $[0, \ 1]$ and stopband $[1.5, \ \pi]$ for $e(\omega)$ (a) and for $A(\omega)$ (b).

$$
\begin{bmatrix}
1 & \cos \omega_1 & \cos 2\omega_1 & \cos 3\omega_1 & \cos 4\omega_1 & 1/W_t(\omega_1) \\
1 & \cos \omega_2 & \cos 2\omega_2 & \cos 3\omega_2 & \cos 4\omega_2 & -1/W_t(\omega_2) \\
1 & \cos \omega_3 & \cos 2\omega_3 & \cos 3\omega_3 & \cos 4\omega_3 & 1/W_t(\omega_3) \\
1 & \cos \omega_4 & \cos 2\omega_4 & \cos 3\omega_4 & \cos 4\omega_4 & -1/W_t(\omega_4) \\
1 & \cos \omega_5 & \cos 2\omega_5 & \cos 3\omega_5 & \cos 4\omega_5 & 1/W_t(\omega_5) \\
1 & \cos \omega_6 & \cos 2\omega_6 & \cos 3\omega_6 & \cos 4\omega_6 & -1/W_t(\omega_6)
\end{bmatrix}
\begin{bmatrix}
d[0] \\ d[1] \\ d[2] \\ d[3] \\ d[4] \\ e_m
\end{bmatrix}
=
\begin{bmatrix}
D_1 \\ D_2 \\ D_3 \\ D_4 \\ D_5 \\ D_6
\end{bmatrix}
\qquad (7.77)
$$

The number of unknowns in (7.70) is $M + 1$; the maximum error e_m is not known. Thus the total number of unknowns in (7.77) is $M + 2$ and the left-most matrix in (7.77) is square. If all ω_i are distinct, the square matrix is nonsingular, and a unique solution can be obtained. We modify Program 7.2 to solve (7.77). We list the program for $M = 4$.

Program 7.3
```
W=[w1 w2 w3 w4 w5 w6]';
D=[D1 D2 D3 D4 D5 D6]';
WW = W*[0:4]; P=cos(WW);
p=[1/wt1 -1/wt2 1/wt3 -1/wt4 1/wt5 -1/wt6]';Pa=[P p];
d=Pa\D;
h1=[d(5)/2 d(4)/2 d(3)/2 d(2)/2];
h=[h1 d(1) fliplr(h1)];
em=d(6)
[H,w]=freqz(h,1);
plot(w,abs(H))
```

This program forms (7.77), solves its solution, forms a linear-phase FIR filter, and plots its magnitude response.

In the preceding development, we assumed implicitly that the extremal frequencies are available. If this is the case, the e_m computed in (7.77) will be the largest error defined in (7.75). Now if the six frequencies in (7.77) are not all extremal, a unique solution still exists in (7.77). However, the computed e_m will not be the maximum error. This property can be used to search extremal frequencies.

The search will be carried out iteratively by using the *Remez exchange algorithm*. First we select arbitrarily $M + 2$ frequencies ω_i in $[0, \omega_p]$ and $[\omega_s, \pi]$. The band-edge frequencies ω_p and ω_s must be included in the set. Whether or not the other two edge frequencies 0 and π are included is immaterial. One way to select these frequencies is to distribute them evenly in $[0, \omega_p]$ and $[\omega_s, \pi]$. We then use Program 7.3 to solve the FIR filter and e_m. We then check whether or not $|e(\omega)| \leq e_m$ for all ω in Ω. If the answer is affirmative, the selected ω_i are extremal frequencies and the computed FIR filter is optimal in the minimax sense. If the answer is negative, not every frequency in the set is an extremal point. In this case, we select, in addition to ω_p and ω_s, four frequencies at which $e(\omega)$ has zero slope or, equivalently, has peak ripple locally. We then use Program 7.3 to solve a new FIR filter and e_m, and repeat the checking. This iterative process will converge rapidly and is called the Remez exchange algorithm. It is also called the *Parks-McClellan algorithm*.

Now we will use the preceding procedure to find $d[n]$ in (7.70) to approximate the desired magnitude response in (7.71) with $\omega_p = 1$ and $\omega_s = 1.5$. No different weights will be imposed in the two bands. According to the alternation theorem, $e(\omega)$ has at least $M + 2 = 6$ extremal frequencies in $[0, 1]$ and $[1.5, \pi]$. In the design, we will plot only $|A(\omega)|$. In terms of the magnitude response $|A(\omega)|$, the extremal frequencies are those frequencies with $|A(\omega)| = 1 \pm e_m$ in the passband and $|A(\omega)| = |e_m|$ in the stopband. Arbitrarily, we select the six frequencies as w=[0 0.5 1 1.5 2.3 3.14]'. Then we have D=[1 1 1 0 0 0]'. Substituting these into Program 7.3 yields the plot in Fig. 7.18(a) with a solid line. It also yields $e_m = 0.09605$. We also mark with small circles the value of $|A(\omega)|$ at the six frequencies. The small circles all have the same distance e_m from the desired value (1 or 0). Clearly, $e_m = 0.09605$ is not the largest error or the ripples are not bounded by ± 0.09605. Thus not every frequency in w=[0 0.5 1 1.5 2.3 3.14]' is an extremal frequency. An alternative way of checking this is to check the slope of the magnitude response at every frequency other than ω_p and ω_s. If all slopes are zero, then all frequencies are extremal. Clearly, this is not the case at $\omega = 0.5$ and 2.3.

Next we select a different set of six frequencies. We retain the three frequencies $\omega = 0, 1$, and 1.5. The other three are selected at $\omega = 0.75, 1.9$, and 3, where the magnitude response denoted with the solid line in Fig. 7.18(a) has roughly a zero slope. Substituting w=[0 0.75 1 1.5 1.9 3]' into Program 7.3 yields the magnitude response in Fig. 7.18(a) with a dashed line and $e_m = 0.1091$. We also mark with plus signs the values of $|A(\omega)|$ at the six frequencies. Other than the band-edge frequencies 1 and 1.5, $|A(\omega)|$ has zero slope at 0 and 1.9 but not at 0.75 and 3. We replace 0.75 and 3 by 0.65 and 2.6, where the dashed line has zero slope. Substituting w=[0 0.65 1 1.5 1.9 2.6]' into Program 7.3 yields the magnitude response in Fig. 7.18(b) with a solid line and $e_m = 0.1153$. The magnitude response is roughly equiripple. For this problem, we obtain the minimax optimal filter in three iterations.

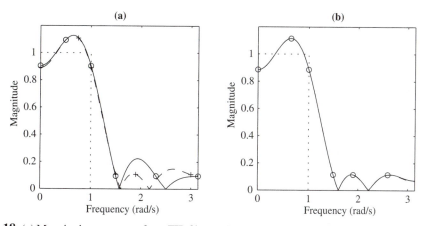

Figure 7.18 (a) Magnitude responses of two FIR filters: First try (solid line) and second try (dashed line). (b) Magnitude responses of FIR filters obtained in the third try (solid line) and using MATLAB function remez (dotted line).

The MATLAB functions h=remez(n,f,d) and h=remez(n,f,d,wt) generate nth-order minimax optimal FIR filters. The vectors f and d are defined as in firls(n,f,d). That is, f consists of pairs of frequencies, in ascending order, in [0, 1]. The desired magnitude of each pair is specified by the corresponding pair in d. The ranges between pairs are "don't care" transition regions. The weight wt has half the length of f and specifies relative weights in all pairs. For our example, the program that follows

```
f=[0 1/pi 1.5/pi 1];d=[1 1 0 0];
h=remez(8,f,d)
[H,w]=freqz(h,1);
plot(w,abs(H))
```

generates the magnitude response in Fig. 7.18(b) with a dotted line. It is indistinguishable from the one denoted by the solid line.

In Fig. 7.19(a), we compare the result of remez(8,f,d) (solid line) and remez(8,f,d,wt) with wt=[2 1] (dashed line). Because of the weight, the stopband tolerance is twice the passband tolerance.

Figure 7.19(b) compares the minimax optimal FIR filter obtained by using remez(8,f,d) (solid line) and the LS optimal filter obtained by using firls(8,f,d) (dashed line). Except in the immediate neighborhoods of $\omega_p = 1$ and $\omega_s = 1.5$, the LS optimal filter has smaller passband and stopband tolerances than the minimax optimal filter.

This section discussed only type I optimal FIR filters. Although the same procedure can be applied to other types of FIR filters, some modifications are needed. The interested reader is referred to Ref. 21.

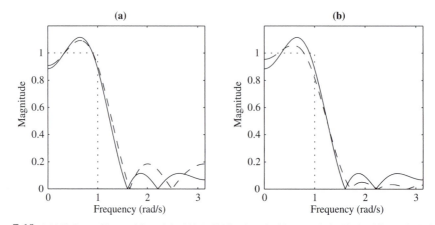

Figure 7.19 (a) Minimax filters with weight 2/1 (solid line) and without weight (dashed line). (b) Minimax (solid line) and LS (dashed line) optimal filters. They all have the "don't care" region $(1, \ 1.5)$.

7.8 Design of Digital Differentiators

This section discusses the design of digital differentiators. Differentiators are used to compute derivatives of signals. This has many applications such as computing the speed from the traveled distance or bringing out boundaries of rapid changes. Differentiation, strictly speaking, is defined only for CT signals. As discussed in (5.87), we have

$$\mathcal{L}\left[\frac{dx(t)}{dt}\right] = sX(s)$$

Thus the transfer function of a CT differentiator is s and its frequency response is $j\omega$ for all ω in $(-\infty, \ \infty)$.

We can carry out differentiation of CT signals digitally as shown in Fig. 6.8. The sampling period T clearly must be selected to avoid frequency aliasing due to time sampling. Once a CT signal is discretized, its digital processing is independent of the sampling period in non-real-time processing. In real-time processing, the only restriction is that the sampling period be large enough to complete the computation of each output in time. Thus in designing digital differentiators, we still can assume $T = 1$. Consequently, the frequency response of an ideal Nth-order FIR differentiator should be

$$H_d(\omega) = j\omega e^{-j0.5N\omega} \quad \text{for } |\omega| \leq \pi \tag{7.78}$$

The linear phase $-0.5N\omega$ is needed for designing good differentiators.

Before proceeding, we compare (7.78) with the difference equation

$$y[n] = x[n] - x[n-1] \tag{7.79}$$

It is a first-order type IV FIR filter. Applying the z-transform to (7.79) yields

$$Y(z) = X(z) - z^{-1}X(z) = (1 - z^{-1})X(z)$$

Thus the transfer function of (7.79) is

$$\bar{H}(z) = \frac{Y(z)}{X(z)} = 1 - z^{-1} = \frac{z-1}{z} \tag{7.80}$$

and its frequency response is

$$\bar{H}(e^{j\omega}) = 1 - e^{-j\omega} = e^{-j0.5\omega} \left(e^{j0.5\omega} - e^{-j0.5\omega} \right)$$

$$= j2 \sin 0.5\omega \times e^{-j0.5\omega} \tag{7.81}$$

Figure 7.20 shows the magnitude responses of (7.78) (dotted line) and (7.81) (solid line). They are very close for ω small. This can also be established directly from (7.81). For ω small, we have $\sin 0.5\omega \approx 0.5\omega$ and $e^{-j0.5\omega} \approx 1$. Thus we have, for ω small,

$$\bar{H}(e^{j\omega}) = j2 \sin 0.5\omega \times e^{-j0.5\omega} \approx j2 \times 0.5\omega = j\omega$$

Thus differentiation of slowly varying CT signals, or CT signals with narrow frequency spectra centered at $\omega = 0$, can be carried out by the difference equation in (7.79). Digital differentiation of CT signals with broad frequency spectra must be implemented with a digital filter with frequency response shown in (7.78). Thus the differentiator in (7.78) is often called a *broadband differentiator*. We discuss its design using two different methods.

Truncation with or without a Window We compute the inverse DTFT of (7.78). Define $M = N/2$. Substituting (7.78) into (3.44) with $T = 1$ yields

$$h_d[n] = \frac{1}{2\pi} \int_{\omega=-\pi}^{\pi} j\omega e^{j(n-M)\omega} d\omega = \frac{1}{2\pi(n-M)} \int_{\omega=-\pi}^{\pi} \omega d e^{j(n-M)\omega}$$

which becomes, using integration by parts,

$$h_d[n] = \frac{1}{2(n-M)\pi} \left(\omega e^{j(n-M)\omega} \Big|_{\omega=-\pi}^{\pi} - \int_{-\pi}^{\pi} e^{j(n-M)\omega} d\omega \right)$$

$$= \frac{1}{2(n-M)\pi} \left(\pi e^{j(n-M)\pi} - (-\pi)e^{-j(n-M)\pi} \right.$$

$$\left. - \frac{1}{j(n-M)} (e^{j(n-M)\pi} - e^{-j(n-M)\pi}) \right)$$

$$= \frac{\cos[(n-M)\pi]}{n-M} - \frac{\sin[(n-M)\pi]}{\pi(n-M)^2}$$

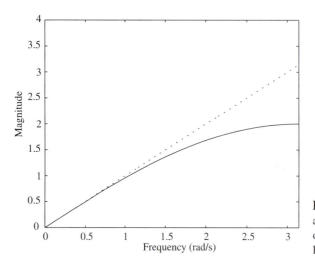

Figure 7.20 Magnitude responses of an ideal differentiator (dotted line) and of a first-order difference equation (solid line).

If N is even, then $M = N/2$ is an integer, and we have $\sin[(n - M)\pi] = 0$ for any integer n. If N is odd, then $\cos[(2n - N)\pi/2] = 0$ for any integer n. Thus we have

$$h_d[n] = \begin{cases} \dfrac{\cos[(n - M)\pi]}{n - M} & \text{for } n \neq M \\ 0 & \text{for } n = M \end{cases} \tag{7.82}$$

for $N = 2M$ even and

$$h_d[n] = \frac{-\sin[(n - M)\pi]}{\pi(n - M)^2} \tag{7.83}$$

for $N = 2M$ odd. Both have the property $h_d[n] = -h_d[N - n]$. Their coefficients are antisymmetric and of infinite length. Truncating them using the windows of length $N + 1$ in (7.37)–(7.40), we obtain

$$h[n] = h_d[n]w_i[n] \tag{7.84}$$

It is a type III FIR filter if N is even and type IV if N is odd. Figure 7.21(a) shows the magnitude responses of the seventh-order type IV differentiators obtained by direct truncation (solid line) and by using a Hamming window (dotted line). The one obtained by direct truncation shows only small ripples. The one obtained using a Hamming window does not have any ripple but has a larger leakage. Figure 7.21(b) shows the magnitude responses of eighth-order type III differentiators obtained by direct truncation (solid line) and by using a Hamming window (dotted line). The one obtained by direct truncation shows significant ripples. These ripples are completely eliminated by using a Hamming window. But the Hamming window also introduces a much larger leakage.

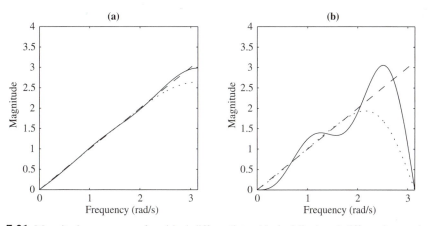

Figure 7.21 Magnitude responses of an ideal differentiator (dashed line) and differentiators obtained using rectangular (solid lines) and Hamming (dotted lines) windows. (a) Seventh order. (b) Eighth order.

Even though the type IV FIR filters in Fig. 7.21(a) have an order one fewer than the type III FIR filters in Fig. 7.21(b), their magnitude responses are much better. The reason is that every type III filter has zeros, as shown in Fig. 7.3(c), at $\omega = 0$ and π. The zero at $\omega = 0$ is desirable; the zero at $\omega = \pi$, however, is detrimental to the magnitude response at high frequencies. Type IV filters have zeros only at $\omega = 0$; thus they are more suitable than type III for designing differentiators. In conclusion, we should use type IV FIR filters to approximate the ideal differentiator in (7.78). Such a filter can be obtained from (7.83) by direct truncation as

```
n=0:N; (% N odd)
h=-sin((n-N/2).*pi)./(pi.*(n-N/2).^ 2);
```

For example, if $N = 7$, then the filter is given by

$$H(z) = -0.0260 + 0.0509z^{-1} - 0.1415z^{-2} + 1.2732z^{-3} - 1.2732z^{-4}$$
$$+ 0.1415z^{-5} - 0.0509z^{-6} + 0.0260z^{-7} \tag{7.85}$$

This filter minimizes

$$E = \int_{-\pi}^{\pi} |H(e^{j\omega}) - j\omega e^{jN\omega/2}|^2 d\omega \tag{7.86}$$

and is LS optimal. This filter will introduce a delay of $M = N/2$ samples. Because M is not an integer, the physical meaning of the delay of a noninteger samples is not clear. However if the filter is used to process CT signals, it will introduce simply a delay of MT seconds.

Discrete Least-Squares Method In this part, we will apply the discrete LS optimization method in Section 7.5 to design linear phase FIR differentiators. We discuss only type IV FIR filters. Consider the seventh-order type IV FIR filter in (7.24)–(7.26):

$$H(e^{j\omega}) = \sum_{n=0}^{7} h[n]e^{-jn\omega} = jA(\omega)e^{-j3.5\omega}$$

with

$$A(\omega) = \sum_{n=0}^{3} d[n]\sin(n+0.5)\omega \tag{7.87}$$

where $h[n] = d[3-n]/2$ and $h[7-n] = -h[n]$, for $n = 0, 1, 2, 3$. This equation has four independent coefficients $d[n]$ and can be used to match exactly four frequency samples of $D(\omega) = \omega$ in $[0, \pi]$. Let us select four frequencies as ω_i for $i = 1, 2, 3, 4$. Then the four equations $A(\omega_i) = \omega_i$ can be expressed in matrix form as

$$\begin{bmatrix} \sin 0.5\omega_1 & \sin 1.5\omega_1 & \sin 2.5\omega_1 & \sin 3.5\omega_1 \\ \sin 0.5\omega_2 & \sin 1.5\omega_2 & \sin 2.5\omega_2 & \sin 3.5\omega_2 \\ \sin 0.5\omega_3 & \sin 1.5\omega_3 & \sin 2.5\omega_3 & \sin 3.5\omega_3 \\ \sin 0.5\omega_4 & \sin 1.5\omega_4 & \sin 2.5\omega_4 & \sin 3.5\omega_4 \end{bmatrix} \begin{bmatrix} d[0] \\ d[1] \\ d[2] \\ d[3] \end{bmatrix} = \begin{bmatrix} \omega_1 \\ \omega_2 \\ \omega_3 \\ \omega_4 \end{bmatrix} \tag{7.88}$$

or

$$\mathbf{Pd} = \mathbf{D}$$

where $\mathbf{P}$ is a 4×4 matrix, $\mathbf{d}$ and $\mathbf{D}$ are 4×1 column vectors. If ω_i are distinct, $\mathbf{P}$ is nonsingular and the solution is unique and can be obtained as

$$\mathbf{d} = \mathbf{P}^{-1}\mathbf{D}$$

The program that follows forms (7.88), solves $\mathbf{d}$, forms an FIR differentiator, and computes its magnitude response.

Program 7.4
```
W=[w1 w2 w3 w4]';
D=W;
WW = W* [0.5 : 3.5]; P = sin(WW);
d=P\D;
k=4:-1:1;
h1=[d(k)/2]'; (% d is a column, thus the need of transpose.)
h=[h1 fliplr(-h1)]
```

```
[H,w]=freqz(h,1);
plot(w,abs(H))
```

If we select W=[0.1 1 2 3]', then the program yields the magnitude response in Fig. 7.22(a) with a solid line. We plot in Fig. 7.22(b) with a solid line the absolute error $|\omega - |H(e^{j\omega})||$. The largest error is about 0.13. Thus the result is quite good. If we select a different set of four frequencies, we will obtain a different differentiator.

Now instead of matching four frequencies, we will match five or more frequencies. In this case, perfect matching is not possible, but we can achieve least-squares matching. Let us replace the first line of Program 7.4 by

```
W=[0.1:0.1:3.1]';
```

In this case, we match 31 frequencies in [0.1 3.1] with increment 0.1. The program yields the filter

$$\bar{H}(z) = -0.0263 + 0.0512z^{-1} - 0.1418z^{-2} + 1.2735z^{-3} - 1.2735z^{-4}$$
$$+ 0.1418z^{-5} - 0.0512z^{-6} + 0.0263z^{-7} \tag{7.89}$$

and the magnitude response in Fig. 7.22(a) and its absolute error in Fig. 7.22(b) with dotted lines. The magnitude response is better than the one obtained by matching four frequencies in some frequency range but worse in another frequency range. The type IV FIR filter in (7.89) minimizes the error

$$\bar{E} = \sum_{i=1}^{31} |A(\omega_i) - \omega_i|^2 = \sum_{i=1}^{31} |A(0.1i) - 0.1i|^2 \tag{7.90}$$

Because the integral squared error in (7.86) roughly equals (7.90) multiplied by 2, the filter in (7.89) should be close to the integral LS optimal filter in (7.85). Indeed, they differ only in the least significant digit.

Now let us introduce the weighting factor $1/\omega^2$ into the discrete squared error in (7.90). That is, (7.90) will be modified as

$$\hat{E} = \sum_{i=0}^{31} \frac{1}{\omega_i^2} |A(\omega_i) - \omega_i|^2 \tag{7.91}$$

As discussed in (7.62)–(7.64), if we modify $\mathbf{Pd} = \mathbf{D}$ as $\bar{\mathbf{P}}\mathbf{d} = \bar{\mathbf{D}}$, where the ith rows of $\bar{\mathbf{P}}$ and $\bar{\mathbf{D}}$ equal the ith rows of $\mathbf{P}$ and $\mathbf{D}$ divided by ω_i, then $\bar{\mathbf{P}}\backslash\bar{\mathbf{D}}$ yields the weighted LS solution. In Program 7.4, if D=W is replaced by D=[ones(1,4)]', and P=sin(WW) by

```
P=sin(WW)./(W*[ones(1,4)]
```

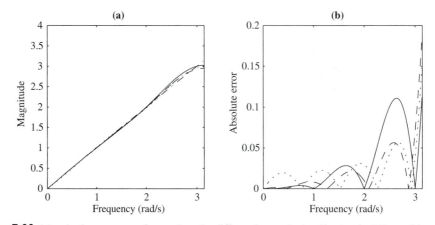

Figure 7.22 Magnitude responses of seventh-order differentiators obtained by 4-point LS matching (solid line), 31-point LS matching (dotted line), and 31-point weighted LS matching (dashed line).

then the program will yield a discrete weighted LS optimal filter. If we select 31 frequency samples as in (7.90), then the result is shown in Figs. 7.22(a) and (b) with dashed line. Because the weighting factor $1/\omega^2$ imposes more weight on low frequencies, the 31-point weighted LS optimal filter has, compared with the corresponding unweighted filter shown in Fig. 7.22 with dotted lines, less error in low frequencies but larger error in high frequencies. The filter obtained by matching the four frequencies at [0.1 1 2 3] actually is the best in low frequencies among the three differentiators.

Next we discuss some MATLAB functions. Let f=[0 1];d=[0 pi]. Then the function h=firls(n, f,d) returns an nth-order integral LS differentiator. We plot in Fig. 7.23(a) the magnitude responses of h=firls(7,f,d) (solid line) and h=firls(8,f,d) (dotted line). The solid line should be better than the discrete LS differentiator in Fig. 7.22. It is not. The dotted line, which is type III, should drop to zero at $\omega = \pi$. But it does not. Thus the MATLAB function firls(n,f,d) with f=[0 1];d=[0 pi] should not be used to design differentiators. However, MATLAB contains

h=firls(n,f,d,'differentiator')

which has the built-in weighting factor $1/f^2$. Typing

h=firls(7,[0 1],[0 pi],'differentiator');
[H,w]=freqz(h,1);
plot(w,abs(H))

yields the solid line in Fig. 7.23(b). Replacing 7 with 8 in the first line yields the dotted line in Fig. 7.23(b). The function indeed yields good differentiators. Thus whenever we use firls to design differentiators, we must use the one with flag "differentiator."

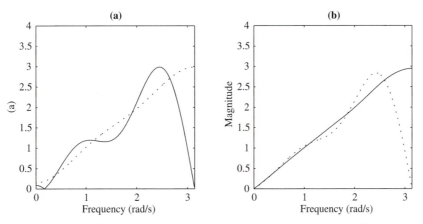

Figure 7.23 (a) Outputs of firls(n,[0 1],[0 pi]) with $n = 7$ (solid line) and $n = 8$ (dotted line). (b) Corresponding outputs of firls(n,[0 1],[0 pi],'differentiator').

The minimax optimization discussed in Section 7.5 can also be applied to design digital differentiators. The MATLAB function

h=remez(n,f,d,'differentiator')

generates an nth-order FIR differentiator. The function uses the built-in weighting factor $1/f$. We compare in Figs. 7.24(a) and (b) the magnitude responses and absolute errors of the seventh-order FIR filters obtained by direct truncation (integral LS optimal) (solid line), by using firls (weighted integral LS optimal) (dotted line), and by using remez (weighted minimax optimal) (dashed line). The one obtained by using firls appears to be the best in low frequencies but the worst at the highest frequency π.

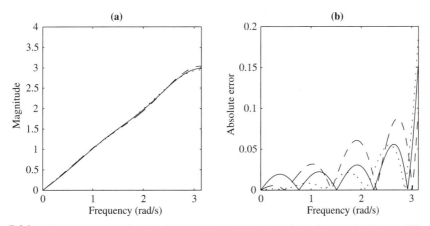

Figure 7.24 Differentiators obtained by integral LS (solid line), weighted integral LS (dotted line), and weighted minimax (dashed line) optimizations. (a) Magnitude responses. (b) Absolute errors.

7.9 Hilbert Transformers

The ideal Hilbert transformer is specified by

$$H_d(e^{j\omega}) = \begin{cases} -j & \text{for } 0 < \omega < \pi \\ j & \text{for } -\pi < \omega < 0 \end{cases} \tag{7.92}$$

Its magnitude response is 1 for all ω; thus it is an all-pass filter. It introduces a phase shift of $-90°$ for $\omega > 0$ and $90°$ for $\omega < 0$. Thus it is also called a 90-degree phase shifter. We first show where it is used.

Consider a DT signal $x[n]$ bandlimited to W as shown in Fig. 7.25(a). In transmission, the signal may be modulated as $x[n] \cos \omega_0 n$, and its frequency spectrum will be as shown in Fig. 7.25(b). The plot is similar to the one in Fig. 3.6 and can be obtained as in (3.24) and (3.25). The bandwidth of the modulated signal is $2W$. If we can cut the bandwidth by half, the number of signals carried by a channel can be doubled. This is highly desirable in communication.

Consider $x[n]$ with frequency spectrum

$$X(e^{j\omega}) = X_r(e^{j\omega}) + jX_i(e^{j\omega})$$

where X_r and X_i denote, respectively, the real and imaginary parts. If $x[n]$ is real-valued, as is always the case in practice, then its spectrum is conjugate symmetric; that is,

$$X_r(-\omega) = X_r(\omega) \quad \text{and} \quad X_i(-\omega) = -X_i(\omega) \tag{7.93}$$

Let $\bar{x}[n]$ be the output of the transformer in (7.92) excited by $x[n]$. Then the spectrum of $\bar{x}[n]$ is

$$\begin{aligned} \bar{X}(e^{j\omega}) &= H_d(e^{j\omega})X(e^{j\omega}) \\ &= \begin{cases} -jX_r(e^{j\omega}) + X_i(e^{j\omega}) & 0 < \omega < \pi \\ jX_r(e^{j\omega}) - X_i(e^{j\omega}) & -\pi < \omega < 0 \end{cases} \end{aligned} \tag{7.94}$$

Now we form the complex-valued signal $y[n] := x[n] + j\bar{x}[n]$. Its frequency spectrum can readily be obtained as

$$Y(e^{j\omega}) = X(e^{j\omega}) + j\bar{X}(e^{j\omega}) = \begin{cases} 2X(e^{j\omega}) & \text{for } 0 < \omega < \pi \\ 0 & \text{for } -\pi < \omega < 0 \end{cases} \tag{7.95}$$

Its frequency spectrum is zero for $\omega < 0$. Its modulated signal or $y[n] \cos \omega_0 n$ has bandwidth W, only half of the bandwidth of the modulated $x[n]$. This cuts the bandwidth of a signal in half. Thus the number of signals transmitted in a channel can be doubled. This is an important application of Hilbert transformers.

We now discuss the design of Nth-order linear-phase FIR filters to approximate the ideal Hilbert transformer denoted by

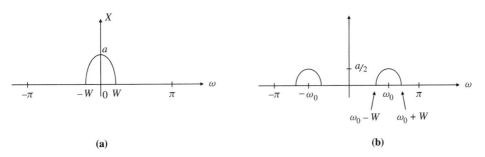

Figure 7.25 (a) Spectrum of $x[n]$. (b) Spectrum of modulated $x[n]$ or $x[n] \cos \omega_o n$.

$$H_d(e^{j\omega}) = \begin{cases} -je^{-j0.5N\omega} & \text{for } 0 < \omega < \pi \\ je^{j0.5N\omega} & \text{for } -\pi < \omega < 0 \end{cases} \tag{7.96}$$

The linear phase $-0.5N\omega$ will introduce a delay of $0.5N$ samples. If its output is to be combined with $x[n]$, as shown in Fig. 7.26, and if the delay $-0.5N$ is not an integer, matching the delays of $\bar{x}[n]$ and $x[n]$ will be difficult. Thus from now on, we consider only N even.

Let us compute the inverse DTFT of (7.96) with $M = M/2$:

$$\begin{aligned}
h_d[n] &= \frac{1}{2\pi} \int_{-\pi}^{\pi} H_d(e^{j\omega}) e^{jn\omega} d\omega \\
&= \frac{1}{2\pi} \left(\int_{-\pi}^{0} j e^{j(n-M)\omega} d\omega - \int_{0}^{\pi} j e^{j(n-M)\omega} d\omega \right) \\
&= \frac{1}{2\pi(n-M)} \left(2 - e^{j(n-M)\pi} - e^{-j(n-M)\pi} \right) \\
&= \frac{1}{\pi(n-M)} [1 - \cos((n-M)\pi)] = \frac{2\sin^2[(n-M)\pi/2]}{(n-M)\pi}
\end{aligned}$$

It is straightforward to show $h_d[M] = 0$ and $h_d[n] = -h_d[N-n]$. For N even, the truncated $h_d[n]$, for $0 \leq n \leq N$, is a type III FIR filter. We plot in Fig. 7.27(a) the impulse response of the fourteenth-order FIR Hilbert transformer and in Fig. 7.27(b) the impulse response of the sixteenth-order FIR Hilbert transformer. The first and last entries of the sixteenth-order transformer are zero. If it is advanced by one sample (it is still causal because $h[0] = 0$), the filter becomes the fourteenth-order filter. Thus in designing Hilbert transformers, we may consider only N even and $M = N/2$ odd. In other words, we consider only $N = 2, 6, 10, 14$, and so forth. For $M = N/2$ odd, we have

$$h_d[n] = \begin{cases} \dfrac{-2}{(M-n)\pi} & \text{for } n = 0, 2, \ldots, M-1 \\ 0 & \text{for } n = 1, 3, \ldots, M \end{cases} \tag{7.97}$$

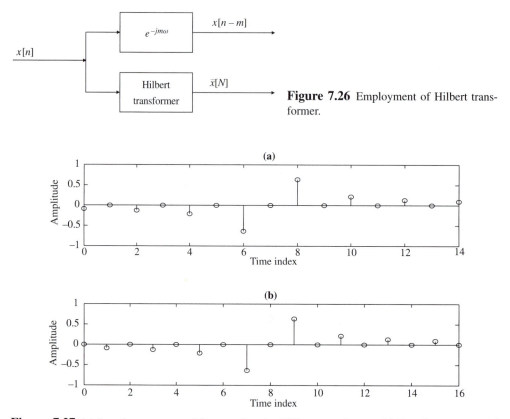

Figure 7.26 Employment of Hilbert transformer.

Figure 7.27 (a) Impulse response of fourteenth-order Hilbert transformer. (b) Impulse response of sixteenth-order Hilbert transformer.

and $h_d[N - n] = -h_d[n]$ for $0 \leq n \leq M$.

A simple way of designing Hilbert transformers is to use the window method. The desired magnitude response of a Hilbert transform is 1 for all ω in $[0, \pi]$. Every type III FIR filter has zeros at $\omega = 0$ and $= \pi$; thus the magnitude response will have sharp rolloff in the neighborhood of 0 and π. Thus direct truncation will introduce ripples. Figure 7.28 shows the magnitude response, with a solid line, of the fourteenth-order filter obtained by direct truncation. It is obtained in MATLAB as

```
h1=[2/(7*pi) 0 2/(5*pi) 0 2/(3*pi) 0 2/pi];
h=[h1 0 fliplr(-h1)];
[H,w]=freqz(h,1);
plot(w,abs(H))
```

The fourteenth-order filter has length 15; **h1** lists the first seven coefficients. The eighth coefficient is $h[7] = 0$; the remaining seven coefficients are the left right flip of **h1** with

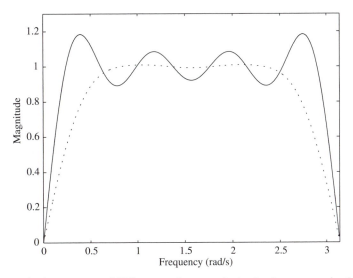

Figure 7.28 Magnitude responses of Hilbert transformers obtained using rectangular (solid line) and Hamming (dotted line) windows.

signs reversed. Indeed, the magnitude response shows significant ripples. The ripples can be eliminated by using, for example, a Hamming window. Replacing h by h=h.*hamming(15) yields the plot in Fig. 7.28 with a dotted line. The magnitude response does not have any ripple but has a wider leakage. Indeed, the design method is simple and straight-forward.

LS Matching Next we discuss discrete least-squares matching of the desired frequency response. As derived in (7.19)–(7.20), an Nth-order type III FIR filter can be expressed as

$$H(e^{j\omega}) = \sum_{n=0}^{N} h[n]e^{-jn\omega} = A(\omega)je^{-jM\omega} \tag{7.98}$$

with

$$A(\omega) = \sum_{n=1}^{M} d[n]\sin n\omega \tag{7.99}$$

and $d[n] = 2h[M - n]$ for $1 \le n \le M$. As discussed earlier, we consider only $M = N/2$ odd. Because the ideal Hilbert transformer has $d[n] = 0$, for n even, as implied by $h[n] = 0$ for n odd in (7.97), we will impose the condition $d[n] = 0$ for n even or

$$d[2k] = 0 \qquad k = 0, 1, \cdots \tag{7.100}$$

Thus in designing Hilbert transformers, (7.99) can be modified as

$$A(\omega) = \sum_{n=1}^{\bar{M}} d[2n-1]\sin[(2n-1)\omega] \tag{7.101}$$

where $\bar{M} = \overline{M/2}$, and the overline rounds a number upward to an integer.

The amplitude response $A(\omega)$ in (7.101) has the following symmetric property

$$A(\omega) = A(\pi - \omega) \quad \text{or} \quad A(0.5\pi + \omega) = A(0.5\pi - \omega) \tag{7.102}$$

for $0 < \omega < 0.5\omega$, which follows directly from

$$\sin[(2n-1)(\pi - \omega)] = \sin[(2n-1)\omega]$$

We call this *midfrequency symmetry*. The amplitude response in (7.101) is a special case of type III FIR filters, which have zeros at $\omega = 0$ and π. Thus it is not possible to design a Hilbert transformer to have

$$A(\omega) = 1 \quad \text{for} \quad 0 \le \omega \le \pi$$

The best we can achieve is

$$A(\omega) = 1 \qquad 0 < \omega_l \le \omega \le \omega_u < \pi \tag{7.103}$$

with $\omega_l = \pi - \omega_u$ (because of midfrequency symmetry) for some ω_l. Thus in design, we need to match a desired magnitude response only in $(0, \pi/2]$ or $[\omega_l, \pi/2]$.

Let us design (7.101) with $N = 14$ and $\bar{M} = \overline{14/4} = \overline{3.5} = 4$ to match a desired Hilbert transformer with amplitude 1 in $(0, \pi/2)$. Such a filter has only four independent coefficients $d[2n-1]$, for $n = 1, 2, 3, 4$. Matching $A(\omega_i) = D_i = 1$ for four ω_i yields the equation

$$\begin{bmatrix} \sin \omega_1 & \sin 3\omega_1 & \sin 5\omega_1 & \sin 7\omega_1 \\ \sin \omega_2 & \sin 3\omega_2 & \sin 5\omega_2 & \sin 7\omega_2 \\ \sin \omega_3 & \sin 3\omega_3 & \sin 5\omega_3 & \sin 7\omega_3 \\ \sin \omega_4 & \sin 3\omega_4 & \sin 5\omega_4 & \sin \omega_4 \end{bmatrix} \begin{bmatrix} d[1] \\ d[3] \\ d[5] \\ d[7] \end{bmatrix} = \begin{bmatrix} 1 \\ 1 \\ 1 \\ 1 \end{bmatrix} \tag{7.104}$$

or

$$\mathbf{Pd} = \mathbf{D}$$

If ω_i are distinct, $\mathbf{P}$ is nonsingular, and the solution is unique and can be obtained as

$$\mathbf{d} = \mathbf{P}^{-1}\mathbf{D}$$

The program that follows forms (7.104), solves **d**, forms a Hilbert transformer, and computes its magnitude response.

Program 7.5
```
W=[w1 w2 w3 w4]';
D=[ones(1,4)]';
WW = W* [1 : 2 : 7]; P = sin (WW);
d=P\D;
h1=[d(4)/2 0 d(3)/2 0 d(2)/2 0 d(1)/2];
h=[h1 0 fliplr(-h1)];
[H,w]=freqz(h,1);
plot(w,abs(H))
```

If we select W=[0.2 0.6 1 1.4]', then the program yields the magnitude response in Fig. 7.29 with a solid line.

Next we match, in the discrete LS sense, 15 equally spaced frequencies in [0.1, 1.5]. Note that we need to match only in the frequency range $(0, \pi/2] = (0, 1.57]$ due to midfrequency symmetry. Program 7.5 with its first two lines replaced by

```
W=[0.1:0.1:1.5]';
D=[ones(1,15)]';
```

yields the magnitude response in Fig. 7.29 with a dotted line. The result is better than the one obtained by matching only four frequencies.

The MATLAB function h=firls(14,[0 1]],[1 1],'hilbert') returns a fourteenth-order FIR Hilbert transformer that minimizes

$$E = \int_0^\pi [A(\omega) - 1]^2 d\omega$$

Its magnitude response is shown in Fig. 7.29 with a dashed line. It is almost indistinguishable from the discrete LS optimal filter shown with a dotted line.

The MATLAB function h=remez(n,f,d,'hilbert') returns a minimax optimal FIR Hilbert transformer. We compare in Fig. 7.30(a) the magnitude responses obtained by using firls (solid line) and remez (dotted line), both using n=14,f=[0.05 0.95],d=[1 1]. Clearly, the integral LS optimal filter is better. We show in Fig. 7.30(b) the corresponding magnitude responses with f=[0.1 0.9]. For this f, the minimax optimal filter is better. We see that the results of firls and remez depend highly on the f used. One f may yield a good result in firls but a poor result in remez and vice versa. It is not clear how to select the best f. We mention that the magnitude responses of firls(14,[0.05 0.95],[1 1],'hilbert') and remez(14,[0.1 0.9],[1 1],'hilbert') are comparable.

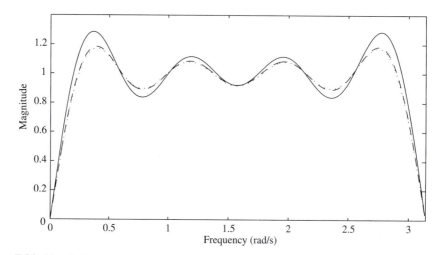

Figure 7.29 Magnitude responses of Hilbert transformers obtained by matching 4 points (solid line), 15 points (dotted line), and infinitely many points (dash-dotted line) in $(0, \pi/2]$.

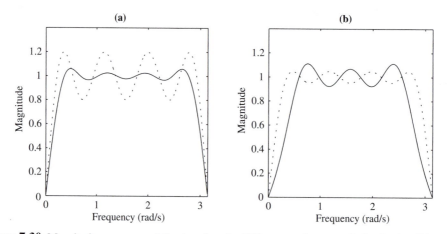

Figure 7.30 Magnitude responses of fourteenth-order Hilbert transformers obtained using firls (solid lines) and remez (dotted lines) with two different f: (a) f=[0.05 0.95] and (b) f=[0.1 0.9].

7.9.1 From FIR Low-Pass Filters to Hilbert Transformers

Consider a type I low-pass FIR filter

$$H(z) = \sum_{n=0}^{N} h[n]z^{-n} \tag{7.105}$$

Suppose its frequency response $H(e^{j\omega})$ is as shown in Fig. 7.31(a). Then $H(e^{j(\omega-\pi/2)})$ is the shifting of $H(e^{j\omega})$ to the right by $\pi/2$ as shown in Fig. 7.31(b) with a solid line. Likewise, $H(e^{j(\omega+\pi/2)})$ is the shifting of $H(e^{j\omega})$ to the left by $\pi/2$, as shown in Fig. 7.31(b) with a dotted line. Their sum yields the frequency response of a Hilbert transformer. Note that the frequency response $H(e^{j(\omega-\pi/2)})$ by itself is not conjugate symmetric, as is required for all filters with real coefficients. Thus we need both $H(e^{j(\omega-\pi/2)})$ and $H(e^{j(\omega+\pi/2)})$ to design a Hilbert transformer with real coefficients.

Let us compute

$$
\begin{aligned}
\bar{H}(e^{j\omega}) &:= H(e^{j(\omega+\pi/2)}) + H(e^{j(\omega-\pi/2)}) \\
&= \sum_{n=0}^{N} h[n]e^{-jn(\omega+\pi/2)} + \sum_{n=0}^{N} h[n]e^{-jn(\omega-\pi/2)} \\
&= \sum_{n=0}^{N} h[n]e^{-jn\omega}\left(e^{-jn\pi/2} + e^{jn\pi/2}\right) \\
&= \sum_{n=0}^{N} 2h[n]e^{-jn\omega}\cos\left(\frac{n\pi}{2}\right)
\end{aligned}
\tag{7.106}
$$

Because $\cos(n\pi/2) = 0$ for n odd, we can drop all odd terms in (7.106). Let us define $M = N/2$ with N even and $\bar{n} = 2n$. Then we can simplify (7.106) as

$$
\bar{H}(e^{j\omega}) = \sum_{\bar{n}=0}^{M} 2h[2\bar{n}]\cos(\bar{n}\pi)e^{-j2\bar{n}\omega}
\tag{7.107}
$$

or, dropping the overbar on n,

$$
\bar{H}(z) = \sum_{n=0}^{M} 2h[2n]\cos(n\pi)z^{-2n} = \sum_{n=0}^{M}(-1)^n 2h[2n]z^{-2n}
\tag{7.108}
$$

We write this out explicitly for $M = 14/2 = 7$ as

$$
\begin{aligned}
\bar{H}(z) = \; &2h[0] - 2h[2]z^{-2} + 2h[4]z^{-4} - 2h[6]z^{-6} + 2h[8]z^{-8} \\
&- 2h[10]z^{-10} + 2h[12]z^{-12} - 2h[14]z^{-14}
\end{aligned}
\tag{7.109}
$$

Now if the fourteenth-order low-pass filter is type I, then we have $h[n] = h[14 - n]$, for $n = 0, 1, \ldots, 7$. From (7.109) we see immediately that the coefficients of $\bar{H}(z)$ are antisymmetric. This is consistent with the discussion in (7.97). In order for this property to hold, we require M to be odd. For example, if M is 6, and if $h[n] = h[12 - n]$ for $n = 0, 1, \ldots, 6$, then the coefficients of $\bar{H}(z)$ are symmetric and cannot be used as a Hilbert transformer. In conclusion, if N is even and $M = N/2$ is odd, we can obtain a Hilbert transformer from an Nth-order type

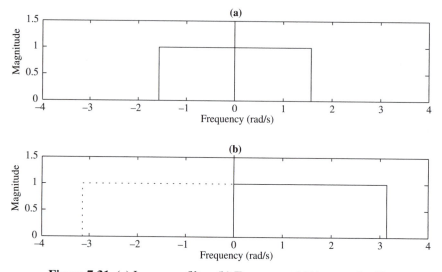

Figure 7.31 (a) Low-pass filter. (b) Frequency shiftings to $\pm\pi/2$.

I FIR low-pass filter.

We summarize the design procedure in the following. Let $h[n]$, $n = 0, 1, \ldots, N$, with $M = N/2$ odd, be an Nth-order type I low-pass filter with cutoff frequency $\omega_c = \pi/2$. We form

$$\bar{h}[n] = (-1)^n 2h[2n] \quad \text{for } 0 \leq n \leq M \tag{7.110}$$

Then the filter

$$\bar{H}(z) = \sum_{n=0}^{M} \bar{h}[n] z^{-2n} \tag{7.111}$$

is a Hilbert transformer.

We give an example. We design a fourteenth-order low-pass filter with cutoff frequency $\omega_c = \pi/2$ by calling h=firls. Its magnitude response is shown in Fig. 7.32(a). We then form (7.110). Note that indices in MATLAB start from 1 rather 0. The program that follows

```
Program 7.6
f=[0 0.5 0.5 1];d=[1 1 0 0];
h=firls(14,f,d);
h1=[2*h(1) 0 -2*h(3) 0 2*h(5) 0 -2*h(7)];
hb=[h1 0 fliplr(-h1)];
[Hb,w]=freqz(hb,1);
plot(w,abs(Hb))
```

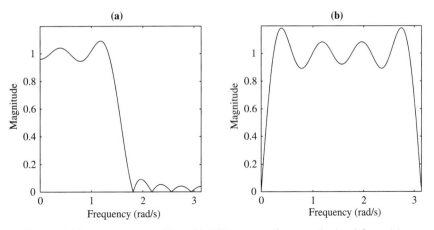

Figure 7.32 (a) Low-pass filter. (b) Hilbert transformer obtained from (a).

generates a Hilbert transformer with magnitude response shown in Fig. 7.32(b). This is a novel way of designing Hilbert transformers. However, it is probably not as useful as the direct methods discussed in the preceding section.

7.10 A Design Example

As discussed in Section 6.3.1, a desired magnitude response is often specified in terms of passband and stopband edge frequencies and tolerances. We discuss these specifications further. Let a_1 and a_2 be the largest and smallest magnitudes in the passband. In other words, the magnitude response in the passband is bounded by a_1 and a_2. Then the *passband tolerance* (peak-to-peak) R_p in dB is defined as

$$R_p = 20 \log_{10} a_1 - 20 \log_{10} a_2 = 20 \log_{10} \left(\frac{a_1}{a_2} \right) \tag{7.112}$$

For example, if $a_1 = 1$ and $a_2 = 0.707$, then the *passband tolerance* is 3 dB. If $a_1 = 1.05$ and $a_2 = 0.9$, then we have

$$R_p = 20 \log_{10}(1.05/0.9) = 1.34 \text{ dB}$$

Let b be the largest magnitude in the stopband. Then the stopband attenuation in dB is defined as

$$R_s = -20 \log_{10} b \tag{7.113}$$

Because b is less than 1, the attenuation is always a positive number. For example, if $b = 0.05$, then $R_s = 26$ dB.

In practice, a low-pass filter is often specified to have passband edge frequency ω_p, passband tolerance *at most* R_p dB, stopband edge frequency ω_s, and stopband attenuation *at least* R_s dB. Usually no filter order is specified. In the design, we then try to find a filter with order as small as possible to meet the specification. Because there is no simple relationship among N, ω_p, ω_s, R_p, and R_s, the search of a filter with a minimum order often requires trial and error. The minimum order also depends on the type of filters used, as we will show next.

We discuss the design of a linear-phase FIR low-pass filter with passband edge frequency 500 Hz, passband tolerance at most 1 dB, stopband edge frequency 600 Hz, and stopband attenuation 20 dB or more. The sampling frequency is 2000 Hz. First we will use the MATLAB function remez(N,f,d,wt) to design a minimax optimal filter to meet the specifications. The sampling frequency is 2000 Hz; thus the frequency range is $(-1000, 1000)$ in Hz or $[0, 1000]$ if we consider only the positive frequency part. The frequency in f is limited to $[0, 1]$. Thus for our problem we have

```
f=[0 500/1000 600/1000 1000/1000]=[0 0.5 0.6 1];
d=[1 1 0 0];
```

The minimax optimal filter is equiripple in the passband and stopband. The magnitude response in the passband is limited to $1 + e_p$ and $1 - e_p$. The passband tolerance $R_p = 1$ dB requires

$$20 \log_{10} \left(\frac{1 + e_p}{1 - e_p} \right) = 1$$

which implies

$$\frac{1 + e_p}{1 - e_p} = 10^{1/20} = 1.122$$

and $e_p = 0.0575$. The stopband attenuation $R_s = 20$ requires

$$20 \log_{10} e_s = -20 \quad \text{or} \quad e_s = 0.1$$

Thus the weighting factor should be wt=[0.1/0.0575 1]=[1.74 1]. As a first try, we select $N = 10$. Then h=remez(10,f,d,wt) will yield a minimax optimal filter with the magnitude response shown in Fig. 7.33(a) with a solid line. The required passband tolerance and stopband attenuation are also marked with horizontal dotted lines. Clearly, the tenth-order minimax optimal filter does not meet the specifications. Next we try $N = 20$. The result is plotted in Fig. 7.33(a) with a dashed line. The passband tolerance is less than 1 dB, and the stopband attenuation is more than 22 dB. Thus the twentieth-order minimax filter overmeets the specification. Proceeding forward, it is possible to find a minimax optimal filter of least order to meet the specification.

In MATLAB, if we type filtdemo, then a menu will appear in which we can select FIR filter REMEZ, FIRLS, KAISER, and some IIR filters to be introduced in the next chapter. Once we type in the sampling frequency, passband and stopband edge frequencies, all in Hz, and passband

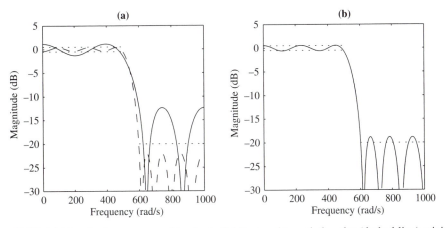

Figure 7.33 (a) Magnitude responses of tenth- (solid line) and twentieth-order (dashed line) minimax filters. (b) Magnitude response of seventeenth-order minimax filter.

ripple R_p and stopband attenuation R_s, both in dB, then the program will automatically find a filter with minimum order to meet the specification. For example, if $f_s = 2000$ Hz, $\omega_p = 500$ Hz, $\omega_s = 600$ Hz, $R_p = 1$ dB, and $R_s = 20$ dB, and if we select REMEZ, then the program will generate a seventeenth-order linear-phase FIR filter with the magnitude response shown in Fig. 7.33(b). The design does not exactly meet the specification. If it is not acceptable, we can increase the filter order until the resulting filter meets the specification.

The design of linear-phase FIR filters using the window and LS methods will be more complex because the band-edge frequencies cannot be specified. Therefore, their design will involve

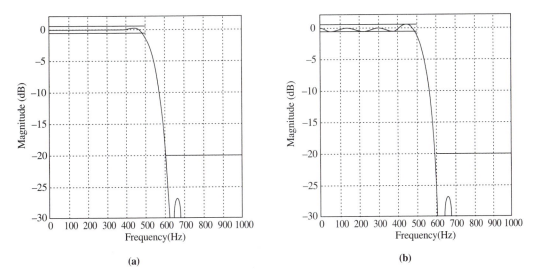

Figure 7.34 (a) LS optimal filter. (b) Kaiser window.

more iterations. For the preceding example, if we select FIRLS, then filtdemo will generate a twentieth-order filter with magnitude response shown in Fig. 7.34(a). Note that the vertical coordinate is limited to -30 dB; thus magnitudes less than -30 dB are not shown. If we select KAISER, then filtdemo will generate a twenty-fourth-order filter with magnitude response shown in Fig. 7.34(b). Thus the design of digital filter using MATLAB is very simple.

For this example, the minimax optimal filter has the lowest order, and the windowed filter has the largest order to meet the same specification. This is true in general.

PROBLEMS

7.1 Use a rectangular and a Hamming window to find fourth-order linear-phase FIR filters to approximate the ideal low-pass filter $e^{-j2\omega}$ for $|\omega| \leq 1$ and 0 for $1 < |\omega| \leq \pi$.

7.2 Use a rectangular and a Hamming window to find fifth-order linear-phase FIR filters to approximate the ideal low-pass filter $e^{-j2.5\omega}$ for $|\omega| \leq 1$ and 0 for $1 < |\omega| \leq \pi$.

7.3 The Hanning window of length $N + 1$ is defined as

$$w[n] = 0.5 - 0.5 \cos\left(\frac{2\pi n}{N}\right)$$

for $0 \leq n \leq N$ and zero otherwise. Compare its magnitude response with the one of the Hamming window for $N = 20$. Plot their magnitude responses using freqz. Also use the procedure of computing frequency spectra discussed in Chapter 4 to compute their magnitude responses in $[0, \pi]$ by using 1024-point FFT.

7.4 Consider the digital ideal low-pass filter in (7.35) for various M. Show that the truncated filter in (7.36) has the smallest error defined in (7.33) if $M = N/2$.

7.5 Show that the convolution of the two frequency responses in Figs. 7.9(b) and (c) yields the response in Fig. 7.9(a).

7.6 What is the result of the convolution of the responses in Figs. 7.9(a) and (c)? What is its inverse DTFT?

7.7 Consider a desired frequency response $H_d(\omega) = D(\omega)e^{-j(N/2)\omega}$ with $D(\omega)$ given in (7.44) and $\omega_p = 1$ and $\omega_s = 1.2$. Find Nth order FIR filters by truncating its DTFT for $N = 8$, 20, and 60. Plot their magnitude responses. Do they show any Gibbs phenomenon?

7.8 Consider a desired frequency response $H_d(\omega) = D(\omega)e^{-j(N/2)\omega}$ with $D(\omega)$ given in (7.44) and $\omega_p = 1$ and $\omega_s = 1.4$. Use the frequency sampling method to design (a) a fourth-order FIR filter and (b) a fifth-order FIR filter. Do they have a lienar phase?

7.9   Consider a desired frequency response $H_d(\omega) = D(\omega)e^{-j(N/2)\omega}$ with $D(\omega)$ given in
(7.44) and $\omega_p = 1$ and $\omega_s = 1.4$. Use the discrete LS method to design (a) a fourth-
order type I FIR filter to match the three frequencies in $[0, \pi]$ used in Problem 7.8(a),
and (b) a fifth-order type II FIR filter to match the three frequencies in $[0, \pi]$ used
in Problem 7.8(b). Are the results the same as those obtained in Problem 7.8? Can
you conclude that the discrete LS method is preferable to the frequency sampmling
method?

7.10   Design an eighth-order type I FIR filter to match, in the least squares sense, thirteen
frequencies of the $D(\omega)$ in Proboem 7.9. Select five frequencies in the passband,
two frequencies in the transition band, and six frequencies in the stopband. Design a
different eighth-order type I FIR filter to match six frequencies in the passband and
seven frequencies in the stopband; the transition band is disregarded. Which filter is
better?

7.11   Use the discrete LS method to design eighth-order type I and type III filters to
approximate the desired frequency response $H_d(\omega) = D(\omega)e^{-j(N/2)\omega}$ with

$$D(\omega) = \begin{cases} 1 & \text{for } 0 \leq \omega \leq 1 \\ 0 & \text{for } 1.4 \leq \omega \leq \pi \end{cases}$$

where $1 \leq \omega \leq 1.4$ is the "don't care" region. Match frequencies with increment 0.1
in the passband and stopband. Can you use type III filters to design low-pass filters?

7.12   Repeat Problem 7.11 by designing ninth-order type II and type IV FIR filtes.

7.13   Design eighth-order discrete LS optimal type I and type III FIR filters to approximate
the desired high-pass filter $H_d(\omega) = D(\omega)e^{-j4\omega}$ with

$$D(\omega) = \begin{cases} 0 & \text{for } 0 \leq \omega \leq 2 \\ 1 & \text{for } 2.5 \leq \omega \leq \pi \end{cases}$$

Which type yields a better magnitude response? Compare your results with the one
generated using the MATLAB function firls.

7.14   Repeat Problem 7.13 by designing ninth order type II and type IV FIR filters.

7.15   Design eighth-order discrete LS optimal type I and type III FIR filters to approximate
the desired bandpass filter $H_d(\omega) = D(\omega)e^{-j4\omega}$ with

$$D(\omega) = \begin{cases} 0 & \text{for } 0 \leq \omega \leq 1 \\ 1 & \text{for } 1.5 \leq \omega \leq 2 \\ 0 & \text{for } 2.5 \leq \omega \leq \pi \end{cases}$$

Compare your results with the one generated using the MATLAB function firls.

7.16   Repeat Problem 7.15 by designing ninth-order type II and type IV FIR filters.

7.17   Design an eighth-order type I FIR filter to approximate the desired bandstop filter
$H_d(\omega) = D(\omega)e^{-j4\omega}$ with

$$D(\omega) = \begin{cases} 1 & \text{for } 0 \le \omega \le 1 \\ 0 & \text{for } 1.5 \le \omega \le 2 \\ 1 & \text{for } 2.5 \le \omega \le \pi \end{cases}$$

Compare your result with the one generated using the MATLAB function firls. Can
type II, III, and IV filters be used in this design?

7.18   Design a minimax optimal linear-phase FIR filter of order 2 to approximate an ideal
low-pass filter with $\omega_p = 1$ and $\omega_s = 1.2$. Select a weighting factor so that the passband
tolerance is half of the stopband tolerance.

7.19   Repeat Problem 7.18 for a third-order filter.

7.20   Design a third-order differentiator using a rectangular window. Repeat the design using
a Hamming window.

7.21   Design a third-order differentiator by matching $D(\omega) = \omega$ at $\omega = \pi/4, \pi/2, 3\pi/4$.
Can you match them exactly? If not, find the LS matching.

7.22   Design a sixth-order Hilbert transformer using a rectangular window. Repeat the design
using a Hamming window.

7.23   Design a sixth-order Hilbert transformer by matching three frequency samples. Can
you match them exactly? Give your reasons. What is the transformer obtained by
matching at $\omega = 0.1, \pi/2, \pi - 0.1$? What is the transformer obtained by matching at
$\omega = 0.1, 0.8, 1.5$? Which is better? Give your reasons.

7.24   Design a sixth-order low-pass filter with $\omega_c = \pi/2$ and then transform it into a Hilbert
transformer.

7.25   Use MATLAB to design a linear-phase FIR high-pass filter of least order to meet $\omega_s = 2$
rad/s, $\omega_p = 2.6$ rad/s, $R_s = 25$ dB, and $R_p = 2$ dB.

7.26   Select a type of FIR filter, type II or IV, and then use the discrete LS method to design
a fifteenth-order linear-phase FIR filter to approximate the desired magnitude response
shown in Fig. 7.16. Compare your result with the ones generated using firls with and
without the flag "differentiator."

CHAPTER

Design of IIR Filters

8.1 Introduction

This chapter discusses the design of infinite-impulse-response (IIR) digital filters. We study only the subclass of IIR filters that can be described by difference equations of the form shown in (5.17) or (5.18) or, equivalently, proper rational transfer functions of z. Design of digital filters with irrational transfer functions will be difficult, and their implementation will also be costly. Thus discrete transfer functions to be designed are restricted to the form

$$H(z) = \frac{b_1 + b_2 z^{-1} + \cdots + b_{N+1} z^{-N}}{a_1 + a_2 z^{-1} + \cdots + a_{N+1} z^{-N}} \tag{8.1}$$

$$= \frac{b_1 z^N + b_2 z^{N-1} + \cdots + b_{N+1}}{a_1 z^N + a_2 z^{N-1} + \cdots + a_{N+1}} \tag{8.2}$$

In this equation, we require $a_1 \neq 0$ so that $H(z)$ is proper and the resulting filter is causal. We also require $H(z)$ to be stable or, equivalently, all its poles to lie inside the unit circle.

The design problem is then to find an $H(z)$ to meet a given frequency response. If we specify both magnitude and phase responses, it will be difficult, if not impossible, to find an $H(z)$ to meet both responses. Thus in practice, we design an $H(z)$ to meet only one of them, usually the magnitude response. The way to carry out the design is to find an analog filter to meet the corresponding magnitude response and then transform it into a digital filter. We first discuss why we take this approach. Because we will encounter both DT and CT transfer functions, we use $H(z)$ to denote the former and $G(s)$, the latter. Thus the frequency response of analog $G(s)$ is $G(j\omega)$ with $-\infty < \omega < \infty$. As discussed earlier, we may assume $T = 1$ in designing digital filters. Thus the frequency response of digital $H(z)$ is $H(e^{j\omega})$ with $-\pi < \omega \leq \pi$.

8.2 Difficulties in Direct IIR Filter Design

One way to design an IIR digital filter is to find a set of a_i and b_i in (8.1) to meet a desired magnitude response. A simpler method is to find an $H(e^{j\omega})$ or

$$
\begin{aligned}
M(\omega) := |H(e^{j\omega})|^2 &= H(e^{j\omega})H^*(e^{j\omega}) \\
&= H(e^{j\omega})H(e^{-j\omega}) = H(z)H(z^{-1})\big|_{z=e^{j\omega}}
\end{aligned}
\tag{8.3}
$$

to meet the desired squared magnitude response. The search of $M(\omega)$ is relatively simple. Unfortunately, that is not the end of the design. The next step is to find a rational proper and stable $H(z)$ to meet (8.3). Clearly not every $M(\omega)$ can be so factored. Thus the search of $M(\omega)$ must be limited to those that can be factored as in (8.3). The condition for such $M(\omega)$ is complex, and the process of carrying out the factorization is again difficult. Thus the direct design of IIR digital filters is difficult.

Next we discuss the analog counterpart of the preceding design. Consider the analog rational transfer function

$$
G(s) = \frac{B(s)}{A(s)} = \frac{b_1 s^N + b_2 s^{N-1} + \cdots + b_{N+1}}{a_1 s^N + a_2 s^{N-1} + \cdots + a_{N+1}}
\tag{8.4}
$$

The problem is to find a proper stable $G(s)$ to meet a desired squared magnitude response or

$$
\begin{aligned}
\bar{M}(\omega) = |G(j\omega)|^2 &= G(j\omega)G^*(j\omega) = G(j\omega)G(-j\omega) \\
&= G(s)G(-s)\big|_{s=j\omega} = \frac{B(s)B(-s)}{A(s)A(-s)}\bigg|_{s=j\omega}
\end{aligned}
\tag{8.5}
$$

The conditions for $\bar{M}(\omega)$ to be factorizable as in (8.5) turn out to be very simple. A necessary condition is that $\bar{M}(\omega)$ be a proper rational function of ω^2 such as

$$
\bar{M}(\omega) =: M(\omega^2) = \frac{D(\omega^2)}{C(\omega^2)} = \frac{d_0\omega^{2N} + d_1\omega^{2N-2} + \cdots + d_N}{c_0\omega^{2N} + c_1\omega^{2N-2} + \cdots + c_N}
\tag{8.6}
$$

We also require some additional conditions, as we will mention shortly. Because $s = j\omega$, we have $s^2 = -\omega^2$. Let us replace ω^2 by $-s^2$. Then $D(-s^2)$ and $C(-s^2)$ become two even polynomials of s. Clearly, if λ is a root of $D(-s^2)$ or $C(-s^2)$, so is $-\lambda$. Because all coefficients are required to be real, if λ is a root, so is its complex conjugate λ^*. Thus the poles and zeros of $M(-s^2)$ are symmetric with respect to the real and imaginary axes, as shown in Fig. 8.1. Now we require that $M(-s^2)$ has no poles on the $j\omega$ axis and that its zeros on the $j\omega$ axis are of even multiplicities. These can be met if we require $M(\omega^2)$ to be finite and non-negative (zero or positive) for all ω. Under these conditions. we can factor $M(\omega^2)$ as in (8.5). There are many possible ways of grouping $A(s)$ and $B(s)$. However, because the resulting $G(s)$ is required to be stable, $A(s)$ must consist of all the left-half plane poles of $\bar{M}(\omega)$. If we require $G(s)$ to be minimum phase, and if $\bar{M}(\omega)$ has no imaginary

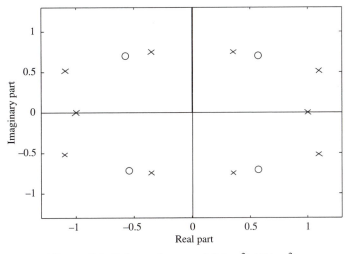

Figure 8.1 Poles and zeros of $D(-s^2)/C(-s^2)$.

zeros, then $B(s)$ must consist of all the left-half plane zeros. In conclusion, the design of an analog filter to meet a desired magnitude response is comparatively straightforward. First we search a rational proper $\bar{M}(\omega)$ whose numerator and denominator are even functions of ω. We compute their roots. By taking their left-half plane poles and zeros, we will obtain a stable analog filter. Once an analog filter is obtained, we can then transform it into a digital filter.

Before proceeding, we mention that the optimization methods discussed for FIR filters in Chapter 7 can also be applied to desigh IIR filters. Let $h_d[n]$, for $n = 0, 1, \cdots, \infty$, be the impulse response of a desired transfer function $H_d(z)$. The transfer function could be irrational or a rational function of a very large degree N_d. The problem is to find a rational function $H(z)$ of degree $N < N_d$ to approximate $H_d(z)$. Let us express $H(z)$ as

$$H(z) = \frac{b_1 z^N + b_2 z^{N-1} + \cdots + b_N z + b_{N+1}}{z^N + a_2 z^{N-1} + \cdots + a_N z + a_{N+1}} = \sum_{n=0}^{\infty} h[n] z^{-n}$$

with $a_1 = 1$. We may try to find a_i and b_i to minimize the squared error

$$\bar{E} = \sum_{n=0}^{\infty} (h_d[n] - h[n])^2 = \frac{1}{2\pi} \int_{\omega=-\pi}^{\pi} \left| H_d(e^{j\omega}) - H(e^{j\omega}) \right|^2 d\omega \qquad (8.7)$$

where we have used (3.53) with $T = 1$. Clearly, the optimal a_i and b_i must meet $\partial \bar{E}/\partial a_i = 0$ and $\partial \bar{E}/\partial b_i = 0$. Unfortunately, the resulting equations are nonlinear and cannot be easily solved. Furthermore, the resulting IIR filter may not be stable. Thus the optimization method cannot be directly applied.

One way to overcome the difficulty is to select a_i and b_i so that $h[n] = h_d[n]$ for $n = 0, 1, \ldots 2N$. Although this can be easily achieved, the resulting IIR filter has the property $h[n] \neq h_d[n]$ for $n > 2N$ and its frequency response may be very different from the frequency response of the desired $H_d(z)$. Furthermore, the resulting IIR flter again may not be stable. Thus this method is generally not acceptable. Another way is to solve a_i, in the least squares sense, from a set of linear algebraic equations, and then to compute b_i. This method, called the *Shanks' method* or *Yule-Walker method* is outside the scope of this text and is extensively discussed in Ref. 12. MATLAB contains the function yulewalk that can be used to design IIR filters. The resulting filters however are not necessarily least squares optimal in the sense of (8.7).

In conclusion, the simplest and most popular method of designing IIR filters is by transforming analog filters, as we will introduce in this chapter.

8.3 Design of Analog Prototype Filters

An analog filter is called a *prototype filter* if it is low-pass with 1 rad/s as its passband, or stopband edge frequency. Once a prototype filter is obtained, low-pass, high-pass, band pass, and bandstop analog filters with any band-edge frequencies can be obtained by frequency transformations. In fact, all frequency selective digital filters can also be obtained directly from analog prototype filters. Thus analog prototype filters are fundamental in filter design.

We will introduce four analog prototype filters: Butterworth, type I Chebyshev, type II Chebyshev, and elliptic. They are available in MATLAB as buttap, cheb1ap, cheb2ap, and ellipap. The last two characters ap stand for *analog prototype*.

Butterworth Analog Prototype Filter The squared magnitude response of the Butterworth analog prototype filter is given by

$$M(\omega^2) = G(j\omega)G(-j\omega) = \frac{1}{1 + \omega^{2N}} \tag{8.8}$$

Its numerator and denominator are even functions of ω. Thus $M(\omega^2)$ can be factored as $G(j\omega)G(-j\omega) = |G(j\omega)|^2$, as we will show shortly, with

$$|G(j\omega)| = \frac{1}{\sqrt{1 + \omega^{2N}}}$$

We plot in Fig. 8.2(a) the function $M(\omega^2)$ and in Fig. 8.2(b) the magnitude of $G(j\omega)$ for $N = 1$ (solid line), 4 (dotted line), and 10 (dashed line). Clearly we have $M(0) = 1$, $M(1) = 1/2$, and $G(0) = 1 = 0$ dB, and $|G(j1)| = 1/\sqrt{2} = 0.707 = -3$ dB. For Butterworth filters, the passband is defined, unless stated otherwise, as the frequency range in which the magnitude lies between 0 and -3 dB. Thus the -3-dB or, more often, 3-dB passband edge frequency is $\omega_p = 1$.

We compute the first derivative of $M(\omega^2)$:

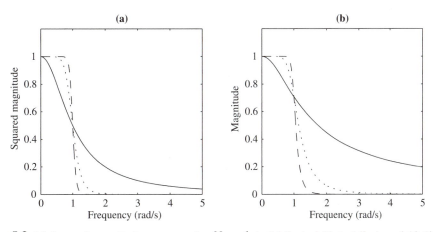

Figure 8.2 (a) Squared magnitude responses for $N = 1$ (solid line), 4 (dotted line), and 10 (dashed line). (b) Corresponding magnitude responses.

$$M^{(1)}(\omega^2) := \frac{dM(\omega^2)}{d\omega} = \frac{-2N\omega^{2N-1}}{(1 + \omega^{2N})^2}$$

which implies $M^{(1)}(0) = 0$; that is, the first derivative of $M(\omega^2)$ is zero at $\omega = 0$. Proceeding forward, we can show $M^{(k)}(0) = 0$, for $k = 1, 2, \ldots, 2N - 1$. The larger k in $M^{(k)}(0) = 0$ is, the flatter the plot of the function $M(\omega^2)$ in the neighborhood of $\omega = 0$. The function in (8.8) has the largest k to have $M^{(k)} = 0$; thus the Butterworth filter is said to be *maximally flat*.

In order to factor $M(-s^2) = 1/[1 + (-s^2)^N]$ as $G(s)G(-s)$, we compute the roots of $1 + (-s^2)^N = 0$ or $(-1)^N s^{2N} = -1$. Multiplying its both sides by $(-1)^N$ yields

$$s^{2N} = -(-1)^N = (-1)^{N+1} = e^{j[(N+1)\pi + 2\pi k]} \quad k = 0, 1, \ldots, 2N - 1$$

Thus the $2N$ roots are located at

$$p_k = e^{j[(N+1)\pi + 2\pi k]/2N} \qquad k = 0, 1, \ldots, 2N - 1 \tag{8.9}$$

as shown in Fig. 8.3(a) for $N = 8$. All of them are located on the unit circle with half of them in the left-half plane and the other half in the right-half plane. Taking those in the left-half plane, we obtain

$$G(s) = \frac{1}{(s - p_0)(s - p_1) \cdots (s - p_{N-1})} \tag{8.10}$$

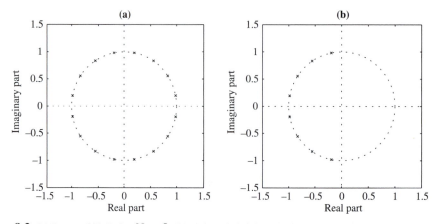

Figure 8.3 (a) Roots of (8.9) for $N = 8$. (b) Poles of eighth-order Butterworth prototype filter.

with its poles shown in Fig. 8.3(b). This is the Nth-order Butterworth low-pass filter with 3-dB passband edge frequency $\omega_p = 1$. Its magnitude responses are shown in Fig. 8.2(b). The MATLAB function [z,p,k]=buttap(N) generates the zeros, poles, and gain of the Nth-order Butterworth prototype filter. Because the filter has no zero, z is empty. The gain k is always 1. The transfer function in (8.10) is said to be in the *zero-pole-gain form*. Typing [b,a]=zp2tf(z,p,k), where zp2tf is an acronym for *zero-pole to transfer function*, yields the transfer function b/a, where b is the numerator and a the denominator. The numerator of a Butterworth filter is always 1; its denominators for $N = 1, 2, 3, 4$ are listed in Table 8.1. We show how the third line is obtained. For $N = 3$, the three roots of (8.9) for $k = 0, 1, 2$ are

$$e^{j4\pi/6} = -0.5 + 0.866j, \quad e^{j6\pi/6} = -1, \quad e^{j8\pi/6} = -0.5 - 0.866j$$

Thus the denominator of the third-order Butterworth prototype filter is

$$(s + 0.5 - 0.866j)(s + 1)(s + 0.5 + 0.866j) = (s + 1)(s^2 + s + 1) = s^3 + 2s^2 + 2s + 1$$

This is the third polynomial in Table 8.1.

Type I Chebyshev Analog Prototype Filter The Nth-order Chebyshev polynomial $V_N(v)$ is defined recursively as

$$V_N(v) = 2v V_{N-1}(v) - V_{N-2}(v)$$

for $N > 2$ with

$$V_1(v) = v \quad \text{and} \quad V_2(v) = 2v^2 - 1$$

From the equation, we can compute

Table 8.1 Butterworth Prototype Filters

$$s + 1$$
$$s^2 + 1.414s + 1$$
$$s^3 + 2.000s^2 + 2.000s + 1$$
$$s^4 + 2.613s^3 + 3.414s^2 + 2.613s + 1$$

$$V_3(v) = 4v^3 - 3v$$
$$V_4(v) = 8v^4 - 8v^2 + 1$$

and so forth. The polynominal $V_N(v)$ is an even function of v if N is even and an odd function if N is odd. Its square, however, is always even. The polynomials have the property $V_N(1) = 1$ for all N. Its value at $v = 0$ is 0 for N odd, and 1 or -1 for N even. The polynomial oscillates between 1 and -1 for $-1 < v < 1$ and grows monotonically to $\pm\infty$ as $|v| \to \infty$. Using this property, we can design a filter that is equiripple in the passband and monotonic in the stopband.

The squared magnitude response of type I Chebyshev prototype filter is given by

$$M(\omega^2) := G(j\omega)G(-j\omega) = \frac{1}{1 + \epsilon^2 V_N^2(\omega)} \tag{8.11}$$

with

$$|G(j\omega)| = \frac{1}{\sqrt{1 + \epsilon^2 V_N^2(\omega)}}$$

The magnitude response (not squared) of $G(j\omega)$ is plotted in Fig. 8.4(a) for $N = 4$ and $\epsilon = 0.776$. Its value at $\omega = 1$ always equals $1/\sqrt{1 + \epsilon^2}$, and its value at $\omega = 0$ is 1 for N odd and $1/\sqrt{1 + \epsilon^2}$ for N even. We see that it is equiripple bounded by 1 and $1/\sqrt{1 + \epsilon^2} = 0.79$ for ω in $[0, 1]$ and decreases monotonically to zero as ω increases from 1 to ∞. Because the denominator of $M(\omega^2)$ is an even function of ω, the poles of $M(-s^2)$ are symmetric with respect to the imaginary axis. Taking those in the left-half plane, we can readily obtain a type I Chebyshev low-pass filter transfer function $G(s)$. For $N = 4$, $G(s)$ has the poles shown in Fig. 8.4(b); it has no zero. Its gain at $\omega = 0$ is 1 for N odd, and can be any value for N even, depending on ϵ.

The passband edge frequency of type I Chebyshev low-pass filters is defined as the largest frequency ω_p at which we have

$$|G(j\omega_p)| = \frac{1}{\sqrt{1 + \epsilon^2}} =: r_p \tag{8.12}$$

Thus the passband is defined as the frequency range in which the magnitude response is bounded between 1 and r_p or between 0 dB and $20 \log_{10} r_p$ dB. We define

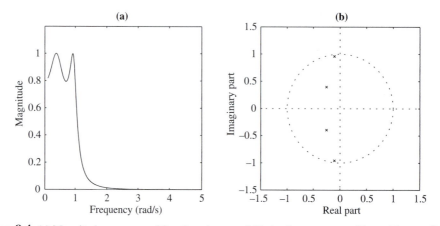

Figure 8.4 (a) Magnitude response of fourth-order type I Chebyshev prototype filter with $\epsilon = 0.776$. (b) Its poles.

$$R_p := -20 \log_{10} r_p \quad \text{or} \quad r_p = 10^{-R_p/20} \tag{8.13}$$

where R_p has the decibel (dB) as its unit and is called the *passband tolerance*. For example, if $R_p = 2$ dB, then $r_p = 10^{-2/20} = 0.79$. If $\omega_p = 1$, it is called a prototype.

The MATLAB function [z,p,k]=cheb1ap(N,Rp) generates the zeros, poles, and gain of an Nth-order type I Chebyshev low-pass filter with $\omega_p = 1$ and passband tolerance R_p dB. Typing [num,den]=zp2tf(z,p,k) yields the transfer function of the filter. As an example, Fig. 8.4 is generated by the program that follows

Program 8.1
```
[z,p,k]=cheb1ap(4,2);
[b,a]=zp2tf(z,p,k);
[H,w]=freqs(b,a);
subplot(1,2,1)
plot(w,abs(H)),title('(a)')
axis([0 5 0 1.2])
axis square
subplot(1,2,2)
splane(b,a),title('(b)')
axis([-1.5 1.5 -1.5 1.5])
axis square
```

The CT frequency response is generated by freqs(b,a) (not freqz), which automatically selects a frequency range in $[0, \infty)$. However, we plot only the frequency range in $[0, 5]$ rad/s using the axis command axis([0 5 0 1.2]). Note that the function splane plots poles and zeros on the s plane and is defined by

```
function splane(b,a)
z=roots(b);
p=roots(a);
plot(real(z),imag(z),'o',real(p),imag(p),'x')
hold on
plot([-5 5],[0 0],[-5 0 0 0 0 5],[0 0 5 -5 0 0])
```
(% plots horizontal and vertical coordinates)

It is not available as an M-file in MATLAB and can be easily created by the user.

Type II Chebyshev Analog Prototype Filter A type I Chebyshev filter is equiripple in the passband and monotonic in the stopband. Now we will design a filter that is monotonic in the passband but equiripple in the stopband. Such a filter is called a type II Chebyshev and can be developed from a type I filter.

Consider the squared magnitude response in (8.11). If ω is replaced by $1/\omega$, then the magnitude response will be as shown in Fig. 8.5(a). Note that the magnitudes at $\omega = 0$, 1, and ∞ in Fig. 8.4(a) become the magnitudes, respectively, at $\omega = \infty$, 1, and 0 in Fig. 8.5(a). The frequency range [0, 1] in Fig. 8.4(a) is stretched to $[1, \infty)$ in Fig. 8.5(a). Subtracting from 1 the magnitude response in Fig. 8.5(a) yields the magnitude response in Fig. 8.5(b). It is monotonic in the passband and equiripple in the stopband. Thus the type II Chebyshev low-pass filter is specified by

$$M(\omega^2) = 1 - \frac{1}{1 + \epsilon^2 V_N^2(1/\omega)} = \frac{\epsilon^2 V_N^2(1/\omega)}{1 + \epsilon^2 V_N^2(1/\omega)} \qquad (8.14)$$

Its numerator and denominator are even functions of ω. Thus $M(-s^2)$ can be factored as $G(s)G(-s)$. Such $G(s)$ has poles and zeros. An Nth-order $G(s)$ has N zeros on the imaginary axis as shown in Fig. 8.5(c) for $N = 4$.

A type II Chebyshev low-pass filter is called a prototype filter if its stopband edge frequency ω_s, defined to be the smallest positive frequency with magnitude

$$|G(j\omega_s)| = \sqrt{\frac{\epsilon^2}{1 + \epsilon^2}} =: r_s$$

equals 1 rad/s. The stopband attenuation is defined as

$$R_s = -20 \log_{10} r_s$$

The MATLAB function [z,p,k]=cheb2ap(N,Rs) generates the zeros, poles, and gain of an Nth-order type II Chebyshev low-pass filter with stopband edge frequency $\omega_s = 1$ and attenuation R_s. Replacing the first line of Program 8.1 by [z,p,k]=cheb2ap(4,15) generates the plots in Figs. 8.5(b) and (c).

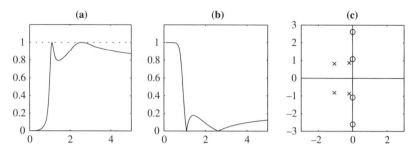

Figure 8.5 (a) Magnitude response. (b) 1-(a). (c) Poles and zeros of fourth-order type II Chebyshev prototype filter.

Elliptic Analog Prototype Filter An elliptic or Cauer analog low-pass filter is equiripple in the passband as well as in the stopband. The squared magnitude response of an elliptic low-pass filter is specified by

$$M(\omega^2) = G(j\omega)G(-j\omega) = \frac{1}{1 + \epsilon^2 Q_N^2(\omega)}$$

where $Q_N(v)$ is a Chebyshev rational function. Its theory is very involved and will not be discussed. We discuss here only the use of the MATLAB function [z,p,k]=ellipap(N,Rp,Rs).

An elliptic low-pass filter is specified by its order N, passband edge frequency ω_p, passband tolerance R_p, stopband edge frequency ω_s, and stopband attenuation R_s. The passband edge frequency is the largest frequency at which the magnitude equals $-R_p$ dB. The stopband edge frequency is the smallest positive frequency at which the magnitude equals $-R_s$ dB. It is a prototype if its passband edge frequency equals 1 or $\omega_p = 1$. For an Nth-order prototype ($\omega_p = 1$), we can specify only two of the remaining three parameters ω_s, R_p, and R_s. If we specify any two, then the resulting elliptic filter will minimize the remaining one. For example, the filter ellipap(N,Rp,Rs) will minimize the stopband edge frequency ω_s or, equivalently, the transition band $\omega_s - \omega_p = \omega_s - 1$. The prototype filter ellipap(4,2,20) has its poles and zeros shown in Fig. 8.6(a) and its magnitude response shown in Fig. 8.6(b). Figures 8.6(c) and (d) show the corresponding plots of ellipap(4,2,40).

To recapitulate, a Butterworth analog low-pass filter with (-3)-dB passband edge frequency $\omega_p = 1$, in rad/s, is called a prototype. A type 1 Chebyshev or elliptic low-pass filter is a prototype if its $-R_p$-dB passband edge frequency equals 1 rad/s. A type II Chebyshev low-pass filter is a prototype if its $-R_s$-dB stopband edge frequency equals 1 rad/s.

8.4 Analog Frequency Transformations

We discussed in the preceding section the design of analog prototype filters. Although the same procedure can be used to design nonprototype filters, it is simpler to do so by using frequency transformations, as we will discuss in this section.

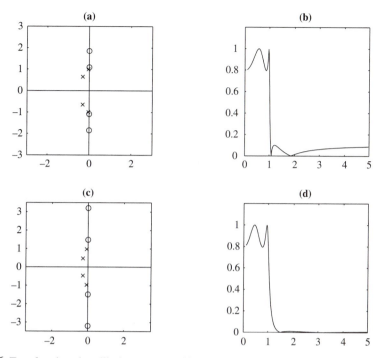

Figure 8.6 Two fourth-order elliptic prototype filters with the same passband tolerance but different stopband tolerances. (a,b) $R_p = 2$ dB, $R_s = 20$ dB. (c,d) $R_p = 2$ dB, $R_s = 40$ dB.

Consider the transformation

$$s = f(\bar{s}) \quad \text{or} \quad j\omega = f(j\bar{\omega}) \tag{8.15}$$

where $f(\bar{s})$ is a rational function of $\bar{s}$ and has the property that $f(j\bar{\omega})$ is pure imaginary for all $\bar{\omega}$. This function maps the $j\bar{\omega}$ axis on the $\bar{s}$ plane into the $j\omega$ axis on the s plane. Because $f(j\bar{\omega})$ is pure imaginary, $\omega = f(j\bar{\omega})/j$ is a real-valued function. Thus a real $\bar{\omega}$ maps into a real ω, and the mapping of the $\bar{\omega}$ axis (horizontal line) into the ω axis (vertical line) can be as shown in Fig. 8.7. This type of mapping need not be a one-to-one mapping.

Suppose we have an analog low-pass filter $G(s)$ with its magnitude response shown on the left-hand side of Fig. 8.7. Substituting $s = f(\bar{s})$ into $G(s)$ yields a new transfer function as

$$\bar{G}(\bar{s}) := G(s)|_{s=f(\bar{s})} = G(f(\bar{s})) \tag{8.16}$$

Its magnitude response can be obtained graphically from that of $G(s)$ as shown. For example, the magnitude of $\bar{G}(j\bar{\omega})$ at $\bar{\omega}_1$ and $\bar{\omega}_1'$ equals the magnitude of $G(j\omega)$ at ω_1 that is measured as a. By this process, the magnitude response of $\bar{G}(\bar{s})$ can be readily obtained. We see that this transformation converts a low-pass filter into a bandpass filter. Note that the form is twisted

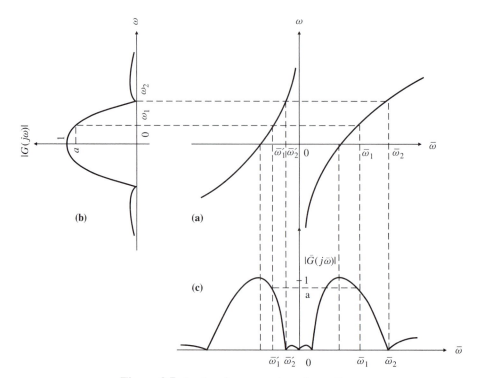

Figure 8.7 Analog frequency transformation.

in the transformation. This type of transformation that changes the coordinate of frequency is called a *frequency transformation*.

Not every rational function can be used as a frequency transformation. If $f(\bar{s})$ is of the form

$$s = f(\bar{s}) = \frac{k(\bar{s}^2 + b_1^2)(\bar{s}^2 + b_2^2) \cdots}{\bar{s}(\bar{s}^2 + a_1^2)(\bar{s}^2 + a_2^2) \cdots} \tag{8.17}$$

with real k, a_i, and b_i, then $f(j\bar{\omega})$ is pure imaginary and can be used as a frequency transformation. Table 8.2 lists a number of frequency transformations. Some of them will be developed next.

Prototype-to-Low-Pass Transformation Consider the two low-pass filters with passband edge frequency $\omega_p = 1$ and $\bar{\omega}_p$ shown in Fig. 8.8. The transformation that achieves this change of bandwidth must transform $\bar{\omega} = 0$ to $\omega = 0$ and $\bar{\omega} = \pm\infty$ to $\omega = \pm\infty$. As a first try, we consider $s = k/\bar{s}$. Clearly, it does not perform the required transformations and cannot be used. Next we try $s = k\bar{s}$ [that is, $s = k(\bar{s}^2 + b_1^2)/\bar{s}$ with $b_1 = 0$], or $\omega = k\bar{\omega}$. It transforms $\bar{\omega} = 0$ to $\omega = 0$ and $\bar{\omega} = \pm\infty$ to $\omega = \pm\infty$ for any positive k. Thus it is a possible candidate for the prototype-to-low-pass transformation. The constant k is to be determined by transforming $\bar{\omega}_p$ to $\omega_p = 1$ or $1 = k\bar{\omega}_p$, which implies $k = 1/\bar{\omega}_p$. This establishes the first transformation in Table 8.2. This transformation maintains the order of the filter. It is possible to find a prototype-to-low-pass

Table 8.2 Analog Frequency Transformations

Prototype filter ($\omega_p = 1$)	$G(s)$
Low-pass with $\bar{\omega}_p$	$s = \bar{s}/\bar{\omega}_p$
High-pass with $\bar{\omega}_p$	$s = \bar{\omega}_p/\bar{s}$
Bandpass with $\bar{\omega}_{pl}$ and $\bar{\omega}_{pu}$	$s = \dfrac{\bar{s}^2 + \bar{\omega}_{pl}\bar{\omega}_{pu}}{\bar{s}(\bar{\omega}_{pu} - \bar{\omega}_{pl})}$
Bandstop with $\bar{\omega}_{sl}$ and $\bar{\omega}_{su}$	$s = \dfrac{\bar{s}(\bar{\omega}_{su} - \bar{\omega}_{sl})\omega_s}{\bar{s}^2 + \bar{\omega}_{sl}\bar{\omega}_{su}}$
Bandstop with $\bar{\omega}_{pl}$ and $\bar{\omega}_{pu}$	$s = \dfrac{\bar{s}(\bar{\omega}_{pu} - \bar{\omega}_{pl})}{\bar{s}^2 + \bar{\omega}_{pu}\bar{\omega}_{pl}}$

transformation that transforms $\bar{\omega}_p$ to $\omega_p = 1$ and $\bar{\omega}_s$ to ω_s, but the transformation will have a degree higher than one. Figure 8.8(c) shows the transformation $\bar{\omega} = \bar{\omega}_p\omega$.

Prototype-to-Bandpass Transformation   Consider the prototype-to-bandpass transformation shown in Fig. 8.9. The transformation must transform $\bar{\omega} = \infty$, $\bar{\omega}_{pu}$, $\bar{\omega}_{pl}$, and 0, respectively, to $\omega = \infty, 1, -1$, and $-\infty$. There are four transformation pairs. We select

$$s = f(\bar{s}) = \frac{k\bar{s}(\bar{s}^2 + b_1^2)}{\bar{s}^2 + a_1^2}$$

or

$$\omega = \frac{k\bar{\omega}(-\bar{\omega}^2 + b_1^2)}{-\bar{\omega}^2 + a_1^2} \tag{8.18}$$

This transforms $\bar{\omega} = \infty$ to $\omega = \infty$ for any parameters $k > 0$, a_1, and b_1, because its numerator has a higher degree in $\bar{\omega}$ than its denominator. The three parameters are to be determined by the remaining three transformation pairs. The transformation from $\bar{\omega} = 0$ to $\omega = -\infty$ requires $a_1 = 0$. Thus the function reduces to

$$s = f(\bar{s}) = \frac{k(\bar{s}^2 + b_1^2)}{\bar{s}} \tag{8.19}$$

or

$$\omega = \frac{-k(-\bar{\omega}^2 + b_1^2)}{\bar{\omega}}$$

Now the transformations from $\bar{\omega}_{pu}$ to 1 and $\bar{\omega}_{pl}$ to -1 implies

$$1 = -\frac{k(-\bar{\omega}_{pu} + b_1^2)}{\bar{\omega}_{pu}} \quad \text{and} \quad -1 = -\frac{k(-\bar{\omega}_{pl} + b_1^2)}{\bar{\omega}_{pl}}$$

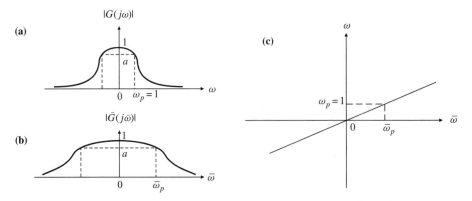

Figure 8.8 Analog prototype-to-low-pass frequency transformation.

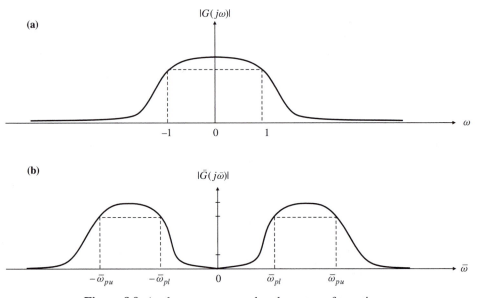

Figure 8.9 Analog prototype-to-bandpass transformation.

Solving these equations yields $b_1^2 = \bar{\omega}_{pu}\bar{\omega}_{pl}$ and $k = 1/(\bar{\omega}_{pu} - \bar{\omega}_{pl})$. Because $\bar{\omega}_{pu} > \bar{\omega}_{pl}$, we have $k > 0$. This establishes the third pair in Table 8.1.

Prototype-to-Bandstop Transformations Consider the prototype-to-bandstop transformation shown in Fig. 8.10. The transformation must transform $\bar{\omega} = 0$ and $\bar{\omega} = \infty$ to $\omega = 0$; thus we select

$$s = f(\bar{s}) = \frac{k\bar{s}}{\bar{s}^2 + a_1^2}$$

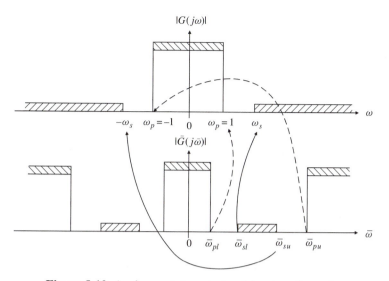

Figure 8.10 Analog prototype-to-bandstop transformation.

as a possible transformation. Next we transform $\bar{\omega}_{sl}$ to ω_s and $\bar{\omega}_{su}$ to $-\omega_s$, as shown in Fig. 8.10 with solid lines. Then we have

$$\omega_s = \frac{k\bar{\omega}_{sl}}{-\bar{\omega}_{sl}^2 + a_1^2} \quad \text{and} \quad -\omega_s = \frac{k\bar{\omega}_{su}}{-\bar{\omega}_{su}^2 + a_1^2}$$

From these we can solve $k = (\bar{\omega}_{su} - \bar{\omega}_{sl})\omega_s$ and $a_1^2 = \bar{\omega}_{su}\bar{\omega}_{sl}$. This establishes

$$s = \frac{\bar{s}(\bar{\omega}_{su} - \bar{\omega}_{sl})\omega_s}{\bar{s}^2 + \bar{\omega}_{sl}\bar{\omega}_{su}} \tag{8.20}$$

This transformation is obtained by transforming stopband edge frequencies to the stopband edge frequencies of a prototype filter. If we select a type II Chebyshev protype filter, then $\omega_s = 1$, and we can use (8.20) directly. If we select a Butterworth prototype filter, then we must compute ω_s. Note that for different prototype filter order, we have different ω_s. Thus we must carry out some computation before using (8.20), as we will discuss in Example 8.1.

A different prototype-to-bandstop transformation can be obtained by transforming $\bar{\omega}_{pl}$ to $\omega_p = 1$ and $\bar{\omega}_{pu}$ to $-\omega_p = -1$, as shown in Fig. 8.10 with dotted lines. The result is

$$s = f(\bar{s}) = \frac{\bar{s}(\bar{\omega}_{pu} - \bar{\omega}_{pl})}{\bar{s}^2 + \bar{\omega}_{pu}\bar{\omega}_{pl}} \tag{8.21}$$

This is the one cited in most analog filter design texts. See, for example, Ref. 31.

Before comparing the two transformations in (8.20) and (8.21), we discuss a number of MATLAB analog filter functions. The function [b,a]=butter(n,wp,'s')

returns the numerator and denominator of an nth-order Butterworth analog low-pass filter with −3 dB passband edge frequency wp in rad/s. Note that the flag "s," which is the Laplace-transform variable, indicates analog.

If wp=[wpl wpu], then [b,a]=butter(n,wp,'s') returns a $2n$-th-order bandpass filter with −3-dB passband edge frequencies wpl and wpu. The function [b,a]=butter(n,wp,'high','s') returns an nth-order analog high-pass filter with −3-dB passband edge frequency wp. The function [b,a]=butter(n,[wpl wpu],'stop','s') returns a $2n$-th-order analog bandstop filter with −3-dB passband edge frequency wpl and wpu.[1] Similar remarks apply to

[b,a]=cheby1(n,Rp,wp,'ftype','s'),
[b,a]=cheby2(n,Rs,ws,'ftype','s'), and
[b,a]=ellip(n,Rp,Rs,wp,'ftype','s').

Note that cheby2 specifies stopband edge frequency and stopband attenuation. In ellip, the stopband edge frequency ws is not specified; the function will yield a minimum ws for the specified n, Rp, Rs, and wp.

◆ **Example 8.1**

Design an analog bandstop filter with attenuation of at least 25 dB between 400 and 600 Hz. The passband tolerance for $\omega \le 100$ Hz and $\omega \ge 900$ Hz must be 3 dB or less.

We first use a MATLAB function to design a Butterworth bandstop filter to meet the specification. Typing

butter(N,[400*2*pi 600*2*pi],'stop','s')

will yield a peculiar filter. See Footnote 1. Typing

[b,a]=butter(N,[100*2*pi 900*2*pi],'stop','s');
[G,w]=freqs(b,a);
plot(w/(2*pi),20*log10(abs(G)))
axis([0 1200 -40 5])

with $N = 3$ generates a sixth-order Butterworth bandstop filter with magnitude response shown in Fig. 8.11(a) with a dotted line. In Fig. 8.11, we also plot with

---

[1] According to MATLAB Version 4 User's Guide, the function designs a $2n$-th-order analog bandstop filter with stopband edge frequencies wsl and wsu. However, this writer is not able to obtain such a filter.

dotted horizontal lines the required passband tolerance and stopband attenuation. We see that the sixth-order filter meets the specification on passband tolerance but not on stopband attenuation. The attenuation between roughly 500 and 600 Hz is less than 25 dB. Next we try $N = 4$. The magnitude response of the resulting eighth-order bandstop filter is plotted in Fig. 8.11(a) with a solid line. Although it meets the specification (except in the immediate neighborhood of 600 Hz), signals with frequency spectrum between roughly 200 and 400 Hz will be severely attenuated. Because the passband edge frequencies equal 100 and 900 Hz, it is believed that the MATLAB function uses the transformation in (8.21).

Next we use the transformation in (8.20) to design a bandstop filter. Before applying (8.20), we must compute the stopband edge frequency ω_s of prototype filters that corresponds to the stopband edge frequencies of the bandstop filter to be designed. For this example, the transformation in (8.20) becomes

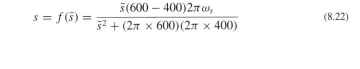

$$s = f(\bar{s}) = \frac{\bar{s}(600 - 400)2\pi\omega_s}{\bar{s}^2 + (2\pi \times 600)(2\pi \times 400)} \tag{8.22}$$

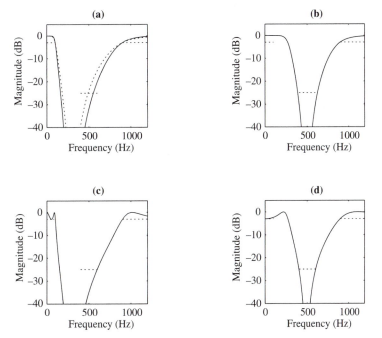

Figure 8.11 (a) Magnitude responses of sixth- (dotted line) and eighth- (solid line) order analog Butterworth bandstop filters obtained using MATLAB. (b) Magnitude responses of sixth-order analog Butterworth bandstop filter obtained using (8.20). (c) Magnitude response of sixth-order type I analog Chebyshev bandstop filter obtained using MATLAB. (d) Magnitude response of fourth-order type I analog Chebyshev bandstop filter obtained using (8.20).

or

$$\omega = f(j\bar{\omega}) = \frac{\bar{\omega}(400\pi)\omega_s}{-\bar{\omega}^2 + 960000\pi^2}$$

where ω_s is the bandstop edge frequency of a Butterworth prototype filter ($\omega_p = 1$). We compute

$$\omega_p' = f(j900 \times 2\pi) = \frac{(900 \times 2\pi)(400\pi)\omega_s}{-(900 \times 2\pi)^2 + 960000\pi^2} = -0.316\omega_s$$

and

$$\omega_p'' = f(j100 \times 2\pi) = \frac{(100 \times 2\pi)(400\pi)\omega_s}{-(100 \times 2\pi)^2 + 960000\pi^2} = 0.087\omega_s$$

Because the passband tolerance is required to be 3 dB or less for $\omega \geq -0.316\omega_s$ and $\omega \leq 0.087\omega_s$ and because the passband of a low-pass filter is symmetric with respect to $\omega = 0$, the passband edge frequency of the prototype filter must be selected as $\omega_p = 0.316\omega_s$. If $\omega_p = 1$, then we have $\omega_s = 1/0.316 = 3.16$ rad/s. In other words, in order for the bandstop filter to meet $R_p = 3$ dB and $R_s = 25$ dB, the Butterworth prototype filter must have attenuation 25 dB or more at the stopband edge frequency $\omega_s = 3.16$. In order to search for such a Butterworth prototype filter, we use (8.8) to find an N to meet

$$20\log_{10}|G(j3.16)| = 10\log_{10}\left|\frac{1}{1 + 3.16^{2N}}\right| \leq -25$$

We compute the left-hand-side term for $N = 1, 2, \ldots$ and find that the smallest N to meet the inequality is $N = 3$. Thus we must use a third- or higher-order Butterworth prototype filter in the transformation.

The least filter order N can also be obtained graphically as shown in Fig. 8.12. Typing

```
[z,p,k]=buttap(N);
[b,a]=zp2tf(z,p,k);
[G,w]=freqs(b,a);
plot(w,20*log10(abs(G)))
hold on
plot([3.16 10],[-25 -25],':')
```

yields the magnitude response for $N = 1$ in Fig. 8.12(a), in which we also plot the -25-dB line for $\omega \geq 3.16$ (dotted horizontal line). We see that the magnitude lies above the horizontal dotted line. Thus $N = 1$ is not large enough. We repeat the plot for $N = 2$ in Fig. 8.12(b). A small section of the magnitude plot lies above the line; thus $N = 2$ is not large enough. We repeat the plot of $N = 3$ in Fig. 8.12(c).

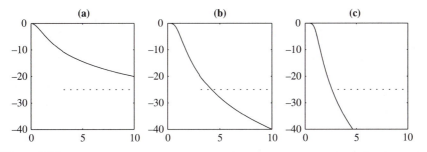

Figure 8.12 Magnitude responses of analog Butterworth prototype filters of order (a) 1, (b) 2, and (c) 3.

The magnitude for $\omega \geq 3.16$ is less than -25 dB; thus the smallest N to meet the specifications is $N = 3$.

The third-order Butterworth prototype filter has the transfer function

$$G(s) = \frac{1}{s^3 + 2s^2 + 2s + 1} \qquad (8.23)$$

Let us write (8.22) as, with $\omega_s = 3.16$,

$$s = f(\bar{s}) = \frac{1264\pi\bar{s}}{\bar{s}^2 + 960000\pi^2} =: \frac{c_1}{c_2}$$

Then the sixth-order Butterworth bandstop filter is given by

$$\bar{G}(\bar{s}) = \frac{c_2^3}{c_1^3 + 2c_1^2 c_2 + 2c_1 c_2^2 + c_2^3}$$

Multiplication of two polynomials is the same as the convolution of their coefficients (see Problem 8.15). Thus the MATLAB program[2] that follows

```
format long
c1=[0 1264*pi 0];c2=[1 0 960000*pi*pi];
c12=conv(c1,c1);c22=conv(c2,c2);
b=conv(c22,c2);
a=conv(c12,c1)+2*conv(c12,c2)+2*conv(c1,c22)+conv(c22,c2);
[Gb,w]=freqs(b,a);
plot(w/(2*pi),20*log10(abs(Gb)))
axis([0 1200 -40 5])
```

---

[2] We use **format long** as the number display format because the coefficients of the filter may spread widely.

computes the numerator and denominator of $\bar{G}(\bar{s})$, computes its frequency response, and plots its magnitude response (in dB) in Fig. 8.11(b) for frequencies between 0 and 1200 Hz. Note that c1 and c2 in the program must be of the same length; otherwise, the four terms in a will be of different lengths, and their sum is not defined. The bandstop filter generated by the program has a degree smaller than the one in Fig. 8.11(a) and yet has a better magnitude response.

We repeat the design by using a type I Chebyshev filter. The MATLAB function

cheby1(3,[100*2*pi 900*2*pi],'stop','s')

returns a sixth-order filter with magnitude response shown in Fig. 8.11(c). Using the transformation in (8.20), we can obtain a fourth-order type I Chebyshev filter with magnitude response shown in Fig. 8.11(d) (Problem 8.4). This shows once again that (8.20) yields a better result than (8.21).

Type II Chebyshev prototype filters are defined to have $\omega_s = 1$ by using stopband attenuation. Thus the MATLAB function

cheby2(3,[400*2*pi 600*2*pi],'stop','s')

will return a satisfactory bandstop filter (Problem 8.5). In conclusion, the transformation in (8.20) will yield a better result than (8.21) unless type II Chebyshev filters are used. However, (8.20) requires the computation of ω_s.

To conclude this section, we mention that the analog frequency transformations are available in MATLAB as lp2lp, lp2hp, lp2bp, and lp2bs. Reading their M-files, we find that they are all carried out using state-space equations. The computation required in state-space equations is fairly complex. For example, lp2hp requires the inversion of square matrices; whereas it can easily be implemented using transfer functions. See Problem 8.7. Thus it is not clear to this writer that state-space equations are definitely better computationally than transfer functions in frequency transformations.

8.5 Impulse Invariance Method

This section introduces the impulse-invariant method that transforms an analog low-pass filter to a digital low-pass filter. Let $G(s)$ be the transfer function of an analog low-pass filter and let $g(t)$ be its impulse response. In other words, $g(t)$ is the inverse Laplace transform of $G(s)$. Let us sample $g(t)$ with sampling period T to yield

$$h[n] = h(nT) = g(t)|_{t=nT} = g(nT) \qquad (8.24)$$

Here we cannot assume $T = 1$. The z-transform of (8.24) is

$$\bar{H}(z) = \sum_{n=0}^{\infty} h[n]z^{-n} = \sum_{n=0}^{\infty} g(nT)z^{-n} \tag{8.25}$$

We will show that if the sampling period T is sufficiently small, then $\bar{H}(z)$ yields a digital low-pass filter whose frequency response $\bar{H}(e^{j\omega T})$ is close to the frequency response of $G(j\omega)$ divided by T in the frequency range $(-\pi/T, \pi/T]$.

Every analog filter is required to be stable; thus its impulse response $g(t)$ is absolutely integrable. Let $F(\omega)$ be the CTFT of $g(t)$ and let $F_d(\omega)$ be the DTFT of $g(nT)$. Then they are related, as derived in (3.67), by

$$F_d(\omega) = \frac{1}{T} \sum_{m=-\infty}^{\infty} F\left(\omega - \frac{2\pi m}{T}\right) \tag{8.26}$$

For stable analog and digital filters, we have $G(j\omega) = F(\omega)$ and $\bar{H}(e^{j\omega T}) = F_d(\omega)$, as shown in (5.97) and (5.73). Note the inclusion of T because we do not assume $T = 1$ here. Thus (8.26) can be replaced by

$$\bar{H}(e^{j\omega T}) = \frac{1}{T} \sum_{m=-\infty}^{\infty} G\left(j\left(\omega - \frac{2\pi m}{T}\right)\right) \tag{8.27}$$

and the discussion in Sections 3.8 and 3.8.1 regarding frequency aliasing due to time sampling is applicable here. Now if T is chosen to be sufficiently small so that

$$|G(j\omega)| \approx 0 \quad \text{for } |\omega| \geq \pi/T \tag{8.28}$$

then we have

$$\bar{H}(e^{j\omega T}) \approx \frac{1}{T} G(j\omega)$$

for $|\omega| \leq \pi/T$. Now let us define

$$H(z) = T\bar{H}(z) \tag{8.29}$$

Then we have $H(e^{j\omega T}) \approx G(j\omega)$, for $|\omega| \leq \pi/T$, and $H(z)$ is called an *impulse-invariant digital filter*. Note that the impulse response sequence of $H(z)$ is not the sample of the impulse response of $G(s)$. It is the sample of the impulse response of $TG(s)$.

Before discussing how to obtain $H(z)$ from $G(s)$, we need the following transform pair

$$\mathcal{L}[e^{\lambda t}] = \frac{1}{s-\lambda} \Leftrightarrow Z[e^{\lambda nT}] = \frac{1}{1-e^{\lambda T}z^{-1}} = \frac{z}{z-e^{\lambda T}} \tag{8.30}$$

where λ can be real or complex. Indeed, we have

$$Z[e^{\lambda nT}] = \sum_{n=0}^{\infty} e^{\lambda nT} z^{-n} = \sum_{n=0}^{\infty} (e^{\lambda T} z^{-1})^n = \frac{1}{1 - e^{\lambda T} z^{-1}}$$

This establishes the pair in (8.30).

The way to obtain an impulse-invariant digital filter from an analog filter $G(s)$ is as follows. Let p_i be the poles of $G(s)$. We assume all p_i to be distinct, as is often the case in practice. We then expand $G(s)$, using partial fraction expansion, as

$$G(s) = \sum_i \frac{r_i}{s - p_i} + k$$

where r_i are called *residues*. The constant k is 0 because $G(s)$ must be strictly proper in order to meet (8.28). Thus the impulse response of $G(s)$ is

$$g(t) = \sum_i r_i e^{p_i t}$$

and its sampled sequence with sampling period T is

$$h[n] = g(nT) = \sum_i r_i e^{p_i nT}$$

The z-transform of $h[n]$ is

$$\bar{H}(z) = Z\left[\sum_i r_i e^{p_i nT}\right] = \sum_i \frac{r_i}{1 - e^{p_i T} z^{-1}}$$

Thus the transfer function of the impulse-invariant digital filter is

$$H(z) = T\left[\sum_i \frac{r_i}{1 - e^{p_i T} z^{-1}}\right] \tag{8.31}$$

The preceding steps can be carried out in MATLAB as follows:

```
[r,p,k]=residue(b,a); [% Perform partial fraction expansion of G(s) = B(s)/A(s)]
[bz,az]=residuez(r,exp(p/Fs),k); [% Compute H̄(z)]
bz=bz/Fs; [% Multiply H̄(z) by T = 1/fₛ]
```

Then the impulse-invariant digital filter is given by $H(z) = B(z)/A(z)$. The preceding three steps are combined in MATLAB to form the function

```
[bz,az]=impinvar(b,a,Fs)
```

Its employment is very simple.

◆ **Example 8.2**

Transform, by using the impulse invariance method, a third-order Butterworth analog low-pass filter to a digital low-pass filter with 3-dB passband edge frequency 5 Hz. The sampling frequency is assumed to be 50 Hz. We use two methods to carry out the design.

Method I

The positive frequency range of the digital filter, for $T = 1/50$, is $[0, \pi/T] = [0, 50\pi]$ in rad/s. The passband edge frequency is $5 \times 2\pi = 10\pi$ rad/s. If we transform a Butterworth analog filter with passband edge frequency 10π, then we will obtain a digital low-pass filter with passband edge frequency 10π in $[0, 50\pi]$. A Butterworth analog low-pass filter with $\bar{\omega}_p = 10\pi$ can be obtained from the following prototype filter

$$G(s) = \frac{1}{s^3 + 2s^2 + 2s + 1} \tag{8.32}$$

by the transformation $s = \bar{s}/\bar{\omega}_p = \bar{s}/10\pi$. Thus we have

$$\bar{G}(\bar{s}) = \frac{1}{\left(\frac{\bar{s}}{10\pi}\right)^3 + 2 \times \left(\frac{\bar{s}}{10\pi}\right)^2 + 2 \times \left(\frac{\bar{s}}{10\pi}\right) + 1}$$

$$= \frac{1000\pi^3}{\bar{s}^3 + 20\pi\bar{s}^2 + 200\pi^2\bar{s} + 1000\pi^3}$$

Typing

b=1000*pi^3;a=[1 20*pi 200*pi^2 1000*pi^3];
[bz,az]=impinvar(b,a,50)

yields the impulse-invariant digital low-pass filter

$$H(z) = \frac{0.0797z^{-1} + 0.0526z^{-2}}{1 - 1.7833z^{-1} + 1.2003z^{-2} - 0.2846z^{-3}} \tag{8.33}$$

Method II

In this method, the frequency range of the digital filter will be normalized to $[0, \pi]$. The original frequency range is $[0, \pi/T] = [0, 50\pi]$ in rad/s, for $T = 1/50$, with passband edge frequency $5 \times 2\pi = 10\pi$ rad/s. Dividing all frequencies by 50, we obtain the frequency range $[0, \pi]$ with passband edge frequency $10\pi/50 = \pi/5$. This is the same as normalizing the sampling period to 1.

Now we need a Butterworth analog low-pass filter with passband edge frequency $\pi/5$. This can be obtained by using the transformation $s = 5\bar{s}/\pi$ as

$$\bar{G}(\bar{s}) = \frac{1}{\left(\frac{5\bar{s}}{\pi}\right)^3 + 2 \times \left(\frac{5\bar{s}}{\pi}\right)^2 + 2 \times \left(\frac{5\bar{s}}{\pi}\right) + 1}$$

$$= \frac{\pi^3}{125\bar{s}^3 + 50\pi\bar{s}^2 + 10\pi^2\bar{s} + \pi^3}$$

Typing

```
b=pi^3;a=[125 50*pi 10*pi^2 pi^3];
[bz,az]=impinvar(b,a,1)
```

yields the same digital filter in (8.33). In fact, if we normalize the frequency range to [0, 1], we will still obtain the same digital filter.

Typing [H,w]=freqz(bz,az);plot(w,abs(H)), we will obtain the magnitude response of the digital filter in the frequency range [0, π]. Typing plot(w./T,abs(H)), we will obtain the magnitude response in [0, π/T].

From the preceding example, we see that we will obtain the same digital filter whether or not the sampling period is normalized to 1. Thus, from now on, we will normalize the specification of digital filters to the frequency range [0, π] before carrying out the design. This is consistent with our standing assumption of $T = 1$ after Chapter 5.

The condition in (8.28) is needed to avoid frequency aliasing in the impulse-invariant method. Thus the method cannot be used to design high-pass and bandstop filters. Type II Chebyshev and elliptic low-pass filters have ripples in the stopband that do not approach zero as $\omega \to \infty$. Thus the impulse invariance method should be applied only to Butterworth and type I Chebyshev low-pass filters.

The impulse invariance method transforms poles of analog filters into poles of digital filters by the transformation

$$z = e^{sT} \tag{8.34}$$

See (8.30) with $T = 1$. This transformation maps the left-half s plane into the interior of the unit circle on the z plane, as shown in Fig. 5.11 with $T = 1$. Thus all stable poles of $G(s)$ are mapped into stable poles on the z plane. Thus $\bar{H}(z)$ and $H(z)$ are always stable. Although the poles of $H(z)$ are the poles of $G(s)$ transformed by $z = e^{sT}$, the zeros of $H(z)$ have no relationship with the zeros of $G(s)$. For example, $G(s)$ in (8.32) has no zero, but its transformed $H(z)$ has two zeros, as shown in (8.33).

To conclude this section, we mention two related methods of transforming an analog filter to a digital one. The first method maps the poles and zeros of $G(s)$, using $z = e^{sT}$, to yield a digital filter. For example, if

$$G(s) = \frac{\prod_i (s - z_i)}{\prod_j (s - p_j)}$$

Then the transformed digital filter is given by

$$H(z) = \frac{\prod_i(z - e^{z_i T})}{\prod_j(z - e^{p_j T})}$$

This is called the *pole/zero mapping* or *matched-z transformation*. The second method is to design a digital filter so that its step response equals the sample of the step response of $TG(s)$. The design can be carried out by computing the impulse-invariant digital filter of $G(s)/s$. Dividing the resulting digital filter by $(1 - z^{-1})$ yields the step-invariant digital filter. This is called the *step-invariance* method. This and the pole/zero mapping may not yield satisfactory results and are rarely used in practice.

8.5.1 Digital Frequency Transformations[3]

The impulse-invariant method can be used to design digital low-pass filters but not digital high-pass or bandstop filters. However, once having a digital low-pass filter, we can obtain other types of digital filters by digital frequency transformations. This is the digital counterpart of Section 8.4.

Consider the transformation

$$z = f(\bar{z}) \quad \text{or} \quad e^{j\omega} = f(e^{j\bar{\omega}})$$

The function $f(\bar{z})$ is required to have the property $|f(e^{j\bar{\omega}})| = 1$ for all $\bar{\omega}$. Such a function is called a *unit function*. A unit function maps the unit circle on the $\bar{z}$ plane into the unit circle on the z plane or $-\pi < \bar{\omega} \le \pi$ into $-\pi < \omega \le \pi$, as shown Fig. 8.13(a). It is a two-to-one mapping. We see that both $\bar{\omega}_{pl}$ and $-\bar{\omega}_{pu}$ on the $\bar{\omega}$ axis are mapped into ω_p on the ω axis. Both $\bar{\omega} = 0$ and $\bar{\omega} = \pi$ are mapped into $\omega = 0$.

Consider the digital transfer function $H(z)$ with the magnitude response shown in Fig. 8.13(b). Define

$$\bar{H}(\bar{z}) := H(z)|_{z=f(\bar{z})} = H(f(\bar{z}))$$

or

$$\bar{H}(e^{j\bar{\omega}}) = H(f(e^{j\bar{\omega}}))$$

Then the magnitude response of $\bar{H}(\bar{z})$ can be obtained from that of $H(z)$, as shown in Fig. 8.13(c). It is obtained point by point. We see that the transformation converts a digital low-pass filter into a digital bandstop filter. This type of transformation is called a *digital frequency transformation*.

A unit function has the property $|f(e^{j\bar{\omega}})| = 1$, for all $\bar{\omega}$. The all-pass filter in (6.27) or (6.28) has the same property. The difference is that all-pass filters must be stable. Here we do not

---

[3] This subsection may be skipped without loss of continuity.

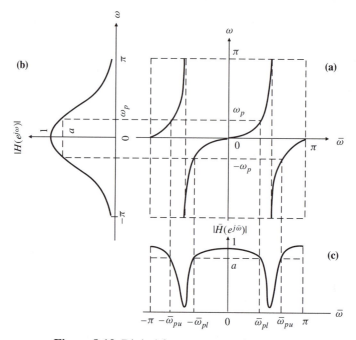

Figure 8.13 Digital frequency transformation.

require $f(\bar{z})$ to be stable, but we do require the resulting $H(f(\bar{z}))$ to be stable. Now we will use some special cases of (6.27) to develop digital frequency transformations.

Low-pass-to-Low-Pass Transformation Consider the two low-pass digital filters shown in Figs. 8.14(a) and (b): one with bandwidth ω_p, the other $\bar{\omega}_p$. The problem is to find a unit function to achieve this change of passband edge frequencies.

Consider the transformation

$$z = \frac{\bar{z} + c}{c\bar{z} + 1} := f_l(\bar{z}) \tag{8.35}$$

or

$$e^{j\omega} = \frac{e^{j\bar{\omega}} + c}{ce^{j\bar{\omega}} + 1}$$

It is straightforward to verify that f_l maps $\bar{\omega} = 0$ into $\omega = 0$, and $\bar{\omega} = \pm\pi$ into $\omega = \pm\pi$. Now if c is chosen so that f_l maps $\bar{\omega}_p$ into ω_p, then $f_l(\bar{z})$ is the desired transformation. We equate

$$e^{j\omega_p} = \frac{e^{j\bar{\omega}_p} + c}{ce^{j\bar{\omega}_p} + 1}$$

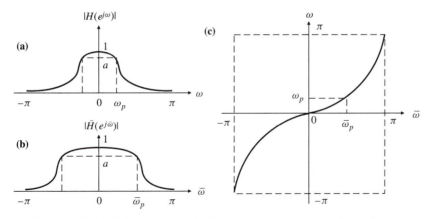

Figure 8.14 Digital low-pass-to-low-pass frequency transformation.

which implies

$$c = \frac{e^{j\bar{\omega}_p} - e^{\omega_p}}{e^{j(\omega_p + \bar{\omega}_p)} - 1}$$

$$= \frac{e^{j(\omega_p + \bar{\omega}_p)/2}(e^{j(\bar{\omega}_p - \omega_p)/2} - e^{-j(\bar{\omega}_p - \omega_p)/2})}{e^{j(\omega_p + \bar{\omega}_p)/2}(e^{j(\bar{\omega}_p + \omega_p)/2} - e^{-j(\bar{\omega}_p + \omega_p)/2})}$$

$$= \frac{\sin[(\bar{\omega}_p - \omega_p)/2]}{\sin[(\bar{\omega}_p + \omega_p)/2]} \tag{8.36}$$

With this c, the unit function $z = f_l(\bar{z})$ will transform a digital low-pass filter with passband edge frequency ω_p into a different digital low-pass filter with passband edge frequency $\bar{\omega}_p$. Figure 8.14(c) shows the transformation between ω and $\bar{\omega}$.

Low-Pass-to-High-Pass Transformation Consider the unit function

$$z = -\frac{\bar{z} + c}{c\bar{z} + 1} := f_h(\bar{z}) \tag{8.37}$$

or

$$e^{j\omega} = -\frac{e^{j\bar{\omega}} + c}{ce^{j\bar{\omega}} + 1} \tag{8.38}$$

This function maps $\bar{\omega} = 0$ into $\omega = \pm\pi$, and $\bar{\omega} = \pm\pi$ into $\omega = 0$, as shown in Fig. 8.15. If $\bar{\omega} = \bar{\omega}_p$ is mapped into $\omega = -\omega_p$, then we have

$$e^{-j\omega_p} = -\frac{e^{j\bar{\omega}_p} + c}{ce^{j\bar{\omega}_p} + 1}$$

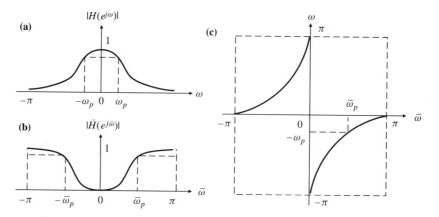

Figure 8.15 Digital low-pass-to-high-pass frequency transformation.

Simple manipulation yields

$$c = -\frac{\cos[(\bar{\omega}_p + \omega_p)/2]}{\cos[(\bar{\omega}_p - \omega_p)/2]} \qquad (8.39)$$

Figure 8.15(c) shows the relationship between ω and $\bar{\omega}$ in (8.38). The transformation in (8.37) with c in (8.39) will transform a digital low-pass filter into a digital high-pass filter.

◆ Example 8.3

Design a digital third-order high-pass filter with 3-dB passband edge frequency 35 Hz and sampling frequency 100 Hz from the digital low-pass filter in (8.33).

The digital low-pass filter in (8.33) has 3-dB passband edge frequency 0.2π in $[0, \pi]$. The digital high-pass filter to be designed has 3-dB passband edge frequency $35 \times 2\pi$ rad/s in the frequency range $[0, \pi/T] = [0, f_s\pi] = [0, 100\pi]$ in rad/s, or $35 \times 2\pi/100 = 0.7\pi$ in $[0, \pi]$. First we compute c in (8.39):

$$c = -\frac{\cos[(\bar{\omega}_p + \omega_p)/2]}{\cos[(\bar{\omega}_p - \omega_p)/2]} = -\frac{\cos[(0.7\pi + 0.2\pi)/2]}{\cos[(0.7\pi - 0.2\pi)/2]} = -0.2212$$

Thus the required frequency transformation is

$$z = \frac{-(\bar{z} - 0.2212)}{-0.2212\bar{z} + 1} = \frac{\bar{z} - 0.2212}{0.2212\bar{z} - 1} =: \frac{d_1}{d_2} \qquad (8.40)$$

Note that this unit function has a pole at $1/0.2212$, which is outside the unit circle; thus it is not stable. As mentioned earlier, the stability of the unit function is immaterial. The only concern is the stability of the resulting high-pass filter.

We write (8.33) in the positive power form by multiplying its numerator and denominator by z^3 to yield

$$H(z) = \frac{0.0797z^2 + 0.0526z}{z^3 - 1.7833z^2 + 1.2003z - 0.2846}$$

Substituting (8.40) into it yields

$$\bar{H}(\bar{z}) = \frac{0.0797(d_1/d_2)^2 + 0.0526(d_1/d_2))}{(d_1/d_2)^3 - 1.7833(d_1/d_2)^2 + 1.2003(d_1/d_2) - 0.2846}$$

$$= \frac{0.0797d_1^2d_2 + 0.0526d_1d_2^2}{d_1^3 - 1.7833d_1^2d_2 + 1.2003d_1d_2^2 - 0.2846d_2^3} \tag{8.41}$$

where we have multiplied its numerator and denominator by d_2^3. The computation of $\bar{H}(\bar{z})$ by hand is formidable. However, if we define

d1=[1 -0.2212];d2=[0.2212 -1];
d12=conv(d1,d1);d22=conv(d2,d2);

then the denominator of $\bar{H}(\bar{z})$ can be obtained as

a=conv(d12,d1)-1.7833*conv(d12,d2)+1.2003*conv(d1,d22)···
    -0.2846*conv(d22,d2)

The numerator can be similarly obtained. The final result is

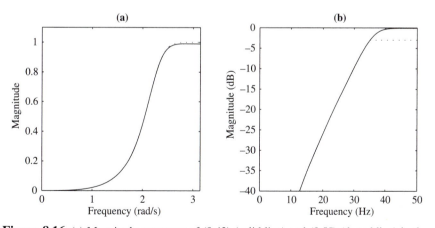

Figure 8.16 (a) Magnitude responses of (8.42) (solid line) and (8.55) (dotted line) in the frequency range $[0, \pi]$. (b) Magnitude response (in dB) of (8.42) in the frequency range $[0, 50]$ (in Hz).

$$\bar{H}(\bar{z}) = \frac{0.0202\bar{z}^3 - 0.1113\bar{z}^2 + 0.0939\bar{z} - 0.0155}{0.6612\bar{z}^3 + 0.7920\bar{z}^2 + 0.4675\bar{z} + 0.0955} =: \frac{\bar{B}(\bar{z})}{\bar{A}(\bar{z})} \tag{8.42}$$

Typing **roots(a)** yields the poles of $\bar{H}(\bar{z})$ as $-0.3538, -0.4228 \pm 0.4799i$. They all lie inside the unit circle. Thus the filter is stable and is the desired digital high-pass filter.

Typing **[H,w]=freqz(b,a);plot(w,abs(H))** yields the magnitude response in Fig. 8.16(a) with a solid line. The function **freqz** selects automatically 512 frequencies in the range $[0, \pi)$. However, the range can be changed to any frequency range, for example, to $[0, 100\pi)$ in rad/s or $[0, 50)$ in Hz. Typing **plot(w*50/pi, 20*log10(abs(H)))** yields the plot in Fig. 8.16(b) with the frequency range in $[0, 50)$ in Hz and with the magnitude in dB. We plot also the -3-dB line from 35 to 50 Hz. The filter indeed meets the specification.

Low-pass-to-Bandpass Transformation The unit function required in the low-pass-to-bandpass transformation must map $\bar{\omega}_{pu}$ into ω_p and $\bar{\omega}_{pl}$ into $-\omega_p$ as shown in Fig. 8.17(c). Thus the unit function must have two parameters or more. Consider

$$z = f_b(\bar{z}) = -\frac{\bar{z}^2 + c_1\bar{z} + c_2}{c_2\bar{z}^2 + c_1\bar{z} + 1} \tag{8.43}$$

or

$$e^{j\omega} = -\frac{e^{j2\bar{\omega}} + c_1 e^{j\bar{\omega}} + c_2}{c_2 e^{j2\bar{\omega}} + c_1 e^{j\bar{\omega}} + 1} \tag{8.44}$$

This function maps $\bar{\omega} = \pi$ into $\omega = \pi$ and $\bar{\omega} = 0$ into $\omega = -\pi$. If the function maps $\bar{\omega}_{pu}$ into ω_p and $\bar{\omega}_{pl}$ into $-\omega_p$, then we have

$$e^{j\omega_p} = -\frac{e^{j2\bar{\omega}_{pu}} + c_1 e^{j\bar{\omega}_{pu}} + c_2}{c_2 e^{j2\bar{\omega}_{pu}} + c_1 e^{j\bar{\omega}_{pu}} + 1}$$

and

$$e^{-j\omega_p} = -\frac{e^{j2\bar{\omega}_{pl}} + c_1 e^{j\bar{\omega}_{pl}} + c_2}{c_2 e^{j2\bar{\omega}_{pl}} + c_1 e^{j\bar{\omega}_{pl}} + 1}$$

From these equations, we can solve the two unknowns c_1 and c_2 as

$$c_1 = -2d\frac{e}{e+1} \qquad c_2 = \frac{e-1}{e+1} \tag{8.45}$$

with

$$e = \cot\left[(\bar{\omega}_{pu} - \bar{\omega}_{pl})/2\right]\tan(\omega_p/2); \qquad d = \frac{\cos[(\bar{\omega}_{pu} + \bar{\omega}_{pl})/2]}{\cos[(\bar{\omega}_{pu} - \bar{\omega}_{pl})/2]} \tag{8.46}$$

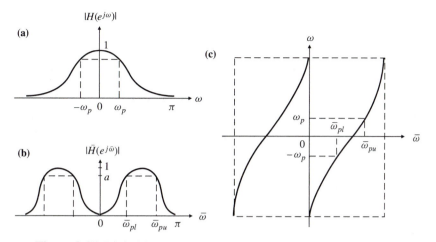

Figure 8.17 Digital low-pass-to-bandpass frequency transformation.

The mapping of $\bar{\omega}$ into ω by (8.44) with c_1 and c_2 in (8.45) and (8.46) is plotted in Fig. 8.17(c).

The derivation of the low-pass-to-bandstop transformation is similar and will not be repeated. We list in Table 8.3 various digital frequency transformations. In using the table, we have freedom in choosing ω_p and ω_s. Thus they may be chosen to simplify the computation. For example, in using the low-pass-to-bandpass or low-pass-to-bandstop transformation, we may choose $\omega_p = \pi/2$ so that $\tan(\omega_p/2) = 1$ or choose ω_p so that $e = 1$. We mention that if a filter obtained by a digital frequency transformation is not stable, it can be stabilized by replacing unstable poles by their reciprocal poles. By so doing, the magnitude response will not be affected. See Problem 6.16.

Table 8.3 Digital Frequency Transformations

Low-pass to low-pass	$z = \dfrac{\bar{z}+c}{c\bar{z}+1}$	$c = \dfrac{\sin[(\bar{\omega}_p - \omega_p)/2]}{\sin[(\bar{\omega}_p - \omega_p)/2]}$
Low-pass to high-pass	$z = -\dfrac{(\bar{z}+c)}{c\bar{z}+1}$	$c = -\dfrac{\cos[(\bar{\omega}_p + \omega_p)/2]}{\cos[(\bar{\omega}_p - \omega_p)/2]}$
Low-pass to bandpass	$z = -\dfrac{(\bar{z}^2 + c_1\bar{z} + c_2)}{c_2\bar{z}^2 + c_1\bar{z} + 1}$	$d = \dfrac{\cos[(\bar{\omega}_{pu} + \bar{\omega}_{pl})/2]}{\cos[(\bar{\omega}_{pu} - \bar{\omega}_{pl})/2]}$
		$e = \cot[(\bar{\omega}_{pu} - \bar{\omega}_{pl})/2]\,\tan(\omega_p/2)$
		$c_1 = \dfrac{-2de}{e+1} \quad c_2 = \dfrac{e-1}{e+1}$
Low-pass to bandstop	$z = \dfrac{\bar{z}^2 + c_1\bar{z} + c_2}{c_2\bar{z}^2 + c_1\bar{z} + 1}$	$d = \dfrac{\cos[(\bar{\omega}_{su} + \bar{\omega}_{sl})/2]}{\cos[(\bar{\omega}_{su} - \bar{\omega}_{sl})/2]}$
		$e = \tan[(\bar{\omega}_{su} - \bar{\omega}_{sl})/2]\,\tan(\omega_s/2)$
		$c_1 = \dfrac{-2d}{e+1} \quad c_2 = \dfrac{1-e}{1+e}$

8.6 Bilinear Transformation

This section introduces a different method of transforming an analog filter into a digital filter. Before proceeding, we introduce some notations. Up to this point, we have been using ω to denote frequency in analog and digital filters. The transformation to be introduced in this section will compress the analog frequency in $[0, \infty)$ into the digital frequency in $[0, \pi]$. Thus we must use different notations to denote them. We use ω and $\bar{\omega}$ to denote digital frequencies and Ω and $\bar{\Omega}$ to denote analog frequencies. We also reserve Ω and s to be associated exclusively with analog prototype filters.

Consider the function

$$\bar{s} = k \frac{z-1}{z+1} \tag{8.47}$$

where k is a positive real constant. Equation (8.47) implies $\bar{s}(z+1) = kz - k$ or $z(k-\bar{s}) = k+\bar{s}$. Thus we have

$$z = \frac{k+\bar{s}}{k-\bar{s}} \tag{8.48}$$

This is called a *bilinear transformation*. For each $\bar{s}$, we can compute a unique z from (8.48), and for each z, we can compute a unique $\bar{s}$ from (8.47). Thus the transformation is one to one. We now show that (8.47) or (8.48) maps the imaginary axis of the $\bar{s}$ plane onto the unit circle of the z plane, and the left-half $\bar{s}$ plane onto the interior of the unit circle on the z plane. Indeed, substituting $\bar{s} = \delta + j\bar{\Omega}$ into (8.48) yields

$$z = \frac{k + \delta + j\bar{\Omega}}{k - \delta - j\bar{\Omega}}$$

If $\delta = 0$ (imaginary axis on the $\bar{s}$ plane), then $|z| = \sqrt{k^2 + \bar{\Omega}^2}/\sqrt{k^2 + (-\bar{\Omega})^2} = 1$ (unit circle on the z plane). Because $k > 0$, if $\delta < 0$ (left-half $\bar{s}$ plane), then $|k - \delta| > |k + \delta|$, which implies

$$|(k-\delta)^2 + (-\bar{\Omega})^2| > |(k+\delta)^2 + \bar{\Omega}^2|$$

and $|z| < 1$ (interior of the unit circle on the z plane). This establishes the assertion. Note that if $k < 0$, the left-half $\bar{s}$ plane will be mapped into the exterior of the unit circle on the z plane. Thus we require $k > 0$.

Next we establish the relationship between the frequency in $\bar{s} = j\bar{\Omega}$ and the frequency in $z = e^{j\omega}$. The frequency range of $\bar{\Omega}$ is $(-\infty, \infty)$ and the frequency range of ω is $(-\pi, \pi]$. From (8.47), we have

$$j\bar{\Omega} = k \frac{e^{j\omega} - 1}{e^{j\omega} + 1} = k \frac{e^{j\omega/2}(e^{j\omega/2} - e^{-j\omega/2})}{e^{j\omega/2}(e^{j\omega/2} + e^{-j\omega/2})}$$

$$= k \frac{2j \sin(\omega/2)}{2 \cos(\omega/2)} = jk \tan(\omega/2)$$

which implies

$$\bar{\Omega} = k \tan(\omega/2) \tag{8.49}$$

and

$$\omega = 2 \tan^{-1}(\bar{\Omega}/k) \tag{8.50}$$

We show a typical relationship between ω and $\bar{\Omega}$ in Fig. 8.18(b). We see that it is not a linear mapping. The frequency range from 0 to ∞ in the analog case is compressed into the frequency range from 0 to π in the digital case. This is called the *frequency warping*.

Consider an analog filter with transfer function $\bar{G}(\bar{s})$. If $\bar{s}$ in $\bar{G}(\bar{s})$ is replaced by $\bar{s} = k(z-1)/(z+1)$, then we obtain

$$H(z) := \bar{G}(\bar{s})\big|_{\bar{s}=k(z-1)/(z+1)} = \bar{G}\left(\frac{k(z-1)}{z+1}\right) \tag{8.51}$$

This is a DT transfer function. Because the bilinear transformation maps stable poles of $\bar{G}(\bar{s})$ into stable poles of $H(z)$, if $\bar{G}(\bar{s})$ is stable, so is $H(z)$. Because of (8.49), the frequency response of $H(z)$ is related to the frequency response of $\bar{G}(\bar{s})$ by

$$H(e^{j\omega}) = \bar{G}(j\bar{\Omega})\big|_{\bar{\Omega}=k \tan(\omega/2)}$$

Thus the magnitude response of $H(z)$ can be obtained graphically from the magnitude response of $\bar{G}(\bar{s})$, as shown in Fig. 8.18, in which an analog high-pass filter is mapped into a digital high-pass filter. We see that there is no aliasing problem as in the impulse invariance method. However, the passband $[\bar{\Omega}_p, \infty)$ in the analog filter is compressed or warped into the passband $[\omega_p, \pi]$ in the digital filter as shown. Because of frequency warping, before applying a bilinear transformation, we must unwarp band-edge frequencies. That is, to design a digital filter with band-edge frequency ω_p, we must find an analog filter with band-edge frequency $\bar{\Omega}_p$ computed from (8.49).

The bilinear transformation in (8.47) holds for any positive real k. It is natural to ask which k to use. Intuitively, warping and unwarping will cancel out; therefore, we should obtain the same result no matter which k is used. To be more specific, suppose we want to design a digital low-pass filter $H(z)$ with passband edge frequency ω_p. It is to be transformed from an analog filter $\bar{G}(\bar{s})$ with passband edge frequency $\bar{\Omega}_p = k \tan(\omega_p/2)$. The analog filter $\bar{G}(\bar{s})$ in turn will be obtained from an analog prototype filter $G(s)$ by the analog frequency transformation $s = \bar{s}/\bar{\Omega}_p$. Thus we have

$$H(z) = \bar{G}(\bar{s})\big|_{\bar{s}=k(z-1)/(z+1)}$$

$$= G(s)\big|_{s=\bar{s}/\bar{\Omega}_p=(z-1)/[(z+1)\tan(\omega_p/2)]}$$

which is independent of k. In other words, we will obtain the same digital filter no matter which k is used. To simplify computation, we select $k = 1$ and use

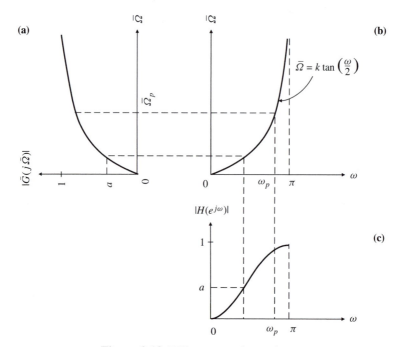

Figure 8.18 Bilinear transformation.

$$\bar{s} = \frac{z - 1}{z + 1} \quad \text{and} \quad \bar{\Omega} = \tan(\omega/2) \tag{8.52}$$

in the design.

◆ Example 8.4

Design a third-order high-pass digital filter with 3-dB passband edge frequency 35 Hz and sampling frequency 100 Hz.

The frequency range for $T = 1/f_s = 1/100$ is $[0, \pi/T] = [0, 100\pi]$. First we normalize it to $[0, \pi]$ by dividing all frequencies by 100. Thus the passband edge frequency becomes $\omega_p = 35 \times 2\pi/100 = 0.7\pi$. We use two approaches to carry out the design.

Method I: Analog Prototype⇒Analog High-Pass ⇒Digital High-Pass
The first arrow uses an analog frequency transformation and the second arrow uses the bilinear transformation. The first step in design is to translate the specification from the digital high-pass to an analog high-pass. First we use (8.52) to compute

$$\bar{\Omega}_p = \tan(\omega_p/2) = \tan(0.7\pi/2) = \tan 0.35\pi = 1.9626$$

Thus we must design an analog high-pass filter with passband edge frequency 1.9626 rad/s, which will be obtained from an analog prototype filter by using an analog frequency transformation.

We select to use Butterworth filters. The third-order Butterworth prototype filter $(\Omega_p = 1)$ is

$$G(s) = \frac{1}{s^3 + 2s^2 + 2s + 1} \tag{8.53}$$

Using the analog frequency transformation $s = \bar{\Omega}/\bar{s} = 1.9626/\bar{s}$ in Table 8.2, we can obtain the following analog high-pass filter with $\bar{\Omega}_p = 1.9626$:

$$\bar{G}(\bar{s}) = \frac{1}{(1.9626/\bar{s})^3 + 2(1.9626/\bar{s})^2 + 2(1.9626/\bar{s}) + 1}$$

$$= \frac{\bar{s}^3}{\bar{s}^3 + 3.9252\bar{s}^2 + 7.7036\bar{s} + 7.5595} \tag{8.54}$$

This can also be obtained in MATLAB as

[b,a]=butter(3,1.9526,'high','s')

Replacing $\bar{s}$ by $(z-1)/(z+1)$ yields the following digital high-pass filter

$$H(z) = \frac{\left(\frac{z-1}{z+1}\right)^3}{\left(\frac{z-1}{z+1}\right)^3 + 3.9252\left(\frac{z-1}{z+1}\right)^2 + 7.7036\left(\frac{z-1}{z+1}\right) + 7.5595}$$

$$= \frac{z^3 - 3z^2 + 3z - 1}{20.1883z^3 + 23.4569z^2 + 14.0497z + 2.7811} \tag{8.55}$$

This filter is stable because the bilinear transformation maps stable poles in analog filters into stable poles in digital filters. The magnitude response of $H(z)$ is shown in Fig. 8.16(a) with a dotted line. The result is almost identical to the one obtained by using the impulse-invariance method in Example 8.3 in the passband but is slightly better in the stopband.

Method II: Analog Prototype$\Rightarrow$Digital Low-Pass $\Rightarrow$Digital High-Pass[4]
The first arrow uses the bilinear transformation, and the second arrow uses a digital frequency transformation. We select the third-order Butterworth prototype filter

$$G(s) = \frac{1}{s^3 + 2s^2 + 2s + 1}$$

---

[4] This part may be skipped without loss of continuity.

with $\Omega_p = 1$. Replacing s by $(z-1)/(z+1)$ yields the digital low-pass filter

$$
\begin{aligned}
H(z) &= \frac{(z+1)^3}{(z-1)^3 + 2(z-1)^2(z+1) + 2(z-1)(z+1)^2 + (z+1)^3} \\
&= \frac{z^3 + 3z^2 + 3z + 1}{6z^3 + 2z}
\end{aligned}
\tag{8.56}
$$

with passband edge frequency, using (8.52),

$$
\omega_p = 2\tan^{-1}\Omega_p = 2\tan^{-1}1 = 2 \cdot 0.785 = 1.57
$$

In order to apply the digital low-pass-to-high-pass frequency transformation in Table 8.3, we compute

$$
c = -\frac{\cos[(\bar{\omega}_p + \omega_p)/2]}{\cos[(\bar{\omega}_p - \omega_p)/2]} = \frac{-\cos[(0.7\pi + 1.57)/2]}{\cos[(0.7\pi - 1.57)/2]} = 0.3246
$$

Let

$$
z = \frac{-(\bar{z} + c)}{c\bar{z} + 1} = \frac{-\bar{z} - 0.3246}{0.3246\bar{z} + 1} =: \frac{d_1}{d_2}
$$

Substituting this into (8.56) yields

$$
\begin{aligned}
\bar{H}(\bar{z}) &= \frac{(d_1/d_2)^3 + 3(d_1/d_2)^2 + 3(d_1/d_2) + 1}{6(d_1/d_2)^3 + 2(d_1/d_2)} \\
&= \frac{d_1^3 + 3d_1^2 d_2 + 3d_1 d_2^2 + d_2^3}{6d_1^3 + 2d_1 d_2^2} \\
&= \frac{\bar{z}^3 - 2.9999\bar{z}^2 + 2.9999\bar{z} - 1}{20.1582\bar{z}^3 + 23.4002\bar{z}^2 + 14.0152\bar{z} + 2.7732}
\end{aligned}
\tag{8.57}
$$

This is the desired digital high-pass filter and is very close to the one in (5.55). Thus the two methods yield roughly the same result.

8.7 Analog-Prototype-to-Digital Transformations

The digital high-pass filter in Example 8.4 was designed using two different methods. One method takes the left-lower route in Fig. 8.19; the other the upper-right route. Each method takes two steps: a bilinear transformation and an analog or digital frequency transformation. In this section, we will combine the two steps in the left-lower route to yield analog-prototype-to-digital frequency transformations. By so doing, the intermediate step of designing a nonprototype analog filter can be bypassed. We first list in Table 8.4 various analog-prototype-to-digital-transformations.

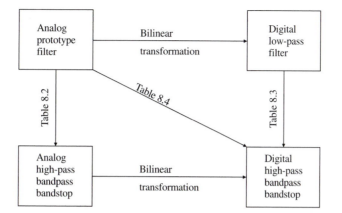

Figure 8.19 Various transformations from analog-prototype-to-digital filters.

Table 8.4 Analog-Prototype-to-Digital Transformations

Analog prototype	Transformation	$\bar{\Omega} = \tan(\omega/2)$
Digital low-pass	$s = \dfrac{z-1}{\bar{\Omega}_p(z+1)}$	$\bar{\Omega}_p = \tan(\omega_p/2)$
Digital high-pass	$s = \dfrac{\bar{\Omega}_p(z+1)}{z-1}$	$\bar{\Omega}_p = \tan(\omega_p/2)$
Digital bandpass	$s = \dfrac{\beta(z^2 + 2\alpha z + 1)}{z^2 - 1}$	$\alpha = \dfrac{\bar{\Omega}_{pl}\bar{\Omega}_{pu} - 1}{\bar{\Omega}_{pl}\bar{\Omega}_{pu} + 1}$
		$\beta = \dfrac{\bar{\Omega}_{pu}\bar{\Omega}_{pl} + 1}{\bar{\Omega}_{pu} - \bar{\Omega}_{pl}}$
Digital bandstop	$s = \dfrac{(z^2 - 1)\Omega_s}{\beta(z^2 + 2\alpha z + 1)}$	$\alpha = \dfrac{\bar{\Omega}_{sl}\bar{\Omega}_{su} - 1}{\bar{\Omega}_{sl}\bar{\Omega}_{su} + 1}$
		$\beta = \dfrac{\bar{\Omega}_{su}\bar{\Omega}_{sl} + 1}{\bar{\Omega}_{su} - \bar{\Omega}_{sl}}$

We develop in the following the third transformation or the analog-prototype-to-digital bandpass transformation. Consider the bilinear transformation

$$\bar{s} = k\frac{z-1}{z+1} \quad \text{and} \quad \bar{\Omega} = k\tan(\omega/2) \tag{8.58}$$

where we do not assume $k = 1$. The analog-prototype-to-analog bandpass transformation is

$$s = \frac{\bar{s}^2 + \bar{\Omega}_{pl}\bar{\Omega}_{pu}}{\bar{s}(\bar{\Omega}_{pu} - \bar{\Omega}_{pl})} \tag{8.59}$$

which is the third transformation in Table 8.2 with $\bar{\omega}$ replaced by $\bar{\Omega}$. Substituting (8.58) into (8.59) yields

$$
s = \frac{k^2 \left(\dfrac{z-1}{z+1}\right)^2 + k \tan\left(\dfrac{\omega_{pl}}{2}\right) k \tan\left(\dfrac{\omega_{pu}}{2}\right)}{k \left(\dfrac{z-1}{z+1}\right) \left[k \tan\left(\dfrac{\omega_{pu}}{2}\right) - k \tan\left(\dfrac{\omega_{pl}}{2}\right)\right]}
$$

$$
= \frac{\left(\dfrac{z-1}{z+1}\right)^2 + \tan\left(\dfrac{\omega_{pl}}{2}\right) \tan\left(\dfrac{\omega_{pu}}{2}\right)}{\left(\dfrac{z-1}{z+1}\right) \left[\tan\left(\dfrac{\omega_{pu}}{2}\right) - \tan\left(\dfrac{\omega_{pl}}{2}\right)\right]}
\tag{8.60}
$$

We see that k is canceled in the preceding equation. Thus the digital filter obtained by the transformation is independent of k, and any positive k can be used. Let us define

$$
\bar{\Omega} = \tan(\omega/2)
$$

Then (8.60) can be simplified as

$$
s = \frac{(z-1)^2 + \bar{\Omega}_{pl}\bar{\Omega}_{pu}(z+1)^2}{(z^2-1)(\bar{\Omega}_{pu} - \bar{\Omega}_{pl})}
$$

$$
= \frac{(1 + \bar{\Omega}_{pl}\bar{\Omega}_{pu})z^2 + 2(\bar{\Omega}_{pl}\bar{\Omega}_{pu} - 1)z + (1 + \bar{\Omega}_{pl}\bar{\Omega}_{pu})}{(\bar{\Omega}_{pu} - \bar{\Omega}_{pl})(z^2 - 1)}
$$

This reduces to the one in the table after defining α and β. The other transformations in the table can be similarly established.

In the last transform in Table 8.4, Ω_s is the stopband edge frequency of an analog prototype filter. Its computation was discussed in Example 8.1. It can also be obtained graphically as shown in Fig. 8.12.

◆ Example 8.5

Repeat the design in Example 8.3 or 8.4. That is, design a third-order high-pass digital filter with 3-dB passband edge frequency 35 Hz and sampling frequency 100 Hz or $\omega_p = 0.7\pi$ in $[0, \pi]$.

We use the second transformation in Table 8.4 to carry out the design. We compute

$$
\bar{\Omega}_p = \tan(\omega_p/2) = \tan(0.7\pi/2) = 1.9626
$$

Substituting

$$s = \frac{1.9625(z+1)}{z-1}$$

into the following third-order Butterworth analog prototype filter

$$G(s) = \frac{1}{s^3 + 2s^2 + 2s + 1}$$

we obtain

$$H(z) = \frac{1}{\left(\frac{1.9625(z+1)}{z-1}\right)^3 + 2\left(\frac{1.9625(z+1)}{z-1}\right)^2 + 2\left(\frac{1.9625(z+1)}{z-1}\right) + 1}$$

$$= \frac{z^3 - 3z^2 + 3z - 1}{20.1883z^3 + 23.4569z^2 + 14.0497z + 2.7811} \tag{8.61}$$

This is identical to the result obtained in Method I of Example 8.4. In Method I of Example 8.4 we must design an analog low-pass filter with passband edge frequency 1.9625. Here the filter is imbedded in the design.

To conclude this section, we discuss the MATLAB functions that generate digital IIR filters. The positive frequency range in these functions is normalized to [0, 1]. Thus a band-edge frequency ω_p in [0, π] must be modified as ω_p/π in [0, 1]. The function [b,a]=butter(n,wp) generates an nth-order IIR digital low-pass filter with -3-dB passband edge frequency wp in [0, 1]. If wp=[wl wu] is a two-element vector, then it returns a $2n$-th-order bandpass filter. The function [b,a]=butter(n,wp,'high') returns a high-pass filter and [b,a]=butter(n,[wl wu],'stop') returns a bandstop filter with -3-dB passband edge frequencies wl and wu. Similar remarks apply to cheby1(n,Rp,wp), cheby2(n,Rs,ws), and ellip(n,Rp,Rs,wp). Note that we specify the passband tolerance R_p in type I Chebyshev and elliptic filters, and the stopband attenuation R_s in type II Chebyshev filters. In ellip, we specify ω_p, R_p, and R_s. Then the function will generate an IIR filter that minimizes the stopband edge frequency ω_s. Thus we do not specify ω_s in using ellip. These functions all take the left-lower route of Fig. 8.19.

The major differences between the MATLAB functions that generate analog and digital filters are as follows: First, the frequency in the analog case is in the unit of radians per second and can assume any value in [0, ∞); the frequency in the digital case is normalized to lie inside [0, 1]. Second, other than prototype filters, a function returns an analog filter if it contains the flag "s" and a digital filter if it does not.

To conclude this section, we mention that the bilinear transformation can be used to design filters with piecewise constant magnitude responses such as low-pass, high-pass, bandpass, and bandstop filters. If a magnitude response is not piecewise constant such as the one of a differentiator, then the magnitude response will be distorted and the bilinear transformation should not be used.

8.8 Comparisons with FIR Filters

Before comparing IIR and FIR filters, we mention that the following MATLAB functions

[N,wn]=buttord(wp,ws,Rp,Rs)
[N,wn]=cheb1ord(wp,ws,Rp,Rs)
[N,wn]=cheb2ord(wp,ws,Rp,Rs)
[N,wn]=ellipord(wp,ws,Rp,Rs)

return filter orders that meet the passband edge frequency ω_p, stopband edge frequency ω_s, passband tolerance at most R_p dB, and stopband attenuation at least R_s dB. All frequencies must be normalized to lie inside [0, 1]. The functions also return −3-dB passband edge frequency wn. In using the filtdemo discussed in Section 7.10 the program will automatically select the required filter order.

We now compare the IIR and FIR filters that meet the passband edge frequency 500 Hz, stopband edge frequency 600 Hz, passband tolerance at most 1 dB and stopband attenuation at least 20 dB, and sampling frequency 2000 Hz. After typing in the specification in filtdemo, the selection of Butterworth, type I and type II Chebyshev, and ellipic will yield, respectively, the magnitude responses shown in Figs. 8.20(a)–(d). Their orders are listed in the following

Filter	Order
Butterworth	10
Type I Chebyshev	5
Type II Chebyshev	5
Elliptic	3
Minimax FIR	17
LS FIR (wdcr)	20
Kaiser FIR	24

Here we also list the orders of the FIR filters obtained in Section 7.10 The orders of the IIR filters are considerably less than those of the FIR filters. This is expected because FIR filters have all poles at $z = 0$ or, equivalently, do not utilize poles in shaping magnitude responses. On the other hand, all FIR filters have linear phases.

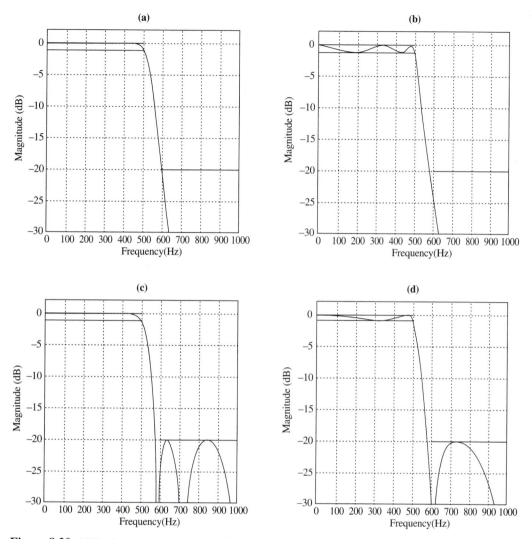

Figure 8.20 (a) Tenth-order Butterworth. (b) Fifth-order type I Chebyshev. (c) Fifth-order type II Chebyshev. (d) Third-order elliptic.

PROBLEMS

8.1 Derive the denominator of the Nth-order Butterworth filter in Table 8.1 for $N = 2$ and 4.

8.2 Design a third-order Butterworth high-pass analog filter with passband edge frequency 100 rad/s.

8.3   In the analog prototype-to-high-pass transformation in Table 8.2, the transformation is obtained by mapping $\bar{\omega}_p$ into $\omega_p = -1$. If you map $\bar{\omega}_p$ into $\omega_p = 1$, what transformation will you obtain? If you use this transformation to design Problem 8.2, will the resulting filter be stable? If not, use the procedure in Problem 6.17 to stabilize the filter. Will the resulting filter be the same as the one in Problem 8.2?

8.4   Repeat Example 8.1 by using a type I Chebyshev filter. Will you obtain the magnitude response in Fig. 8.11(c) if you use a MATLAB function? Will you obtain the magnitude response in Fig. 8.11(d) if you use (8.20)?

8.5   Repeat Example 8.1 by using a type II Chebyshev filter. Use the MATLAB function cheby2 to design such a filter. Also use (8.20) to design such a filter. Compare their transfer functions and magnitude responses.

8.6   Design a fourth-order Butterworth analog bandpass filter with passband edge frequencies 8 and 11 Hz from a prototype filter. Check your result by calling a MATLAB function.

8.7   Let b and a be of the same length and be, respectively, the numerator and denominator coefficients of an Nth-order analog prototype filter. Show that the following MATLAB program

```
n=0:N;w=[wbp. ^ n];
bb=b.*w
ab=a.*w
```

carries out the analog low-pass frequency transformation in Table 8.2. Show that the following program

```
n=0:N;w=[wbp. ^ n];
bb=fliplr(b).*w
ab=fliplr(a).*w
```

carries out the analog high-pass frequency transformation in Table 8.2.

8.8   Transform a second-order Butterworth analog filter, by using the impulse-invariant method, to a digital low-pass filter to meet 3-dB passband edge frequency 30 rad/s. The sampling frequency is 100 Hz. Plot the magnitude responses of the analog and digital filters and compare them.

8.9   Use a digital frequency transformation to transform the digital filter in Problem 8.8 to a digital bandpass filter with 3-dB passband edge frequencies 100 and 180 rad/s and sampling frequency 120 Hz. Note that their sampling frequencies are different.

8.10   Design three digital low-pass filters with 3-dB passband edge frequency 1 rad/s in $[0, \pi]$ by using (8.47) with $k = 0.2, 2, 100$ from first-order Butterworth analog low-pass filters. Are the three digital filters the same?

8.11   Repeat the design of the digital bandpass filter in Problem 8.9 by using a bilinear transformation from an analog bandpass filter. Compare this filter with the one obtained in Problem 8.9. Which is better? Why?

8.12   Repeat the design of the digital bandpass filter in Problem 8.11 using the analog-prototype-to-digital-bandpass transformation in Table 8.4.

8.13   Develop from (8.11) that the second-order type I Chebyshev prototype filter with 1-dB passband tolerance is given by

$$G(s) = \frac{0.9826}{s^2 + 1.0977s + 1.1025}$$

8.14   Design a second-order type I Chebyshev digital high-pass filter with 1-dB passband edge frequency 600 Hz and sampling frequency 3000 Hz. Use the prototype filter in Problem 8.13 to carry out the design using the three routes in Fig. 8.19. Are the three results the same?

8.15   Consider the two polynomials

$$g(x) = a_0x^3 + a_1x^2 + a_2x + a_3; \qquad f(x) = b_0x^2 + b_1x + b_2$$

Show

$$h(x) = g(x)f(x) =: c_0x^5 + c_1x^4 + c_2x^3 + c_3x^2 + c_4x + c_5$$

where c_n, for $n = 0, 1, \ldots, 5$, are given by

$$c_n = \sum_{k=0}^{n} a_{n-k}b_k$$

Use numbers to verify in MATLAB that, if

a=[a0 a1 a2 a3];b=[b0 b1 b2];

then c=conv(a,b).

Structures of Digital Filters

9.1 Introduction

We introduced in the preceding two chapters the design of FIR and IIR digital filters. In this chapter, we discuss their various structures that can be used in actual implementations. Digital filters can be implemented in software or hardware, or on a PC or a DSP processor. They can be implemented using MATLAB as in Program 6.2. They can also be programmed using high-level languages such as C or C++ without using any software package. See Refs. 11 and 17. Actual programming, without using MATLAB, is outside the scope of this text. We discuss in this chapter only structures of digital filters.

Every linear time-invariant, lumped, and causal filter can be described by the convolution summation

$$y[n] = \sum_{k=0}^{n} h[n-k]x[k] = \sum_{k=0}^{n} h[k]x[n-k] \tag{9.1}$$

or the transfer function

$$Y(z) = H(z)X(z) = \frac{b_1 + b_2 z^{-1} + \cdots + b_{N+1} z^{-N}}{1 + a_2 z^{-1} + \cdots + a_{N+1} z^{-N}} X(z) \tag{9.2}$$

where we have normalized a_1 to 1, or the difference equation

$$
\begin{aligned}
y[n] = &-a_2 y[n-1] - \cdots - a_{N+1} y[n-N] \\
&+ b_1 x[n] + b_2 x[n-1] + \cdots + b_{N+1} x[n-N]
\end{aligned}
\tag{9.3}
$$

Equations (9.1) and (9.3) are in the time domain and (9.2) is in the transform domain. They are, however, all equivalent, and any one can be obtained from the other two. All of them can be used to compute the output of a filter excited by any input.

Actual computation, however, involves many issues, as listed in the following.

- Efficiency in terms of the number of arithmetic operations (multiplications and additions)
- Filter coefficient quantization
- Storage requirement
- Ease of programming
- Quantization of signals
- Numerical errors due to rounding or truncation
- Overflow (saturation)

A complete discussion of these topics is beyond the scope of this text. In this chapter, we will touch upon only some of them. For a more detailed treatment, the reader is referred to Ref. 21. Before proceeding, we mention that these issues are important in specialized hardware that uses a small number of bits and fixed-point representation. If digital filters are implemented using a general-purpose computer, these issues can often be disregarded.

In this chapter, we first develop block diagrams for FIR filters. Block diagrams provide structures or schematic diagrams for software and hardware implementations, and can be used to compare different implementations. We then discuss FFT implementation of FIR filters. It is followed by a discussion of block diagrams for IIR filters. We then develop a set of first-order difference equations for their software implementation. Some examples are introduced to demonstrate some of the issues listed previously. These provide reasons for introducing cascade and parallel implementations. We discuss the implementation of second-order sections to conclude this chapter.

9.2 Direct Form of FIR Filters

Consider an FIR filter with impulse response $h[n]$, for $n = 0, 1, \ldots, N$. Then its imput $x[n]$ and output $y[n]$ are related by

$$y[n] = \sum_{k=0}^{n} h[k]x[n-k]$$

for all $n \geq 0$. This equation reduces to, because $h[k] = 0$ for $k > N$,

$$y[n] = \sum_{k=0}^{N} h[k]x[n-k] \tag{9.4}$$

This is actually, as discussed in Section 5.3.1, a nonrecursive difference equation of order N. Thus the filter has length $N + 1$ and order N.

Let us write (9.4) explicitly for $N = 4$ as

$$y[n] = h[0]x[n] + h[1]x[n-1] + h[2]x[n-2] + h[3]x[n-3] + h[4]x[n-4]$$

This equation can be readily constructed, using the basic elements discussed in Fig. 5.3(a). The equation contains $x[n-4]$. Thus we need four unit-delay elements as shown in Fig. 9.1. If the input of the left-most element is assigned as $x[n]$, then the outputs of the four unit-delay elements are, respectively, $x[n-i]$, for $i = 1, 2, 3, 4$. Multiplying $x[n-i]$ by $h[i]$, for $i = 0, 1, \ldots, 4$, and then summing them, we obtain the block diagram in Fig. 9.1. This is called the *direct-form* block diagram of the FIR filter. We see that the output is simply a weighted sum of current and past four inputs. Because of the series of unit-delay elements, the structure in Fig. 9.1 is also called a *tapped delay line* or *transversal filter*.

The structure in Fig. 9.1 has an important feature: It has *no loop*. A loop is defined as a connection of directed branches that, starting from any point, can return along the loop to the same point. Because of no loop, the errors due to quantization of $h[k]$ and due to rounding or truncation in multiplications and additions have no effect on other parts of the diagram and will propagate directly to the output. Thus the direct form is not only simple but also a good structure for implementing FIR filters.

All FIR filters designed in Chapter 7 have linear phase or have the property $h[n] = \pm h[N-n]$. Using the property, we can combine the outputs of some unit-delay elements as shown in Fig. 9.2. By so doing, the number of multiplications can be cut almost in half.

9.2.1 Cascade Form

Consider a high-order FIR filter with transfer function

$$H(z) = \sum_{k=0}^{N} h[k]z^{-k} = \sum_{k=0}^{N} b_{k+1}z^{-k}$$

If we factor it as

$$H(z) = \prod_i (b_{1i} + b_{2i}z^{-1} + b_{3i}z^{-2}) =: \prod_i H_i(z)$$

then we can implement the FIR filter as a cascade of second-order sections as shown in Fig. 9.3. Each section can be individually implemented in the direct form. Thus every FIR filter can be implemented using either the direct form or cascade form.

If all coefficients can be exactly implemented, then the direct and cascade forms will yield the same frequency response. However, exact implementation of filter coefficients is not possible because of finite number of bits used, thus the two forms may yield different frequency responses in practice. We compare them in the following.

Let c_1 and c_2 be any two real numbers. Then generally we have

$$|Q(c_1c_2) - c_1c_2| \leq |Q[Q(c_1)Q(c_2)] - c_1c_2| \tag{9.5}$$

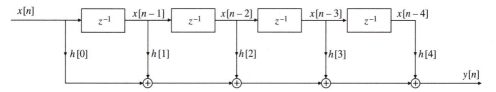

Figure 9.1 Direct form of a fourth-order FIR filter.

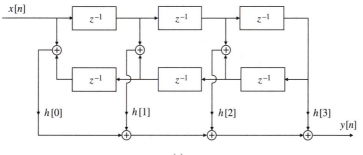

(a)

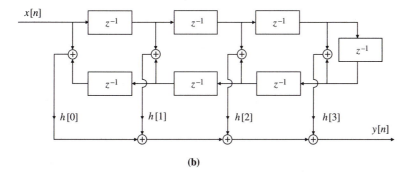

(b)

Figure 9.2 (a) Sixth-order type I FIR filter. (b) Seventh-order type II.

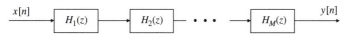

Figure 9.3 Cascade implementation.

where Q stands for quantization either by rounding or truncation. Because of (9.5), the direct-form implementation generally will yield a frequency response closer to the original frequency response, as we will demonstrate in the next example.

◆ **Example 9.1**

Consider the fourth order minimax optimal FIR filter

$$H(z) = 0.069735 + 0.388726z^{-1} + 0.360530z^{-2} + 0.388726z^{-3} + 0.069735z^{-4}$$

It is obtained by typing h=remez(4,[0 1/pi 1.5/pi 1],[1 1 0 0]). It is a type I FIR filter. It has two real zeros and a pair of complex-conjugate zeros on the unit circle. By grouping the two real zeros and grouping the complex-conjugate zeros, we obtain

$$H(z) = H_1(z)H_2(z) = (1 + 4.931530z^{-1} + z^{-2})$$
$$\times (0.069735 + 0.044826z^{-1} + 0.069735z^{-2}) \tag{9.6}$$

Note that the coefficients of $H_1(z)$ and $H_2(z)$ are symmetric. If $H(z)$ has complex-conjugate zeros not on the unit circle, then $H_i(z)$ needs at least degree 4 to have symmetric coefficients.

Now we will round every coefficient to three decimal digits. Clearly we have

$$Q[H(z)] = 0.070 + 0.389z^{-1} + 0.361z^{-2} + 0.389z^{-3} + 0.070z^{-4}$$

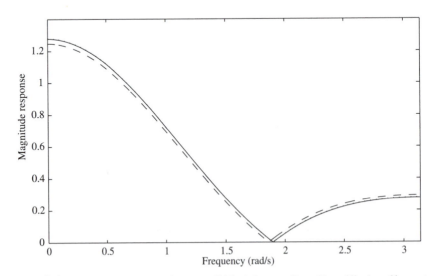

Figure 9.4 Magnitude responses of exact (solid line), inexact direct (dotted line), and inexact cascade (dashed line) implementations.

and

$$Q[Q[H_1(z)]Q[H_2(z)]] = 0.070 + 0.390z^{-1} + 0.362z^{-2} + 0.390z^{-3} + 0.070z^{-4}$$

We see that the coefficients of $Q[Q[H_1(z)]Q[H_2(z)]]$ deviate more from the corresponding coefficients of $H(z)$ than those of $Q[H(z)]$. This confirms the assertion in (9.5). We plot in Fig. 9.4 the magnitude responses of $H(z)$ (solid line), $Q[H(z)]$ (dotted line), and $Q[H_1(z)] \times Q[H_2(z)]$ (dashed line). If we round all coefficients to three decimal digits, then the three plots are indistinguishable. If we round all coefficients to two decimal digits, then the plots are as shown. The direct form has a closer magnitude response to the exact magnitude response than the cascade form.

From the preceding discussion, we conclude that the direct form is better than the cascade form in implementing FIR filters. In addition to the direct form, FIR filters can also be implemented in the lattice structure. Its employment is mostly used in adaptive filtering. The interested reader is referred to Ref. 21.

9.3 DFT of Periodic Convolutions

FIR filters can be directly implemented in the time domain as in (9.4); they can also be implemented using FFT. To develop such an implementation, we first discuss periodic convolutions and their DFT.

Consider two finite sequences $x_i[n]$, for $n = 0, 1, \ldots, N$, of length $N + 1$, and $i = 1, 2$. If they have different lengths, we can pad trailing zeros to the shorter one to make them of the same length. Let us define $W = e^{-j2\pi/(N+1)}$. Then their discrete Fourier transforms (DFTs) are

$$X_i[m] = \mathcal{D}[x_i[n]] = \sum_{n=0}^{N} x_i[n]W^{nm} = \sum_{n=k}^{N+k} \tilde{x}_i[n]W^{nm} \tag{9.7}$$

for any integer k, where $\tilde{x}_i[n]$ is the periodic extension of $x_i[n]$ with period $N + 1$ as defined in (4.13). The inverse DFT of $X_i[m]$ is

$$\tilde{x}_i[n] = \mathcal{D}^{-1}[X_i[m]] = \frac{1}{N} \sum_{m=0}^{N} X_i[m]W^{-nm} \tag{9.8}$$

As discussed in Section 4.2, although DFT is defined for finite sequences, it can also be considered to be defined for periodic infinite sequences, and their DFTs are also periodic with the same period.

Let us use $\tilde{x}_i[n]$ to define a new sequence $\tilde{x}[n]$ as

$$\tilde{x}[n] = \sum_{k=0}^{N} \tilde{x}_1[n-k]\tilde{x}_2[k] = \sum_{k=0}^{N} \tilde{x}_2[n-k]\tilde{x}_1[k] \tag{9.9}$$

This has the convolution form in (9.1) and (9.4). Because $\tilde{x}_i[n]$, $i = 1, 2$, are periodic with period $N + 1$ and because the summation is carried out over one period, the resulting $\tilde{x}[n]$ is also periodic with period $N + 1$. Thus (9.9) is called the *periodic convolution*.

Let us apply the DFT to $\tilde{x}[n]$. Then we have

$$
\begin{aligned}
X[m] = \mathcal{D}[\tilde{x}[n]] &= \sum_{n=0}^{N} \tilde{x}[n] W^{nm} \\
&= \sum_{n=0}^{N} \left(\sum_{k=0}^{N} \tilde{x}_1[n-k]\tilde{x}_2[k] \right) W^{(n-k)m} W^{km} \\
&= \sum_{k=0}^{N} \left(\sum_{n=0}^{N} \tilde{x}_1[n-k] W^{(n-k)m} \right) \tilde{x}_2[k] W^{km} \\
&= \sum_{k=0}^{N} \left(\sum_{\bar{n}=-k}^{N-k} \tilde{x}_1[\bar{n}] W^{\bar{n}m} \right) \tilde{x}_2[k] W^{km} \\
&= X_1[m] \sum_{k=0}^{N} \tilde{x}_2[k] W^{km} = X_1[m] X_2[m]
\end{aligned}
\tag{9.10}
$$

where we have interchanged the order of summations, introduced a new index $\bar{n} = n - k$, and used the periodicity of $\tilde{x}_1$ and W. We see that the DFT transforms the periodic convolution of two sequences into the multiplication of their DFTs.

An important implication of (9.10) is

$$
\tilde{x}[n] = \mathcal{D}^{-1} \left[\mathcal{D}[\tilde{x}_1[n]] \times \mathcal{D}[\tilde{x}_2[n]] \right]
\tag{9.11}
$$

and

$$
\tilde{x}[n] = \mathsf{ifft} \left(\mathsf{fft}(\tilde{x}_1[n]). * \mathsf{fft}(\tilde{x}_2[n]) \right)
\tag{9.12}
$$

It means that the periodic convolution can also be computed using FFT. This is similar to that the response of a digital filter can be computed directly from (9.1) or using the z-transform as

$$
y[n] = Z^{-1} \left[Z[h[n]] \times Z[x[n]] \right]
$$

We will apply (9.12) to compute the response of FIR filters in the next subsection.

9.3.1 FFT Computation of FIR Filters

Every digital filter can be described by a discrete convolution as in (9.1). Before proceeding, we discuss in more detail the convolution. Consider the sequence $h[n]$ of length $N + 1$ and the sequence $x[n]$ of length $P + 1$ shown in Fig. 9.5(a). Because the two sequences are assumed to be zero outside the range, their convolution can be written as

$$y[n] = \sum_{k=0}^{n} h[k]x[n-k] = \sum_{k=-\infty}^{\infty} h[k]x[n-k] \qquad (9.13)$$

for all n. To compute $y[n]$ at $n = n_0$, we flip x and then shift it to n_0 as shown in Fig. 9.5(c). Multiplying $x[k - n_0]$ and $h[k]$ for all k, and then adding them, we obtain the value of $y[n_0]$. Note that for $n_0 < 0$ and $n_0 > P + N$, because there is no overlapping of nonzero parts of $x[k - n_0]$ and $h[k]$, the products of $x[k - n_0]$ and $h[k]$ are zero for all k, and $y[n_0]$ is zero. For $0 \le n_0 \le P + N$, there are at most n_0 number of nonzero products of $x[k - n_0]$ and $h[k]$. Their sum yields the output $y[n_0]$, as shown in Fig. 9.5(d). The sequence has length at most $P + N + 1$ as shown. We call the convolution in (9.13) a *linear convolution* or, simply, a *convolution*.

Next we discuss their periodic convolution. First we assume $P \le N$ and extend $x[n]$ to length $N + 1$ by padding trailing zeros. They are then extended periodically to all n as shown in Figs. 9.5(e) and (f). Their periodic convolution will be as shown in Fig. 9.5(h). The process of computing the convolution in Fig. 9.5(d) is directly applicable to compute the periodic convolution in Fig. 9.5(h). Because of periodic extensions, the result in Fig. 9.5(h) is different from the one in Fig. 9.5(d). Thus periodic convolutions cannot be directly applied to compute linear convolutions.

Now we will modify the periodic convolution in (9.9) so that it can be used to compute the linear convolution in (9.13). Let

$$L \ge P + N + 1$$

We define

$$\bar{h}[n] = \begin{cases} h[n] & \text{for } n = 0, 1, \ldots, N \\ 0 & \text{for } n = N + 1, N + 2, \ldots, L \end{cases} \qquad (9.14)$$

and

$$\bar{x}[n] = \begin{cases} x[n] & \text{for } n = 0, 1, \ldots, P \\ 0 & \text{for } n = P + 1, P + 2, \ldots, L \end{cases} \qquad (9.15)$$

Let $\bar{H}[m]$ and $\bar{X}[m]$ be respectively the $(L + 1)$-point DFT of $\bar{h}[n]$ and $\bar{x}[n]$. We claim that

$$\bar{y}[n] := \mathcal{D}^{-1}[\bar{H}[m]\bar{X}[m]] = \sum_{k=0}^{L} \tilde{\bar{h}}[k]\tilde{\bar{x}}[n-k]$$

$$= \sum_{k=0}^{n} h[n-k]x[k] = y[n] \qquad (9.16)$$

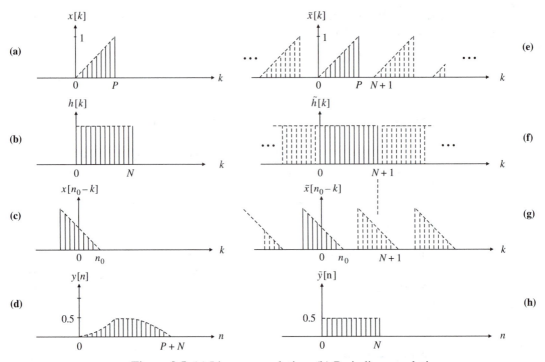

Figure 9.5 (a) Linear convolution. (b) Periodic convolution.

for $n = 0, 1, \ldots, L$, where $\tilde{\bar{h}}[n]$ and $\tilde{\bar{x}}[n]$, are periodic extension of $\bar{h}[n]$ and $\bar{x}[n]$ with period $L + 1$. The validity of (9.16) can be seen from Fig. 9.6, which is redrawn from Figs. 9.5(e)–(h). Because of the padded zeros, the product of Figs. 9.6(b) and (c) is the same as the product of Figs. 9.5(b) and (c). Thus the periodic convolution in Fig. 9.6 equals the linear convolution in Figs. 9.5(a)–(d). It is important to mention that (9.16) holds only for $n = 0, 1, \ldots, L$, for $y[n]$ is zero outside the range, whereas $\bar{y}[n]$ can be extended periodically for all n.

Equation (9.16) states that if two finite sequences are extended by trailing zeros and if the extensions are sufficiently long, the linear convolution of $h[n]$ and $x[n]$ equals the periodic convolution of $\tilde{\bar{h}}[n]$ and $\tilde{\bar{x}}[n]$, which, in turn, can be computed using FFT. This provides an alternative way of computing linear convolutions.

Now we have two ways of computing the convolution of two finite sequences or two ways of implementing FIR filters.

1. Direct computation in the time domain: For each n_0, compute the products of $h[k]$ and $x[n_0 - k]$ for $0 \le k \le n_0$. Their sum yields $y[n_0]$. We carry out the computation from $n_0 = 0$ to $n_0 = P + N$.
2. Using FFT: Compute the FFTs of $\bar{h}[n]$ and $\bar{x}[n]$. Multiply the two FFTs. The convolution equals the inverse FFT of the product.

We now compare these two methods. A detailed and precise comparison will be complicated

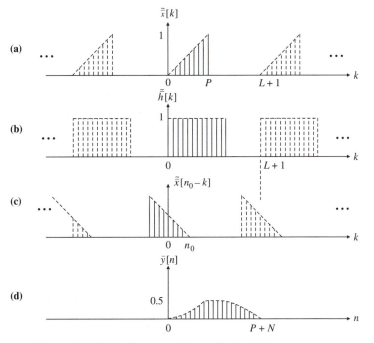

Figure 9.6 Periodic convolution of extended sequences.

because it depends on how each method is actually implemented. Therefore, we will show only roughly that the FFT method requires fewer multiplications than the direct method for N and P large.

Consider the Nth-order FIR filter described by (9.4). Let us compute its output excited by an input sequence of length $P + 1$. In direct computation of (9.4), each $y[n]$ requires $N + 1$ real multiplications. Thus to compute $y[n]$ from $n = 0$ to $P + N$, we need a total of

$$M_1 = (P + N + 1)(N + 1)$$

real multiplications. To use the FFT method, we must extend, by padding trailing zeros, the impulse sequence $h[n]$ and the input $x[n]$ to length $L + 1$ with

$$L \geq P + N + 1$$

Because the $(L + 1)$-point FFT fft(h,L+1) automatically pads trailing zeros to $h[n]$, the program that follows

Program 9.1
```
H=fft(h,L+1);X=fft(x,L+1);
y=ifft(H.*X)
```

generates the output $y[n]$ from $n = 0$ to L. Because all sequences are positive-time, there is no need to rearrange the inputs of fft and the output of ifft. See Section 4.9. Thus the use of FFT to compute the response of an FIR filter is very simple.

Now we count the number of real multiplications required in computing Program 9.1. We assume $\bar{L} = L + 1$ to be a power of 2. Then each FFT requires roughly $0.5\bar{L} \log_2 \bar{L}$ complex multiplications. See Section 4.4. Each complex multiplication requires four real multiplications. Thus each FFT requires $2\bar{L} \log_2 \bar{L}$ real multiplications. Program 9.1 requires two FFTs and one inverse FFT. Because an inverse FFT requires about the same amount of operations as an FFT, we conclude that Program 9.1 requires a total of

$$M_2 = 3 \times 2\bar{L} \log_2 \bar{L} = 6\bar{L} \log_2 \bar{L} \tag{9.17}$$

real multiplications. If $P = 4N$, then we have

$N =$	10	20	30	40	50	60	70
$M_1 =$	561	2121	4681	8241	12801	18361	24921
$M_2 =$	1779	4084	6610	9282	12062	14928	17866

We see that, if $N \leq 40$, direct computation requires less number of real multiplications; for $N \geq 50$, FFT requires less number of real multiplications.

The preceding comparison is a very rough one. For example, the numbers of additions are not compared. The number of $\bar{L}$ multiplications of H.$*$X is not included in M_2. On the other hand, one $\bar{L}$-point FFT can compute the $\bar{L}$-point FFTs of two real sequences and cut the number of operations almost in half (see Section 4.4.2). Direct computation actually requires fewer multiplications than M_1 because the first and last N outputs require less than $(N + 1)$ multiplications. Nevertheless, it is a fact that, for large N and P, FFT is more efficient than direct computation in implementing FIR filters.

Although FFT is more efficient, it requires more memory to store all inputs $x[n]$. Moreover, we can carry out the computation only after all inputs are received. Thus FFT cannot be used in real-time computation. If the input is a very long sequence, then it must be subdivided into subsequences as we will discuss in the next subsection, and its programming will become more complex. As we can see from Fig. 9.1, direct computation requires only $2(N + 1)$ memory locations and is applicable whether the input sequence is of finite or infinite duration. Thus for real-time processing, we must use direct computation. For non-real-time processing and for two long data sequences of finite lengths, Program 9.1 is a possible alternative. This will be discussed further in Section 9.4.1.

9.3.2 Convolution of Finite and Infinite Sequences

The convolution discussed in the preceding section is not applicable if one of the sequence is of infinite duration or exceeds the computer memory. In this case, the following procedure can be employed.

Let $h[n]$ be a finite sequence of length $N + 1$ and let $x[n]$ be an infinite sequence as shown in Fig. 9.7. We divide $x[n]$ into subsequences, each of length $P + 1$, as shown in Fig. 9.7(b).

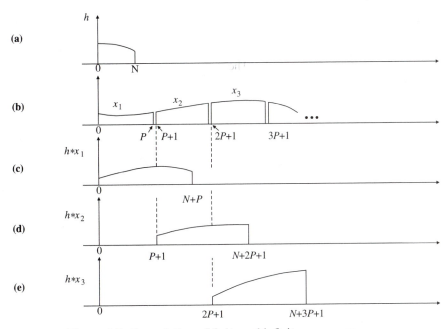

Figure 9.7 Convolution of finite and infinite sequences.

The first subsequence x_1 consists of the first $P + 1$ elements of $x[n]$; the second subsequence $x_2[n]$ consists of the next $P + 1$ elements of $x[n]$; and so forth. Because the convolution is a linear operator, the convolution of $h[n]$ and $x[n]$ equals the sum of the convolutions of $h[n]$ and $x_k[n]$, for $k = 1, 2, \ldots$. The convolution of $h[n]$ and $x_k[n]$ can be obtained as shown in Figs. 9.7(c)–(e), for $k = 1, 2, 3$. Summing the three convolutions yields the convolution of $h[n]$ and $x[n]$. This method is called the *overlap-add* method. There is a different method, called *overlap-save*; the interested reader is referred to Ref. 21.

9.4 Direct and Canonical Forms of IIR Filters

Every IIR filter can be described by a convolution, a transfer function, or a recursive difference equation. Its implementation using a convolution, either by direct computation or using FFT, is complicated in programming, requires infinitely many memory locations, and a large number of operations. Thus convolutions are not used to implement IIR filters. In this section, we discuss only transfer functions or, equivalently, difference equations.

For easy presentation, we discuss the transfer function of degree 4:

$$H(z) = \frac{Y(z)}{X(z)} = \frac{b_1 + b_2 z^{-1} + b_3 z^{-2} + b_4 z^{-3} + b_5 z^{-4}}{1 + a_2 z^{-1} + a_3 z^{-2} + a_4 z^{-3} + a_5 z^{-4}} \tag{9.18}$$

where we have normalized a_1 to 1. We will develop a number of block diagrams for (9.18). First we write (9.18) as

$$\frac{Y(z)}{X(z)} = \frac{Y(z)}{V(z)} \frac{V(z)}{X(z)} = \frac{1}{1 + a_2 z^{-1} + a_3 z^{-2} + a_4 z^{-3} + a_5 z^{-4}}$$
$$\times (b_1 + b_2 z^{-1} + b_3 z^{-2} + b_4 z^{-3} + b_5 z^{-4}) \qquad (9.19)$$

Let us define

$$\frac{Y(z)}{V(z)} = \frac{1}{1 + a_2 z^{-1} + a_3 z^{-2} + a_4 z^{-3} + a_5 z^{-4}} \qquad (9.20)$$

Then (9.19) implies

$$\frac{V(z)}{X(z)} = b_1 + b_2 z^{-1} + b_3 z^{-2} + b_4 z^{-3} + b_5 z^{-4} \qquad (9.21)$$

In the time domain, (9.21) and (9.20) become, respectively,

$$v[n] = b_1 x[n] + b_2 x[n-1] + b_3 x[n-2] + b_4 x[n-3] + b_5 x[n-4] \qquad (9.22)$$

and

$$y[n] + a_2 y[n-1] + a_3 y[n-2] + a_4 y[n-3] + a_5 y[n-4] = v[n]$$

or

$$y[n] = v[n] - a_2 y[n-1] - a_3 y[n-2] - a_4 y[n-3] - a_5 y[n-4] \qquad (9.23)$$

Equation (9.22) contains $x[n-4]$; thus we need four unit-delay elements as shown on the left-hand side of Fig. 9.8. If we assign the input of the top unit-delay element as $x[n]$, then the outputs of the four unit delay elements are $x[n-1]$, $x[n-2]$, $x[n-3]$, and $x[n-4]$. Using these outputs, we can readily build the left-hand-side block diagram for (9.22). Equation (9.23) contains $y[n-4]$; thus we need additional four unit-delay elements as shown on the right-hand side of Fig. 9.8. Using the same argument, we can draw the right-hand-side block diagram for (9.23). Thus Fig. 9.8 is a complete block diagram for (9.18) and is called the *direct form*.

The direct form is obtained by defining $Y(z)/V(z)$ as in (9.20). Now, instead, we will define

$$\frac{Y(z)}{V(z)} = b_1 + b_2 z^{-1} + b_3 z^{-2} + b_4 z^{-3} + b_5 z^{-4} \qquad (9.24)$$

Then (9.19) implies

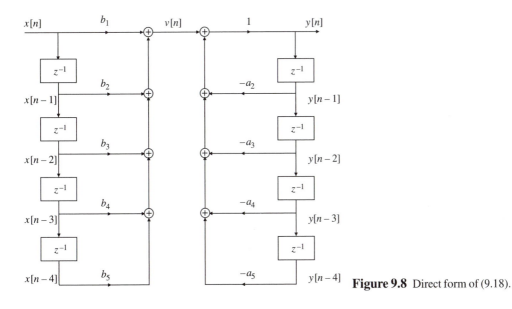

Figure 9.8 Direct form of (9.18).

$$\frac{V(z)}{X(z)} = \frac{1}{1 + a_2 z^{-1} + a_3 z^{-2} + a_4 z^{-3} + a_5 z^{-4}} \tag{9.25}$$

In the time domain, (9.25) and (9.24) become, respectively,

$$v[n] = x[n] - a_2 v[n-1] - a_3 v[n-2] - a_4 v[n-3] - a_5 v[n-4] \tag{9.26}$$

and

$$y[n] = b_1 v[n] + b_2 v[n-1] + b_3 v[n-2] + b_4 v[n-3] + b_5 v[n-4] \tag{9.27}$$

Equation (9.26) contains $v[n-4]$. Thus we need four unit-delay elements as shown in Fig. 9.9. If we assign the input of the top unit-delay element as $v[n]$, then the outputs of the four unit-delay elements are $v[n-1]$, $v[n-2]$, $v[n-3]$, and $v[n-4]$. Using (9.26), we can readily obtain the left-hand-side block diagram in Fig. 9.9. Because $v[n-k]$, for $k = 0, 1, \ldots, 4$, are already available, we can readily obtain (9.27) as shown on the right-hand side of Fig. 9.9. Thus (9.18) can also be implemented as shown in Fig. 9.9. Because the block diagram in Fig. 9.9 uses only four unit-delay elements as opposed to eight in Fig. 9.8, this form is preferred to the direct form. In fact, the block diagram in Fig. 9.9 uses the fewest possible numbers of unit-delay elements, multipliers, and adders; thus it is called a *canonical form*. It is called the *direct form II* in Ref. 21.

We next discuss a block diagram that is dual to the canonical form. By reversing the direction of every branch (without changing b_i, $-a_i$, or z^{-1}), changing every adder into a branch-out point and vice versa, and interchanging the input $x[n]$ and output $y[n]$ of Fig. 9.9, we obtain the

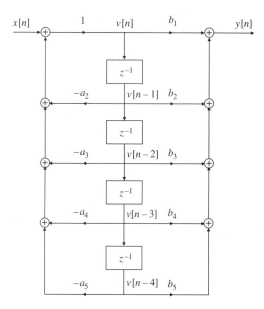

Figure 9.9 Canonical form of (9.18).

block diagram in Fig. 9.10(a). We redraw Fig. 9.10(a) in Fig. 9.10(b) by reversing only the flow of input and output. They are identical but will be assigned with different variables. We use Fig. 9.10(a) to show that it has (9.18) as its transfer function and, thus, is a different block diagram of (9.18).[1]

Let us assign the input of each unit delay element in Fig. 9.10(a) as $v_i[n]$ as shown. Then the corresponding output is $v_i[n-1]$. In computing the transfer function, all variables must be in the z-transform domain. Let $V_i(z)$, $X(z)$, and $Y(z)$ be, respectively, the z-transforms of $v_i[n]$, $x[n]$, and $y[n]$. Then the z-transform of $v_i[n-1]$ is $z^{-1}V_i(z)$. From Fig. 9.10(a), we have

$$Y(z) = z^{-1}V_1(z) + b_1 X(z)$$
$$V_1(z) = z^{-1}V_2(z) + b_2 X(z) - a_2 Y(z)$$
$$V_2(z) = z^{-1}V_3(z) + b_3 X(z) - a_3 Y(z) \qquad (9.28)$$
$$V_3(z) = z^{-1}V_4(z) + b_4 X(z) - a_4 Y(z)$$
$$V_4(z) = b_5 X(z) - a_5 Y(z)$$

Substituting successively $V_{i+1}(z)$ into $V_i(z)$, for $i = 3, 2, 1$, and then into $Y(z)$, we obtain

$$Y(z) = \left[b_1 + b_2 z^{-1} + b_3 z^{-2} + b_4 z^{-3} + b_5 z^{-4} \right] X(z)$$

---

[1] This can be established using graph theory and Mason's formula as in most DSP texts. To be self-contained, we show directly that its transfer function equals (9.18).

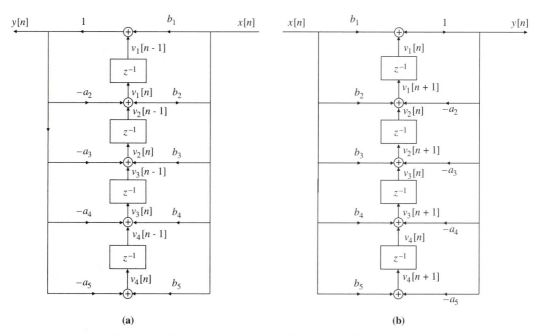

Figure 9.10 (a) Transposition of Fig. 9.9. (b) Redrawn from (a).

$$- \left[a_2 z^{-1} + a_3 z^{-2} + a_4 z^{-3} + a_5 z^{-4} \right] Y(z)$$

which implies

$$\left[1 + a_2 z^{-1} + a_3 z^{-2} + a_4 z^{-3} + a_5 z^{-4} \right] Y(z)$$
$$= \left[b_1 + b_2 z^{-1} + b_3 z^{-2} + b_4 z^{-3} + b_5 z^{-4} \right] X(z)$$

and

$$\frac{Y(z)}{X(z)} = \frac{b_1 + b_2 z^{-1} + b_3 z^{-2} + b_4 z^{-3} + b_5 z^{-4}}{1 + a_2 z^{-1} + a_3 z^{-2} + a_4 z^{-3} + a_5 z^{-4}}$$

This shows that the block diagram in Fig. 9.10 has the transfer function in (9.18). We call the block diagram the *transposed canonical form*. It is called the *direct form II transposed* in Ref. 21 and MATLAB.

It is possible to develop many other block diagrams for (9.18). However, the canonical and transposed canonical forms are most important. Note that if $a_i = 0$ for all i, then (9.18) is an FIR filter and the block diagram in Fig. 9.9 reduces to the one in Fig. 9.1. Thus the preceding discussion is applicable to FIR filters as well.

9.4.1 Implementation Using State-Space Equations

We developed in the preceding section block diagrams for the transfer function in (9.18). We show in this subsection how they can be used in software implementation. Before proceeding, we mention that (9.18) can be readily expressed as

$$y[n] = -a_2 y[n-1] - a_3 y[n-2] - a_4 y[n-3] - a_5 y[n-4]$$
$$+ b_1 x[n] + b_2 x[n-1] + b_3 x[n-2] + b_4 x[n-3] + b_5 x[n-4]$$

This fourth-order difference equation can be directly programmed. See Problem 5.14. However, it is more convenient and more systematic in programming to use a set of first-order difference equations than a single high-order difference equation. We develop in this subsection such a set.

Block diagrams provide a way of developing sets of first-order difference equations and any block diagram in the preceding section can be used. The procedure is very simple: We assign the output of each unit-delay element as a variable such as $v_i[n]$. Then its input is $v_i[n+1]$. Relating all variables, the input, and output, we will obtain a set of first-order difference equations.

Because the MATLAB function filter is based on the transposed canonical form in Fig. 9.10, we use the diagram to illustrate the procedure. If we assign the outputs of the four unit-delay elements in Fig. 9.10(b) as $v_i[n]$ as shown, then the corresponding inputs are $v_i[n+1]$. From Fig. 9.10(b), we can readily obtain

$$y[n] = v_1[n] + b_1 x[n]$$
$$v_1[n+1] = v_2[n] + b_2 x[n] - a_2 y[n]$$
$$v_2[n+1] = v_3[n] + b_3 x[n] - a_3 y[n]$$
$$v_3[n+1] = v_4[n] + b_4 x[n] - a_4 y[n]$$
$$v_4[n+1] = b_5 x[n] - a_5 y[n]$$

Substituting the first equation into the next four equations, we obtain

$$v_1[n+1] = v_2[n] - a_2 v_1[n] + (b_2 - a_2 b_1)x[n]$$
$$v_2[n+1] = v_3[n] - a_3 v_1[n] + (b_3 - a_3 b_1)x[n]$$
$$v_3[n+1] = v_4[n] - a_4 v_1[n] + (b_4 - a_4 b_1)x[n] \tag{9.29}$$
$$v_4[n+1] = \qquad - a_5 v_1[n] + (b_5 - a_5 b_1)x[n]$$
$$y[n] = v_1[n] + b_1 x[n]$$

The first four equations are first-order difference equations, and the last equation is only an algebraic equation. If the initial conditions $v_i[0]$ for $i = 1, 2, 3, 4$ are given, we can compute $y[n]$ and $v_i[n]$ recursively from $n = 0$ on for any input sequence $x[n]$. The set of equations is called a state-space equation. It is usually expressed in matrix form. See Problem 9.8 and Ref. 7.

The set of equations in (9.29) can easily be programmed. For example, if we use C language, after initialization and declaring variables, the main program will be

```
int L=100, n;
for (n=0;n<=L; ++n)
        {
        y[n] = v1[n] + b1*x[n];
        v1[n+1] = v2[n] - a2*v1[n] + (b2 - a2*b1)*x[n];
        v2[n+1] = v3[n] - a3*v1[n] + (b3 - a3*b1)*x[n];
        v3[n+1] = v4[n] - a4*v1[n] + (b4 - a4*b1)*x[n];
        v4[n+1] = - a5*v1[n] + (b5 - a5*b1)*x[n];
        printf("y[%d] = %g\ n", n, y[n]);
        }
```

This program computes and prints $y[n]$ for $n = 0$ to $L = 100$.

The MATLAB function y=filter(bn,an,x) generates the output of a discrete transfer function $H(z) = B(z)/A(z)$ excited by the input $x[n]$. The row vectors **bn** and **an** are the coefficients of $B(z)$ and $A(z)$ in negative-power form. The output has the same length as the input. The function is based on the block diagram in Fig. 9.10 or, equivalently, on the set of equations in (9.29), and is applicable to IIR and FIR filters. For FIR filters, we have **an=1**. We mention that **filter** is not an M-file; it is a built-in function in MATLAB. Thus it is not accessible to the user and is *probably* written in C language as discussed in the preceding.[2]

The MATLAB function **conv** (convolution) is also built on **filter**. Let c_1 and c_2 be two finite sequences of length N_1 and N_2. Then **conv(c1,c2)** yields the linear convolution of c_1 and c_2 and has length $N_1 + N_2 - 1$. If we consider c_1 as the impulse response of an FIR filter and c_2 as its input, then the output of the filter is the convolution of c_1 and c_2. However, the output of the filter has the same length as c_2. In order for the output of the filter to have length $N_1 + N_2 - 1$, we must pad trailing zeros to c_2 as **cb2=[c2 zeros(1,max(size(c1))-1)]**, which has length $N_1 + N_2 - 1$. Then we have

$$\text{conv(c1,c2)=filter(c1,1,cb2)} \qquad (9.30)$$

Generally, we select the shorter of c_1 and c_2 as the impulse response and the longer one as the input. Then the computation will require less time.

Equation (9.30) does not use FFT. If we want to use FFT, then we have, as in Program 9.1,

$$\text{conv(c1,c2)=ifft(fft(c1,N1+N2-1).*fft(c2,N1+N2-1))} \qquad (9.31)$$

---

[2] This writer was not able to obtain its codes from the Math Works, Inc. This is not important in any case because we are interested in only basic ideas and procedures.

Thus there are two ways to implement conv or, equivalently, FIR filters: (1) time domain using (9.30) with $a_i = 0$; (2) frequency domain using FFT. Even though we showed in Section 9.3.1 that FFT requires less time for long sequences, we compare there only computing time. If we include display time, then the conclusion may be different. See Problems 9.15 and 9.16. We mention that (9.30) deals exclusively with real numbers, whereas (9.31) must deal with complex numbers. Thus (9.31) may introduce more numerical errors than (9.30). This may be the reasons that the function conv in MATLAB is based on (9.30) even for long sequences.

9.5 Effects of Filter Coefficient Quantizations

Error analysis is not a simple topic. Errors depend on how a number is represented in binary form: fixed point or floating point; sign-and-magnitude, one's complement, or two's complement. They also depend on how a number is approximated: rounding or truncation. Because errors may occur in all operations, a precise analysis would be difficult. Thus error analysis is usually carried out using statistical methods. In this and the following sections, we will use only examples to demonstrate some effects of coefficient quantizations.

Consider an IIR filter with transfer function

$$H(z) = \frac{z^{-1} + 0.4z^{-2} - 0.03z^{-3} + 0.232z^{-4}}{1 - 3.4890z^{-1} + 4.4805z^{-2} - 2.4899z^{-3} + 0.4985z^{-4}}$$

$$= \frac{z^{-1}(1 + 0.8z^{-1})(1 - 0.4z^{-1} + 0.29z^{-2})}{(1 - 0.999z^{-1})(1 - 0.5z^{-1})(1 - 0.995z^{-1} + 0.998z^{-2})} \tag{9.32}$$

The filter has two real poles at 0.999 and 0.5, and a pair of complex conjugate poles at $0.4975 \pm j0.8663$ with magnitude 0.9990. They all lie inside the unit circle; thus the filter is stable.

In implementation, we assume, for simplicity, that all decimal numbers will be coded in sign-and-magnitude binary coding, as discussed in Section 1.2. For example, the binary coding of 0.4 is 0.01100110011 . . . . If it is truncated to six bits as 0.01100, then the resulting decimal number is 0.375. Thus, if we implement (9.32) in the canonical form in Fig. 9.9, and if we truncate the binary coding of all coefficients to six bits, the implemented IIR filter is actually governed by

$$\bar{H}(z) = \frac{z^{-1} + 0.375z^{-2} + 0.21875z^{-4}}{1 - 3.4375z^{-1} + 4.375z^{-2} - 2.4375z^{-3} + 0.46875z^{-4}} \tag{9.33}$$

rather than by (9.32). Clearly, the frequency response of (9.33) will be different from that of (9.32). Their poles and zeros will also be different. In filter design, we are mainly concerned with frequency responses. However, the filter must remain to be stable, otherwise its frequency response is not defined. Let us check the stability of (9.33). It can be checked using the Jury test discussed in Section 5.6.1 or using MATLAB. Typing

roots([1 -3.4375 4.375 -2.4375 0.46875])

yields 1.3076, 0.8659 ± 0.3883i, and 0.3981. The implemented filter in (9.33) has the pole 1.3076 outside the unit circle; thus it is not stable and cannot be used as a filter. This is one of many effects of imprecise implementation of digital filters.

It is well known that the roots of a polynomial may deviate greatly even when its coefficients change only slightly. This is called root sensitivity due to parameter variations. The higher the degree of a polynomial, the poorer the sensitivity. Before discussing a way of reducing the sensitivity, we give a different interpretation of the problem. From the block diagrams in Figs. 9.9 and 9.10, we see that there are many *loops* in the diagrams. A loop is defined as a connection of directed branches which, starting from any point, can return along the loop to the same point. In Fig. 9.9 or 9.10, there are four loops. An error occurred at a point will propagate, through the loops, to other parts of the block diagram, and its effect will be multiplied. Thus one way to reduce the sensitivity of an IIR filter is to develop a block diagram that has the fewest number of interactive loops.

Consider again the transfer function in (9.32). Now we write it as

$$H(z) = \frac{1 + 0.8z^{-1}}{1 - 0.5z^{-1}} \frac{z^{-1}}{1 - 0.999z^{-1}} \frac{1 - 0.4z^{-1} + 0.29z^{-2}}{1 - 0.995z^{-1} + 0.998z^{-2}} \tag{9.34}$$

It has two transfer functions of degree 1 and one transfer function of degree 2. We implement each one in canonical form and then connect them together as shown in Fig. 9.11. Although it still has four loops, only the two loops on the right-hand side affect each other. Thus this may be a better implementation. To verify this, we implement each decimal coefficient in (9.34) by truncating its sign and magnitude binary coding to six bits to yield

$$H(z) = \frac{1 + 0.78125z^{-1}}{1 - 0.5z^{-1}} \cdot \frac{z^{-1}}{1 - 0.96875z^{-1}} \cdot \frac{1 - 0.375z^{-1} + 0.28125z^{-2}}{1 - 0.96875z^{-1} + 0.96875z^{-2}}$$

The poles of the second-order transfer function are 0.4844 ± 0.8568i whose magnitudes are 0.9843. Thus the implemented filter is stable. This shows that the poles of the cascade implementation in Fig. 9.11 is less sensitive to coefficient quantizations than the canonical form in Fig. 9.9 or 9.10. Thus for high-order IIR filters, we should implement them in cascade form. We will return to this problem after discussing some other phenomena due to finite word length.

9.5.1 Dead Band and Limit Cycle

Consider the following first-order digital filter

$$y[n] = ay[n - 1] \tag{9.35}$$

for $n \geq 0$. It has no input $x[n]$ and its response is excited by a nonzero initial condition $y[-1]$. By direct substitution, we have $y[0] = ay[-1]$, $y[1] = ay[0] = a^2 y[-1]$, and, in general,

$$y[n] = a^{n+1} y[-1] \tag{9.36}$$

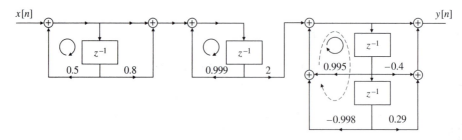

Figure 9.11 Cascade implementation of (9.33).

This solution can also be obtained by applying the z-transform. Applying the z-transform to (9.35) and using Problem 5.7, we obtain

$$Y(z) = a \left\{ z^{-1} Y(z) + y[-1] \right\}$$

which can be written as

$$Y(z) = \frac{a}{1 - az^{-1}} y[-1]$$

Its inverse z-transform is (9.36).

If $|a| < 1$ or the filter is stable, then $a^{n+1} \to 0$ as $n \to \infty$, and $y[n]$ approaches zero for *any* $y[-1]$. This is in fact a general property of any linear time-invariant digital filter. That is, if a digital filter is stable, then its output excited by any initial condition will approach zero as $n \to \infty$.

When we implement such a filter using a finite word length, this property may not hold. For example, consider

$$y[n] = Q[ay[n-1]] \tag{9.37}$$

where the quantization Q rounds a number to its nearest integer. If $a = 0.8$, then we can readily verify the following:

- $1.875 \le y[-1] < \infty \implies y[n] \to 2$
- $0.625 \le y[-1] < 1.875 \implies y[n] = 1$ for $n \ge 0$
- $-0.625 \le y[-1] < 0.625 \implies y[n] = 0$ for $n \ge 0$
- $-1.875 \le y[-1] < -0.625 \implies y[n] = -1$ for $n \ge 0$
- $-\infty < y[-1] < -1.875 \implies y[n] \to -2$

This implies that if a stable digital filter is not exactly implemented, its output excited by a nonzero initial condition may approach a nonzero constant. Furthermore, the nonzero constant depends on the value of the initial condition. The set $\{0, \pm 1, \pm 2\}$ which the responses are trapped to may be called the dead set. If $a = -0.8$ instead of $a = 0.8$, then we can show that the output

$y[n]$ takes alternatively the value 2 and -2 or -1 and 1, depending on the initial condition. This is called a *limit cycle*. The reason for these phenomena is due to the fact that a linear equation becomes a nonlinear equation after quantizations. See Section 1.2.

It is possible to develop a range in which the dead set or limit cycle resides. If (9.37) sustains a constant output or an oscillation with alternative value, then we have

$$y[n] = Q[ay[n-1]] = \begin{cases} y[n-1] & 0 < a < 1 \\ -y[n-1] & -1 < a < 0 \end{cases} \tag{9.38}$$

Furthermore, we have

$$|Q[ay[n-1]] - ay[n-1]| \le \frac{\delta}{2} \tag{9.39}$$

where δ is the quantization step. Substituting (9.38) into (9.39), we obtain

$$|y[n-1]| = |y[n]| \le \frac{\delta}{2(1-|a|)} := k$$

For the preceding example, the quantization step δ is 1 and $|a| = 0.8$; thus we have

$$k = \frac{1}{2(1-0.8)} = \frac{1}{0.4} = 2.25$$

Thus the dead set and the magnitude of the limit cycle lie inside $[-k, \ k] = [-2.25, \ 2.25]$. This is indeed the case. The range $[-k, \ k]$ is called the *dead band*.

If the number of bits used is large, for example, 32 bits, then the quantization step is $\delta = 2^{-32} \approx 2.3 \times 10^{-10}$. If $a = 0.8$, then the dead band will be $[-5 \times 10^{-10}, \ 5 \times 10^{-10}]$. It is very small. Thus if the number of bits used is large, the phenomena of dead set and limit cycle can be simply disregarded.

9.6 Cascade and Parallel Implementations

As discussed in Section 9.5, in implementing high-order IIR filters, it is better to implement it so that the resulting block diagram has a small number of interactive loops. One possible way is to factor a transfer function $H(z)$ as

$$H(z) = H_1(z)H_2(z)H_3(z)\cdots \tag{9.40}$$

where $H_i(z)$ are proper transfer functions of degree either one or two and with real coefficients. In order to standardize the implementation, we will use the following transfer function of degree 2

$$H_i(z) = \frac{b_{1i} + b_{2i}z^{-1} + b_{3i}z^{-2}}{1 + a_{2i}z^{-1} + a_{3i}z^{-2}} \tag{9.41}$$

called a *second-order section (sos)*, as a building block. Thus for an Nth-order transfer function, we will factor it as

$$H(z) = H_1(z)H_2(z)\cdots H_M(z) \tag{9.42}$$

where each $H_i(z)$ is a second-order section. If N is even, then we have $M = N/2$. If N is odd, then we have $M = (N+1)/2$ with one of $H_i(z)$ has degree one or $b_{3i} = 0$ and $a_{3i} = 0$. We can then implement each $H_i(z)$ and connect them in tandem to yield an implementation of $H(z)$. This is called a *tandem* or *cascade* implementation.

◆ **Example 9.2**

Implement the following fifth-order elliptic low-pass filter

$$
\begin{aligned}
H(z) = {} & \frac{0.0606[1 - (0.1920 + 0.9814i)z^{-1}][1 - (0.1920 - 0.9814i)z^{-1}]}{[1 - (0.3080 + 0.9230i)z^{-1}][1 - (0.3080 - 0.9230i)z^{-1}]} \\[2mm]
& \times \frac{[1 - (-0.0903 + 0.9959i)z^{-1}][1 - (-0.0903 - 0.9959i)z^{-1}]}{[1 - (0.4329 + 0.7204i)z^{-1}][1 - (0.4329 - 0.7204i)z^{-1}]} \\[2mm]
& \times \frac{(1 + z^{-1})}{(1 - 0.6181z^{-1})}
\end{aligned}
\tag{9.43}
$$

It is obtained in MATLAB by typing [z,p,k]=ellip(5,2,35,0.4) (fifth-order elliptic filter with passband tolerance 2 dB, stopband attenuation 35 dB, and passband edge frequency 0.4 in the normalized frequency range [0, 1].) We group the poles and zeros as

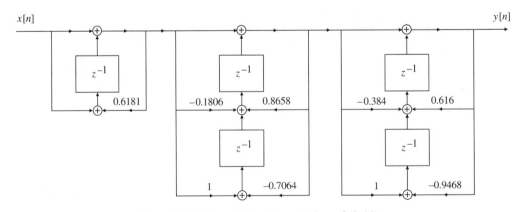

Figure 9.12 Cascade implementation of (9.44).

$$H(z) = \frac{1 + z^{-1}}{1 - 0.6181z^{-1}} \frac{1 - 0.1806z^{-1} + z^{-2}}{1 - 0.8658z^{-1} + 0.7064z^{-2}}$$

$$\times \frac{1 - 0.384z^{-1} + z^{-2}}{1 - 0.616z^{-1} + 0.9468z^{-2}} := H_1(z)H_2(z)H_3(z) \qquad (9.44)$$

It is implemented in Fig. 9.12, in which each $H_i(z)$ is implemented using the transposed canonical form. This block diagram has the smallest number of interactive loops and its poles are less sensitive to coefficient quantizations than the canonical form implementation in Fig. 9.10.

Cascade or tandem implementation actually involves three issues: grouping, pairing, and ordering. If a filter has three or more real poles, then there are many ways of grouping two real poles as a quadratic term. For example, if there are four real poles, then we have $4 \times 3/2 = 6$ ways of grouping two real poles. The same applies to real zeros. If a transfer function has all complex-conjugate poles and zeros except possibly one or two real poles and zeros, then the grouping is unique. This is the case in Example 9.2.

The next issue is to pair poles and zeros to form a second-order section. If both the numerator and denominator of $H(z)$ have M quadratic terms, then there are $M!$ ways of pairing. It is recommended in Ref. 14 that we start with complex-conjugate poles that are closest to the unit circle and pair them with the zeros that are closest to them. We repeat the process until all are paired. For example, consider the filter transfer function in (9.43) whose poles and zeros are plotted in Fig. 9.13. The complex-conjugate poles closest to the unit circle are $0.3080 \pm 0.9230i$. They are paired with the closest zeros $0.1920 \pm 0.9814i$ to form $H_3(z)$ in (9.44). The advantage of this pairing is that its magnitude response has a smaller peak value. A large peak value may cause the filter to saturate or overflow. Next we pair the poles at $0.4329 \pm 0.7204i$ with the zeros at $-0.0903 \pm 0.9959i$, as shown in Fig. 9.13, to form $H_2(z)$ in (9.44). The remaining pole and zero form $H_1(z)$. This process of pairing is straightforward.

The last issue is ordering. We can order $H(z)$ as $H_1(z)H_2(z)H_3(z)$, $H_1(z)H_3(z)H_2(z)$, and so forth. In general, if we have M sections, there are $M!$ possible orderings. In computer computation, it is a well-established practice to postpone, to the latest possible stage, the computation that is most prone to errors. Following this convention, in implementing (9.44), we shall implement $H_1(z)$, then $H_2(z)$, and finally $H_3(z)$, as shown in Fig. 9.12, or order $H(z)$ as $H_3(z)H_2(z)H_1(z)$. For mathematical analysis of these issues, see Ref. 14.

In Professional MATLAB, *Signal Processing Toolbox* includes the function sos=zp2sos(z, p,k), which carried out the pairing and ordering discussed. The student edition, however, does not contain this function.

Parallel Implementation In addition to the cascade implementation, we can also implement an IIR filter in parallel form, which also has a small number of interactive loops. The basic idea is to expand an Nth-order $H(z)$ as

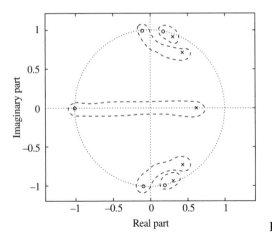

Figure 9.13 Pairing of poles and zeros.

$$H(z) = k + \sum_{i=0}^{M} H_i(z) \tag{9.45}$$

where $H_i(z)$ are proper rational functions of degree 1 or 2. If $H(z)$ has four real poles, then we can have four $H_i(z)$ with degree 1. In order to standardize the procedure, we will require all $H_i(z)$ to be of degree two if N is even and all $H_i(z)$ except one to be of degree two if N is odd. Furthermore, we require $H_i(z)$ to be of the form

$$H_i(z) = \frac{b_{1i} + b_{2i}z^{-1}}{1 + a_{2i}z^{-1} + a_{3i}z^{-2}} \tag{9.46}$$

This is different from the building block in (9.41) for cascade implementation. After completing the expansion in (9.45), if $k \neq 0$, we can add k to any $H_i(z)$ to have its form as in (9.41). By so doing, we have $k = 0$, and there will be one less parallel path. We then implement each $H_i(z)$ and connect them as in Fig. 9.14. This is called a *parallel implementation.*

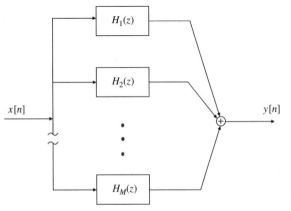

Figure 9.14 Parallel implementation.

◆ **Example 9.3**

Consider the fifth-order elliptic filter studied in Example 9.2. We expand it as

$$H(z) = \frac{(00606 + 0.0483z^{-1} + 0.1047z^{-2} + 0.11047z^{-3} + 0.0483z^{-4} + 0.0606z^{-5})}{(1 - 2.0998z^{-1} + 3.1022z^{-2} - 2.6061z^{-3} + 1.4443z^{-4} - 0.04134z^{-5})}$$

$$= k + \frac{b_{11}}{1 - 0.6181z^{-1}} + \frac{b_{12} + b_{22}z^{-1}}{1 - 0.8658z^{-1} + 0.7064z^{-2}}$$

$$+ \frac{b_{13} + b_{23}z^{-1}}{1 - 0.616z^{-1} + 0.9468z^{-2}} \tag{9.47}$$

Using the program that follows

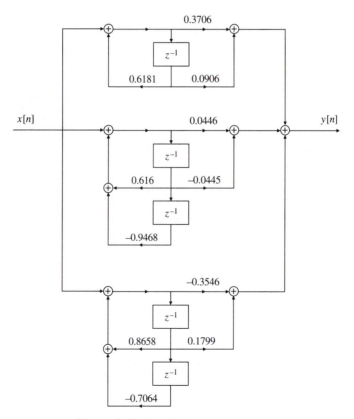

Figure 9.15 Parallel implementation.

```
[b,a]=ellip(5,2,35,0.4);
[r,p,k]=residuez(b,a);
[b1,a1]=residuez([r(1) r(2)],[p(1) p(2)],0);
[b2,a2]=residuez([r(3) r(4)],[p(3) p(4)],0);
```

we can obtain

$$H(z) = \frac{0.0446 - 0.0445z^{-1}}{1 - 0.616z^{-1} + 0.9468z^{-2}} + \frac{-0.3546 + 0.1799z^{-1}}{1 - 0.8658z^{-1} + 0.7064z^{-2}}$$
$$+ \frac{0.5173}{1 - 0.6181z^{-1}} - 0.1467 \tag{9.48}$$

If we combine the last two terms as $(0.3706 + 0.0906z^{-1})/(1 - 0.6181z^{-1})$, then (9.48) can be implemented as in Fig. 9.15, where each $H_i(z)$ is implemented in the canonical form. This is a parallel implementation of the fifth-order elliptic IIR filter.

It is natural to ask at this point which is better: cascade or parallel. This certainly depends on the criterion used. Shifting of poles due to coefficient quantizations are the same in both implementations. In computing the frequency response of (9.48), we must sum its three terms, and the resulting numerator may be severely affected by coefficient quantizations. For example, if $H(z)$ has zeros on the unit circle, it would be difficult to have the same zeros in the parallel implementation. Thus the effect on frequency responses probably is less in cascade implementations. Furthermore, cascade implementations require less computation. Thus cascade implementations are preferable to parallel implementations.

In conclusion, we should implement high-order IIR filters in cascade form if sensitivity is of concern. Recall that high-order FIR filters, as discussed in Section 9.2, should be implemented in direct form rather than cascade form. This is not necessarily inconsistent. For FIR filters, coefficient quantizations will not affect poles, and implemented filters are always stable. For IIR filters, coefficient quantizations will affect poles, and resulting filters may become unstable. Furthermore, the effects on frequency responses by pole variations may be more severe than by zeros variations. Thus there are no reasons to implement high-order IIR and FIR filters in the same form.

9.6.1 Implementation of Second-Order Sections

Because cascade and parallel implementations are based on the second-order section

$$H(z) = \frac{b_1 + b_2z^{-1} + b_3z^{-2}}{1 + a_2z^{-1} + a_3z^{-2}} \tag{9.49}$$

it is important to discuss further its implementations. We discuss two issues: pole sensitivity and overflow. We first discuss the former. Suppose the poles of (9.49) are complex conjugate as

$\alpha \pm j\beta$. Then in addition to the canonical form shown in Fig. 9.16(a), we can also implement it as shown in Fig. 9.16(b). Using a procedure similar to (9.28), we can show that the block diagram in Fig. 9.16(b) has (9.49) as its transfer function (Problem 9.13). The form is called a *coupled form*.

We next compare the canonical and coupled forms in terms of pole shifting due to finite word length. Let the poles of (9.49) be $re^{\pm j\theta}$. Then we have

$$1 + a_2 z^{-1} + a_3 z^{-2} = (1 - re^{j\theta}z^{-1})(1 - re^{-j\theta}z^{-1})$$
$$= 1 - 2r\cos\theta z^{-1} + r^2 z^{-2}$$

and

$$a_2 = -2r\cos\theta, \qquad a_3 = r^2 \qquad\qquad (9.50)$$

Now suppose a_2 and a_3 can assume only three bits or only eight distinct values in $[0, 1)$. Because of $r^2 = a_3$, if a_3 can assume only $k/8$ for $k = 0, 1, \dots, 7$, then the values which r can assume are limited to $\sqrt{k/8}$ for $k = 0, 1, \dots, 7$ as shown in Fig. 9.17(a) with circular curves. Because of $a_2 = -2r\cos\theta$, the values which the real part of a pole can assume are equally spaced as shown in Fig. 9.17(a) with vertical dotted lines. Thus in the canonical form implementation in Fig. 9.16(a), if a_2 and a_3 are limited to three bits, the pole locations which the implementation can assume are limited to those denoted by small circles shown in Fig. 9.17(a). If a pole is not exactly on one of these circles, it must be approximated by the one that is closest to it. We see that for r small or θ small, the small circles spread widely. In these cases, the approximation will be poor. Thus this implementation should be avoided if r or θ is small such as in narrow-band low-pass or high-pass filters.

Next we consider the block diagram in Fig. 9.16(b) with

$$\alpha = r\cos\theta, \qquad \beta = r\sin\theta \qquad\qquad (9.51)$$

If α and β assume one of the eight equally spaced values in $[0, 1)$ as shown in Fig. 9.17(b), then the quantized pole locations are those denoted by small circles shown in Fig. 9.17(b). Clearly, for r and θ small, the implementation in Fig. 9.16(b) is less sensitive than the canonical form in Fig. 9.16(a). However, the coupled form implementation requires, as we can see from Fig. 9.16(b), two more multiplications than the canonical form. Thus the canonical form is more efficient than the coupled form. Thus which form to use depends on which is more important: pole sensitivity or computational efficiency.

The preceding discussion is applicable to the implementation of the sinusoidal generator discussed in Fig. 6.25. The structure in Fig. 6.25 is basically in the canonical form. Because $a_3 = 1$, the poles of the generator are always on the unit circle even if $a_2 = 2\cos\omega_0$ cannot be precisely implemented. If the generator is implemented in the coupled form, then the poles may shift out the unit circle in imprecise implementation. If the poles move inside the unit circle, the amplitude of oscillation will decrease with time and the implementation requires resetting. If

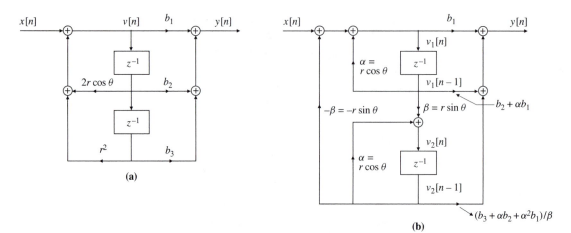

Figure 9.16 (a) Canonical form. (b) Coupled form.

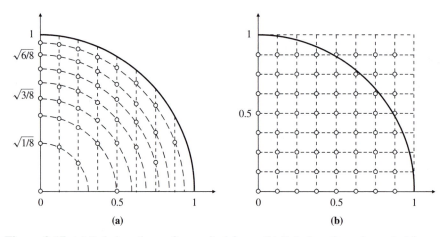

Figure 9.17 (a) Pole locations of canonical form. (b) Pole location of coupled form.

the poles move outside the unit circle, the amplitude will grow and requires limiting. Thus the canonical form is a better implementation.

Next we discuss the problem of overflow. Consider the notch filter in (6.23) repeated in the following[3]

$$H(z) = \frac{1 - 1.4579z^{-1} + z^{-2}}{1 - 1.3850z^{-1} + 0.9025z^{-2}} \qquad (9.52)$$

---

[3] This issue was pointed out to this writer by Professor John Murray.

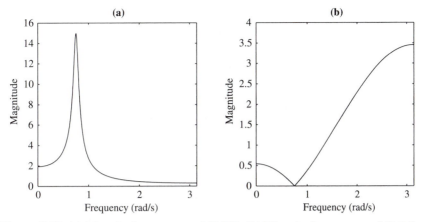

Figure 9.18 (a) Magnitude response of (9.53). (b) Magnitude response of (9.54).

This filter has two zeros on the unit circle at $e^{\pm j0.754}$ and two poles at $0.95e^{\pm j0.754}$. If this filter is implemented in the canonical form shown in Fig. 9.16(a), then the internal variable $v[n]$ shown is related to the input $x[n]$ by

$$V(z) = \frac{1}{1 - 1.3850z^{-1} + 0.9025z^{-2}} X(z) =: H_1(z)X(z) \qquad (9.53)$$

The magnitude response of $H_1(z)$ is plotted in Fig. 9.18(a). We see that it has a peak gain of 15. If the filter is implemented with fixed point arithmetic, then the chance for $v[n]$ to overflow is large. Now let us implement (9.52) in the direct form shown in Fig. 9.8 with $N = 2$. The internal variable $v[n]$ shown is related to the input by

$$V(z) = (1 - 1.4579z^{-1} + z^{-2})X(z) =: H_2(z)X(z)$$

The magnitude response of $H_2(z)$ is plotted in Fig. 9.18(b). Its peak magnitude is slightly less than 3.5. Thus the chance for the direct-form implementation to overflow is much less than the canonical-form implementation. This is also true if we use the transposed canonical form. Thus in fixed-point implementation, the direct-form block diagram should not be lightly dismissed even though it uses twice the number of unit-delay elements. In floating-point implementation, overflow will be less of a problem.

PROBLEMS

9.1 Consider the fourth order minimax optimal FIR filter

$$H(z) = 0.1196 + 0.4105z^{-1} + 0.2656z^{-2} + 0.4105z^{-3} + 0.1196z^{-4}$$

Implement it in the direct form shown in Fig. 9.1. Compute its zeros z_i and plot its magnitude response. Now suppose every coefficient is rounded to two decimal digits after the decimal point. Compute its zeros $\bar{z}_i$ and plot its magnitude response. What is its total zero deviation defined as $\sum |z_i - \bar{z}_i|$?

9.2   Implement the transfer function in Problem 9.1 in cascade form. Repeat Problem 9.1 for this implementation. Which implementation is better in terms of magnitude responses? Which implementation is better in terms of total zero deviations?

9.3   Find the linear and periodic convolutions of the sequences

$$[1\ 1\ 1\ 1] \qquad [2\ -2\ 2]$$

Note that the shorter sequence must be padded by one trailing 0 before computing their periodic convolution. Are the two convolutions the same?

9.4   Find 4-point DFT of the sequences in Problem 9.3. Let $X_1[m]$ and $X_2[m]$, $m = 0, 1, 2, 3$, be their DFT. Compute the inverse DFT of $X_1[m]X_2[m]$. Is the result the same as the periodic convolution in Problem 9.3?

9.5   What is the result of typing

x1 = [1 1 1 1]; x2 = [2 -2 2]; ifft(fft (x1).∗ fft (x2))

in MATLAB? Modify it so that it will yield the periodic convolution in Problem 9.3. Modify it so that it will yield the linear convolution in Problem 9.3.

9.6   Implement the fourth order elliptic filter

$$H(z) = \frac{0.0647 - 0.0106z^{-1} + 0.0997z^{-2} - 0.0106z^{-3} + 0.0647z^{-4}}{1 - 2.28z^{-1} + 2.6543z^{-2} - 1.5624z^{-3} + 0.4215z^{-4}}$$

in the direct form.

9.7   Implement the IIR filter in Problem 9.6 in the canonical form and transposed canonical form.

9.8   Verify that (9.29) can be expressed in matrix form as

$$\begin{bmatrix} v_1[n+1] \\ v_2[n+1] \\ v_3[n+1] \\ v_4[n+1] \end{bmatrix} = \begin{bmatrix} -a_2 & 1 & 0 & 0 \\ -a_3 & 0 & 1 & 0 \\ -a_4 & 0 & 0 & 1 \\ -a_5 & 0 & 0 & 0 \end{bmatrix} \begin{bmatrix} v_1[n] \\ v_2[n] \\ v_3[n] \\ v_4[n] \end{bmatrix} + \begin{bmatrix} b_2 - a_2b_1 \\ b_3 - a_3b_1 \\ b_4 - a_4b_1 \\ b_5 - a_5b_1 \end{bmatrix} x[n]$$

$$y[n] = [1\ 0\ 0\ 0]\mathbf{v}[n] + b_1 x[n]$$

where $\mathbf{v}[n] = [v_1[n]\ v_2[n]\ v_3[n]v_4[n]]'$.

9.9   Express the transfer function (9.18) as

$$H(z) = b_1 + \frac{\bar{b}_2 z^{-1} + \bar{b}_3 z^{-2} + \bar{b}_4 z^{-3} + \bar{b}_5 z^{-4}}{1 + a_2 z^{-1} + a_3 z^{-2} + a_4 z^{-3} + a_5 z^{-4}}$$

and then develop, using the transposed canonical form, a block diagram for it. What is its difference from Fig. 9.10(b)? Develop, by assigning the output of each unit-delay element as $v_i[n]$, a set of first-order difference equations to describe it. Is the set different from the one in Problem 9.8?

9.10   Implement the IIR filter in Problem 9.6 in a cascade form. How many cascade forms can you have? Give reasons of your selection.

9.11   Implement the IIR filter in Problem 9.6 in a parallel form. Is the form unique?

9.12   Implement the IIR filter with impulse sequence

$$h[n] = 0.2^n + 5\,(-0.6)^n + (0.9)^n \sin 0.2n$$

for $n \geq 0$, in the cascade form.

9.13   Verify that the transfer function of Fig. 9.16(b) equals (9.49).

9.14   Let $c_1 = [1\ 2\ 3\ 4]$ and $c_2 = [2\ 3\ 4\ 5\ 6]$. Compute their convolution on your computer by using (9.30) and (9.31). Are the results the same?

9.15   Let $c_1 = [\text{ones}(1, 1024)]$ and $c_2 = [\text{ones}(1, 2048)]$. Compute their convolution on your computer by typing, with $L = N + N_2 - 1$,

tic, (9.30); toc
tic, (9.31) (with L = 1024 + 2048 - 1); toc
tic, (9.31) (with L = 2^(12)); toc

What are their elapsed times? Which has the smallest elapsed time?

9.16   In Problem 9.15, the convolutions are not displayed because of the semicolon after (9.30) and (9.31). (1) Compare their elapsed times if the results are displayed on the monitor by replacing every semicolon by a comma. (2) The convolution has at most 3071 nonzero entries. Compare their elapsed times if only 3071 entries are displayed. (3) Compare their elapsed times if they are plotted on the monitor.

The Impulse

Consider the pulse $\delta_\epsilon(t - t_0)$ shown in Fig. A.1(a). The pulse is centered at $t = t_0$. It has height $1/\epsilon$ and width ϵ. As ϵ decreases, the pulse becomes higher and narrower but its area remains to be 1. We define

$$\delta(t - t_0) := \lim_{\epsilon \to 0} \delta_\epsilon(t - t_0)$$

It is called the Dirac delta function, δ-function, impulse function, or, simply, the *impulse*. We may define the impulse $\delta(\omega - \omega_0)$ in the same way with independent variable ω. Clearly all subsequent discussion remains valid if t is replaced by ω.

The impulse at t_0 has the properties

$$\delta(t - t_0) = \begin{cases} 0 & \text{if } t \neq t_0 \\ \infty & \text{if } t = t_0 \end{cases} \tag{A.1}$$

and

$$\int_{-\infty}^{\infty} \delta(t - t_0)dt = \int_{t_0-}^{t_0+} \delta(t - t_0)dt = 1 \tag{A.2}$$

Because $\delta(t - t_0)$ is zero everywhere except at t_0, the integration interval from $-\infty$ to ∞ in (A.2) can be reduced to the immediate neighborhood of t_0 from t_0- to t_0+. The notation t_0- is used to denote a time instant slightly smaller than t_0 and t_0+ a time instant slightly larger than t_0. The width from t_0- to t_0+ is practically zero but still includes the whole impulse at t_0. If an integration interval includes an impulse wholly, then the integration is 1, such as

$$\int_{-\infty}^{0+} \delta(t - 0)dt = 1, \qquad \int_{0}^{1} \delta(t - 0.5)dt = 1$$

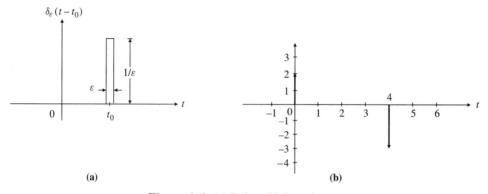

Figure A.1 (a) Pulse. (b) Impulses.

If an integration does not include any part of an impulse, then the integration is 0 such as

$$\int_{-\infty}^{-1} \delta(t-0)dt = 0, \qquad \int_{0.5+}^{1} \delta(t-0.5)dt = 0$$

If integration intervals cover impulses partially such as in

$$\int_{-\infty}^{0} \delta(t-0)dt, \qquad \int_{0.5}^{1} \delta(t-0.5)dt$$

then ambiguity may occur. In these situations, unless stated otherwise, we assume that the integrations cover the whole impulses and the integrations equal 1.

The impulse is not an ordinary function. The integration of an ordinary function such as

$$x_1(t) = t^{1000} + 10^{10}\sin 3t \tag{A.3}$$

or

$$x_2(t) = \begin{cases} 1 & \text{if } t < 4 \\ a & \text{if } t = 4 \\ 1 & \text{if } t > 4 \end{cases} \tag{A.4}$$

for any constant a, for example, $a = 1000^{1000}$, over any zero time interval is 0; that is

$$\int_{t_0-}^{t_0+} x_i(t)dt = 0$$

for $i = 1, 2$, and for any t_0. Nor is there any difference in using t_0, t_0- or t_0+ in integrations of ordinary functions. For example, we have

$$\int_{4-}^{6} x_2(t)dt = \int_{4}^{6} x_2(t)dt = \int_{4+}^{6} x_2(t)dt = 2$$

even though the value of $x_2(t)$ is 1000^{1000} at $t = 4$. Therefore, the value of an ordinary function at an isolated time instant has no effect on integrations. This is, however, not the case for impulses; impulses have zero width but contribute nonzero values in integrations.

Consider the "function"

$$x(t) = 2\delta(t) - 3\delta(t - 4)$$

It has an impulse at $t = 0$ and an impulse at $t = 4$. The impulse at $t = 0$ is said to have *weight* 2 and the one at $t = 4$ has weight -3. Graphically, it is represented as shown in Fig. A.1(b). We see that impulses are denoted by arrows with areas indicated by the vertical scale. Note that their actual heights are infinity.

Let $x(t)$ be an ordinary function. If $x(t)$ is continuous at t_0, then

$$x(t)\delta(t - t_0) = x(t_0)\delta(t - t_0) \tag{A.5}$$

and

$$\int_{-\infty}^{\infty} x(t)\delta(t - t_0)dt = x(t_0)\int_{t_0-}^{t_0+} \delta(t - t_0)dt = x(t_0) \times 1$$

$$= x(t)|_{t=t_0} = x(t_0) \tag{A.6}$$

We require $x(t)$ to be continuous at t_0 to avoid possible ambiguity. Equation (A.5) follows from the property that $\delta(t - t_0)$ is 0 everywhere except at t_0, and (A.6) follows immediately from (A.5) and (A.2). The integration in (A.6) is called the *sifting* property, because the impulse picks or sifts out the value of $x(t)$ at $t = t_0$. For example, we have

$$\int_{t=-\infty}^{\infty} (t^2 + \sin t)\delta(t - 1)dt = (t^2 + \sin t)|_{t=1} = 1^2 + \sin 1$$

$$= 1 + \sin 57.3^o = 1 + 0.841 = 1.841$$

$$\int_{t=-\infty}^{\infty} x(t - 2)\delta(t - \tau + 3)dt = x(t - 2)|_{t=\tau-3}$$

$$= x(\tau - 3 - 2) = x(\tau - 5)$$

and

$$\int_{\omega=-\infty}^{\infty} e^{j\omega t}\delta(\omega - \omega_0)d\omega = e^{j\omega t}|_{\omega=\omega_0} = e^{j\omega_0 t}$$

Thus whenever an integrand is the product of a function and an impulse, we simply move the function out of the integration and replace the integration variable by the one by setting the argument of the impulse to 0. This sifting property will be constantly used in this text.

REFERENCES

[1] Ambardar, A. *Analog and Digital Signal Processing*. Boston: PWS Publishing, 1995.

[2] Burrus, C. S., R. A. Gopinath, and H. Guo. *Introduction to Wavelets and Wavelet Transforms*. Upper Saddle River, N.J.: Prentice-Hall, 1998.

[3] Burrus, C. S., A. W. Soewito, and R. A. Gopinath. "Least Squared Error FIR Filter Design with Transition Bands." *IEEE Trans. Signal Proc.* 40 (1992): 1327–1340.

[4] Cartinhour, J. *Digital Signal Processing*. Upper Saddle River, N.J.: Prentice-Hall, 2000.

[5] Chen, C. T. *One-dimensional Digital Signal Processing*. New York: Dekker, 1979.

[6] Chen, C. T. *System and Signal Analysis*. 2nd ed. New York: Oxford University Press, 1994.

[7] Chen, C. T. *Linear System Theory and Design*. 3rd ed. New York: Oxford University Press, 1999.

[8] Cooley, J. W., and J. W. Tukey. "An Algorithm for the Machine Computation of Complex Fourier Series." *Mathematics of Computation* 19 (1965): 297–301.

[9] Daugherty, K. M. *Analog-to-Digital Conversion*. New York: McGraw-Hill, 1995.

[10] Deller, J. R., Jr., J. G. Proakis, and J. H. L. Hansen. *Discrete-Time Processing of Speech Signals*. New York: Macmillan, 1993.

[11] Embree, P. M., and D. Danieli. *C++ Algorithms for Digital Signal Processing*. 2nd ed. Upper Saddle River, N.J.: Prentice-Hall, 1999.

[12] Hayes, M. H. *Statistical Digital Signal Processing and Modeling*. New York: John Wiley, 1996.

[13] Hubbard, B. B. *The World according to Wavelets*. 2nd ed. Natick, Mass.: A.K. Peters, 1998.

[14] Jackson, L. B. *Digital Filters and Signal Processing*. 3rd ed. Boston: Kluwer, 1996.

[15] Karl, J. H. *An Introduction to Digital Signal Processing*. New York: Academic Press, 1989.

[16] Kuc, R. *Introduction to Digital Signal Processing*. New York: McGraw-Hill, 1988.

[17] Lynn, P. A., and W. Fuerst. *Digital Signal Processing with Computer Applications*. 2nd ed. New York: John Wiley & Sons, 1998.

[18] MATLAB Help Desk, Version 5. The Math Works, 1997.

[19] McClellan, J. H., R. W. Schafer, and M. A. Yoder. *DSP First: A Multimedia Approach*. Upper Saddle River, N.J.: Prentice-Hall, 1998.

[20] Mitra, S. K. *Digital Signal Processing: A Computer-Based Approach*. New York: McGraw-Hill, 1998.

[21] Oppenheim, A. V., and R. W. Schafer. *Discrete-Time Signal Processing*. 2nd ed. Upper Saddle River, N.J.: Prentice-Hall, 1999.

[22] Oppenheim, A. V., A. S. Willsky, with S. H. Nawab. *Signals & Systems*. 2nd ed. Upper Saddle River, N.J.: Prentice-Hall, 1997.

[23] Orfanidis, S. J. *Introduction to Signal Processing*. Englewood Cliffs, N.J.: Prentice-Hall, 1996.

[24] Parks, T. W., and C. S. Burrus. *Digital Filter Design*. New York: John Wiley, 1987.

[25] Proakis, J. G., and D. G. Manolakis. *Digital Signal Processing: Principles, Algorithms, and Applications*. 3rd ed. Upper Saddle River, N.J.: Prentice-Hall, 1996.

[26] Rabiner, L. R., and B. Gold. *Theory and Application of Digital Signal Processing*. Englewood Cliffs, N.J.: Prentice-Hall, 1975.

[27] Selesnick, I. W., M. Lang, and C. S. Burrus. "Constrained Least Squares Design of FIR Filters without Specified Transition Bands." *IEEE Trans. Signal Proc.* 44 (1996): 1879–1892.

[28] Seo, B., and C. T. Chen. "A Relationship between the z-Transform and the Fourier Transform." *IEEE Trans. on Acoustic, Speech, and Signal Processing* 37 (1989): 759–760.

[29] Steiglitz, K. *A Digital Signal Processing Primer*. Reading, Mass.: Addison-Wesley, 1996.

[30] Stanley, W. D., G. R. Dougherty, and R. Dougherty. *Digital Signal Processing*. 2nd ed. Reston, Va.: Reston Publishing, 1984.

[31] Su, K. L. *Analog Filters*. New York: Chapman & Hall, 1996.

[32] Zemanian, A. H. *Distribution Theory and Transform Analysis*. New York: Dover, 1987.

Index